Pipefitting
Level Three

Trainee Guide
Third Edition

PEARSON
Prentice Hall

nccer

Upper Saddle River,
New Jersey
Columbus, Ohio

National Center for Construction Education and Research

President: Don Whyte
Director of Curriculum Revision and Development: Daniele Stacey
Pipefitting Project Manager: Tania Domenech
Production Manager: Jessica Martin
Product Maintenance Supervisor: Debie Ness
Editor: Brendan Coote
Desktop Publisher: James McKay

NCCER would like to acknowledge the contract service provider for this curriculum:
Topaz Publications, Liverpool, New York.

This information is general in nature and intended for training purposes only. Actual performance of activities described in this manual requires compliance with all applicable operating, service, maintenance, and safety procedures under the direction of qualified personnel. References in this manual to patented or proprietary devices do not constitute a recommendation of their use.

1 0 9 8 7 6 5
ISBN 0-13-227284-9

PREFACE

TO THE TRAINEE

There are some who may consider pipefitting synonymous with plumbing, but these are really two very distinct trades. Plumbers install and repair the water, waste disposal, drainage and gas systems in homes and commercial and industrial buildings. Pipefitters, on the other hand, install and repair both high- and low-pressure pipe systems used in manufacturing, in the generation of electricity, and in the heating and cooling of buildings.

If you're trying to imagine a setting involving pipefitters, think of large power plants that create and distribute energy throughout the nation; think of manufacturing plants, chemical plants, and piping systems that carry all kinds of liquids, gaseous, and solid materials.

If you're trying to imagine a job in pipefitting, picture yourself in a job that won't go away for a long time. As the U.S. government reports, the demand for skilled pipefitters continues to outpace the supply of workers trained in this craft. And high demand typically means 'higher pay' making pipefitters among the highest paid construction occupations in the nation.

While pipefitters and plumbers perform different tasks, the aptitudes involved in these crafts are comparable. Attention to detail, spatial and mechanical abilities, and the ability to work efficiently with the tools of their trade are key. If you think you might have what it takes to work in this high-demand occupation, contact your local NCCER Training Sponsor to see if they offer a training program in this craft or contact your local union or non-union training programs. You might make the perfect 'fit'.

We wish you success as you embark on your third year of training in the pipefitting craft and hope that you'll continue your training beyond this textbook. There are more than a half-million people employed in this work in the United States, and as most of them can tell you, there are many opportunities awaiting those with the skills and desire to move forward in the construction industry.

CONTREN® LEARNING SERIES

The National Center for Construction Education and Research (NCCER) is a not-for-profit 501(c)(3) education foundation established in 1995 by the world's largest and most progressive construction companies and national construction associations. It was founded to address the severe workforce shortage facing the industry and to develop a standardized training process and curricula. Today, NCCER is supported by hundreds of leading construction and maintenance companies, manufacturers, and national associations. The Contren® Learning Series was developed by NCCER in partnership with Prentice Hall, the world's largest educational publisher.

Some features of NCCER's Contren® Learning Series are as follows:

- An industry-proven record of success
- Curricula developed by the industry for the industry
- National standardization, providing portability of learned job skills and educational credits
- Compliance with the Office of Apprenticeship requirements for related classroom training (CFR 29:29)
- Well-illustrated, up-to-date, and practical information

NCCER also maintains a National Registry that provides transcripts, certificates, and wallet cards to individuals who have successfully completed modules of NCCER's Contren® Learning Series. *Training programs must be delivered by an NCCER Accredited Training Sponsor in order to receive these credentials.*

Contents

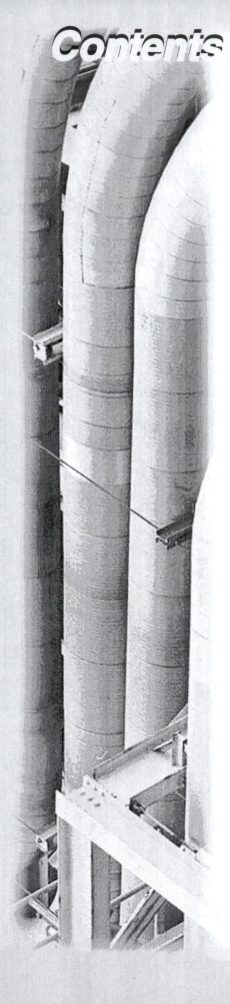

Contren® Curricula

NCCER's training programs comprise more than 40 construction, maintenance, and pipeline areas and include skills assessments, safety training, and management education.

Boilermaking
Carpentry
Cabinetmaking
Careers in Construction
Concrete Finishing
Construction Craft Laborer
Construction Technology
Core Curriculum: Introductory Craft Skills
Currículum Básico
Electrical
Electrical Topics, Advanced
Electronic Systems Technician
Heating, Ventilating, and Air Conditioning
Heavy Equipment Operations
Highway/Heavy Construction
Hydroblasting
Instrumentation
Insulating
Ironworking
Maintenance, Industrial
Masonry
Millwright
Mobile Crane Operations
Painting

Painting, Industrial
Pipefitting
Pipelayer
Plumbing
Reinforcing Ironwork
Rigging
Scaffolding
Sheet Metal
Site Layout
Sprinkler Fitting
Welding

Pipeline

Control Center Operations, Liquid
Corrosion Control
Electrical and Instrumentation
Field Operations, Liquid
Field Operations, Gas
Maintenance
Mechanical

Safety

Field Safety
Orientación de Seguridad
Safety Orientation
Safety Technology

Management

Introductory Skills for the Crew Leader
Project Management
Project Supervision

Acknowledgments

This curriculum was revised as a result of the farsightedness and leadership of the following sponsors:

Becon
Cianbro
Flint Hills Resources/Koch Industries
Fluor Global Craft Services
Kellogg Brown & Root
Lee College
Zachry Construction Corporation

This curriculum would not exist were it not for the dedication and unselfish energy of those volunteers who served on the Authoring Team. A sincere thanks is extended to the following:

Glynn Allbritton
Tom Atkinson
Ned Bush
Adrian Etie
Tina Goode
Ed LePage
Toby Linden

NCCER PARTNERING ASSOCIATIONS

American Fire Sprinkler Association
API
Associated Builders & Contractors, Inc.
Associated General Contractors of America
Association for Career and Technical Education
Association for Skilled and Technical Sciences
Carolinas AGC, Inc.
Carolinas Electrical Contractors Association
Center for the Improvement of Construction
   Management and Processes
Construction Industry Institute
Construction Users Roundtable
Design-Build Institute of America
Electronic Systems Industry Consortium
Merit Contractors Association of Canada
Metal Building Manufacturers Association
NACE International
National Association of Minority Contractors

National Association of Women in Construction
National Insulation Association
National Ready Mixed Concrete Association
National Systems Contractors Association
National Technical Honor Society
National Utility Contractors Association
NAWIC Education Foundation
North American Crane Bureau
North American Technician Excellence
Painting & Decorating Contractors of America
Portland Cement Association
SkillsUSA
Steel Erectors Association of America
Texas Gulf Coast Chapter ABC
U.S. Army Corps of Engineers
University of Florida
Women Construction Owners & Executives, USA

08301-07

Rigging Equipment

08301-07
Rigging Equipment

Topics to be presented in this module include:

Overview

Rigging equipment includes all the tools to safely move materials and equipment from place to place on the work site. It is also the equipment used to raise and support pipe spools while they are assembled. You as a pipefitter must know how all these types of equipment work, and their capabilities and limitations. You will use chains, slings, wire rope, hooks, clamps, and shackles most days of your working life. You must learn to use them safely.

Objectives

When you have completed this module, you will be able to do the following:

1. Identify and describe the uses of common rigging hardware and equipment.
2. Perform a safety inspection on hooks, slings, and other rigging equipment.
3. Describe common slings and determine sling capacities and angles.
4. Select, inspect, use, and maintain special rigging equipment, including:

    - Simple block and tackle
    - Chain hoists
    - Come-alongs
    - Jacks
    - Tuggers

5. Inspect heavy rigging hardware.
6. Tie knots used in rigging.

Trade Terms

Anneal	Hydraulic
Bird caging	Kinking
Equalizer beam	Parts of line
Fixed block	Seizing
Hauling line	Sling angle

Required Trainee Materials

1. Pencil and paper
2. Appropriate personal protective equipment

Prerequisites

Before you begin this module, it is recommended that you successfully complete *Core Curriculum*; *Pipefitting Level One*, and *Pipefitting Level Two*.

This course map shows all of the modules in the third level of the *Pipefitting* curriculum. The suggested training order begins at the bottom and proceeds up. Skill levels increase as you advance on the course map. The local Training Program Sponsor may adjust the training order.

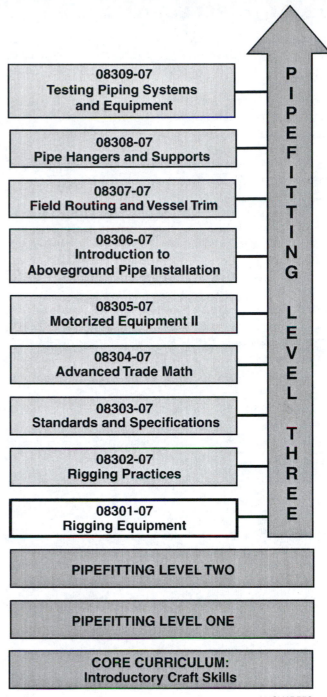

301CMAP.EPS

1.0.0 ◆ INTRODUCTION

Rigging is the planned movement of an object from one place to another (*Figure 1*). However, it does not apply to the actual operation of a crane. It may also include moving an object from one position to another or tying down an object. For example, tools and equipment must often be lifted several floors to get them where they are needed. This moving operation must be done safely and effectively. This module reviews and expands on the hardware and procedural information previously learned in the *Core Curriculum* module, *Basic Rigging*.

The rigger and the crane operator work together in preparing a lift. The rigger must make sure the crane operator understands what is to be lifted, what it weighs, and any complexities involving the lift. At the same time, the rigger should be aware of the crane's capabilities and limitations in order to ensure that the crane is able to safely make the lift. As a team, the rigger and crane operator should examine the area of the lift for any potential safety hazards such as power lines and structures, as well as risks to nearby workers.

301F01.EPS

Figure 1 ◆ A rigging application.

2.0.0 ◆ RIGGING HARDWARE

Hardware used in rigging includes hooks, shackles, eyebolts, spreader and **equalizer beams**, and blocks. These hardware items must be carefully matched to the lifting capability of the crane or hoist and the slings used in each application. Mixing items with different safe working loads makes it impossible to determine if the lift will be safe. Careful inspection and maintenance of all lifting hardware is essential for continuous operation. Hardware should always be inspected before each use.

> **NOTE**
>
> Rigging hooks must be properly marked or tagged with the manufacturer's ratings and application information. If a hook is not properly marked, don't use it. Also, rigging hardware must never be modified because modifications could affect its capacity.

2.1.0 Hooks

A rigging hook can be attached directly to a sling or to a load. As you learned earlier, there are six basic types of rigging hooks, with the eye-type hook being the most commonly used. Rigging hooks are equipped with safety latches to prevent a sling attachment from coming off of the hook when a sling is slackened. The capacity of a rigging hook, in tons, is determined by its size and physical dimensions. Information about a specific hook's capacity is readily available from the hook manufacturer.

The safe working load of the hook is accurate only when the load is suspended from the saddle of the hook. If the working load is applied anywhere between the saddle and the hook tip, the safe working load is considerably reduced, as shown in *Figure 2*. Some manufacturers do not permit their hooks to be used in any other position.

Always inspect hooks before each use. Look for wear in the saddle of the hook. Also, look for cracks, severe corrosion, and twisting of the hook body. The safety latch should be in good working order. Measure the hook throat opening; if a hook has been overloaded or if it is beginning to weaken, the throat will open. OSHA regulations require that the hook be replaced if the throat has opened 15 percent from its original size or if the body has been twisted 10 degrees from its original plane. *Figure 3* shows the location of inspection points for a hook.

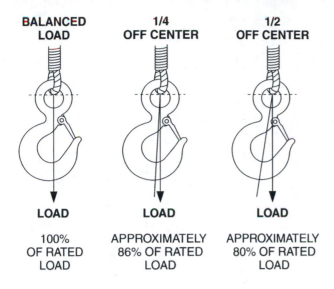

BALANCED LOAD — **LOAD** — 100% OF RATED LOAD

1/4 OFF CENTER — **LOAD** — APPROXIMATELY 86% OF RATED LOAD

1/2 OFF CENTER — **LOAD** — APPROXIMATELY 80% OF RATED LOAD

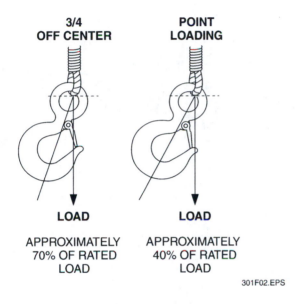

3/4 OFF CENTER — **LOAD** — APPROXIMATELY 70% OF RATED LOAD

POINT LOADING — **LOAD** — APPROXIMATELY 40% OF RATED LOAD

301F02.EPS

Figure 2 ◆ Rigging hook rated load versus load location.

Never use a sling eye over a hook with a body diameter larger than the natural width of the eye. Never force the eye onto the hook. Always use an eye with at least the nominal diameter of the hook.

2.2.0 Shackles

A shackle is used to attach an item to a load or to attach slings together. It can be used to attach the end of a wire rope to an eye fitting, hook, or other type of connector. Shackles are made in several configurations (*Figure 4*).

Shackles used for overhead lifting are made of forged alloy steel. They are sized by the diameter of the steel in the bow section rather than the pin size. When using shackles, be sure that all pins are straight, all screw pins are completely seated, and

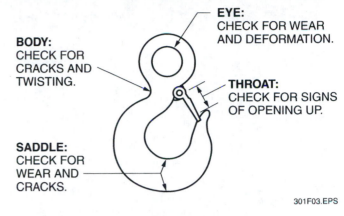

BODY: CHECK FOR CRACKS AND TWISTING.

EYE: CHECK FOR WEAR AND DEFORMATION.

THROAT: CHECK FOR SIGNS OF OPENING UP.

SADDLE: CHECK FOR WEAR AND CRACKS.

301F03.EPS

Figure 3 ◆ Rigging hook inspection points.

cotter pins are used with all round pin shackles. A shackle must not be used in excess of its rated load.

Shackle pins should never be replaced with a bolt. Bolts cannot take the bending force normally applied to a pin. Shackles that are stretched, or that have crowns or pins worn more than 10 percent, should be destroyed and replaced. If a shackle is pulled out at an angle, the capacity is reduced. Use spacers to center the load being hoisted on the pin. Only shackles of suitable load ratings can be used for lifting. When using a shackle on a hook, the pin of the shackle should be hung on the hook. Spacers can be used on the pin to keep the shackle hanging evenly on the hook (*Figure 5*). Never use a screw-pin shackle in a situation where the pin can roll under load, as shown in *Figure 6*.

 WARNING!
Shackles may not be loaded at an angle unless approved by the manufacturer. Contact the manufacturer for any angular loading.

2.3.0 Eyebolts

Eyebolts (*Figure 7*) are often attached to heavy loads by a manufacturer in order to aid in hoisting the load. One type of eyebolt, called a ringbolt, is equipped with an additional movable lifting ring. Eyebolts and ringbolts can be of either the shoulder or shoulderless type. The shoulder type is recommended for use in hoisting applications because it can be used with angular lifting pulls, whereas the shoulderless type is designed only for lifting a load vertically. The safe working load of shoulder eyebolts and ringbolts is reduced with angular loading. Loads should always be applied to the plane of the eye to reduce bending. This procedure is particularly important when bridle slings are used.

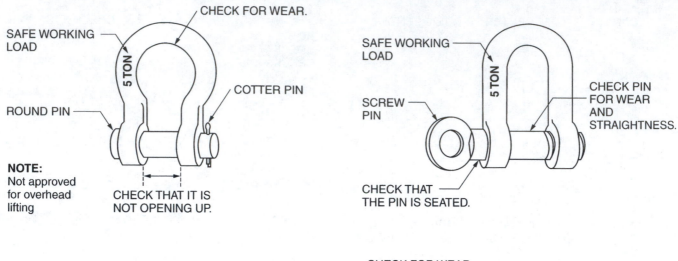

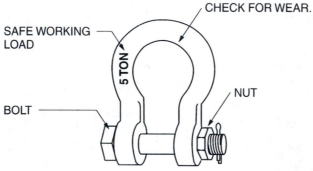

301F04.EPS

Figure 4 ◆ Round pin and screw pin shackle.

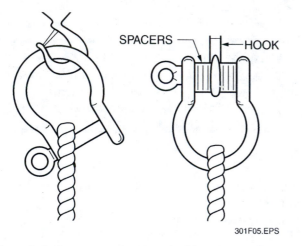

301F05.EPS

Figure 5 ◆ Spacers used with a shackle to keep it hanging evenly.

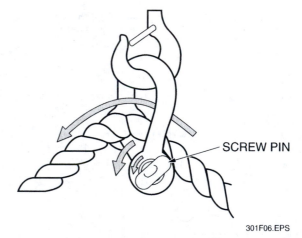

301F06.EPS

Figure 6 ◆ Example of how a rope can cause a shackle screw-pin to roll under load.

USE OF EYEBOLTS

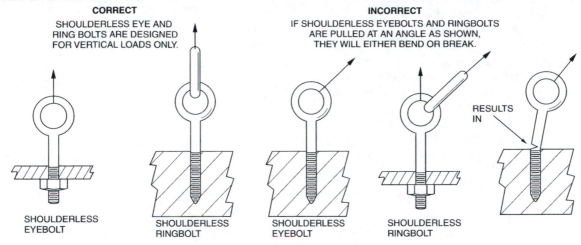

CORRECT
SHOULDERLESS EYE AND RING BOLTS ARE DESIGNED FOR VERTICAL LOADS ONLY.

INCORRECT
IF SHOULDERLESS EYEBOLTS AND RINGBOLTS ARE PULLED AT AN ANGLE AS SHOWN, THEY WILL EITHER BEND OR BREAK.

SHOULDERLESS EYEBOLT

SHOULDERLESS RINGBOLT

SHOULDERLESS EYEBOLT

SHOULDERLESS RINGBOLT

RESULTS IN

USE OF SHOULDER TYPE EYEBOLTS AND RINGBOLTS

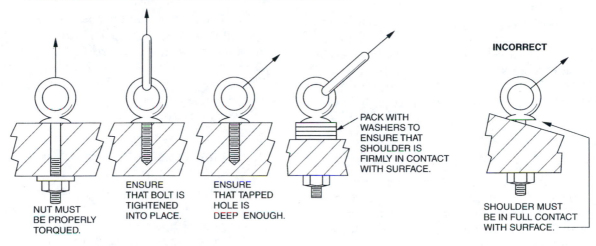

CORRECT
FOR SHOULDER TYPE EYEBOLTS AND RINGBOLTS PROVIDING LOADS ARE REDUCED TO ACCOUNT FOR ANGULAR LOADING.

NUT MUST BE PROPERLY TORQUED.

ENSURE THAT BOLT IS TIGHTENED INTO PLACE.

ENSURE THAT TAPPED HOLE IS DEEP ENOUGH.

PACK WITH WASHERS TO ENSURE THAT SHOULDER IS FIRMLY IN CONTACT WITH SURFACE.

INCORRECT

SHOULDER MUST BE IN FULL CONTACT WITH SURFACE.

ORIENTATION OF EYEBOLTS

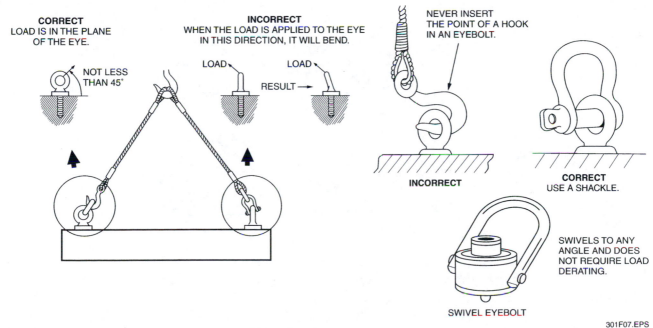

CORRECT
LOAD IS IN THE PLANE OF THE EYE.

NOT LESS THAN 45°

INCORRECT
WHEN THE LOAD IS APPLIED TO THE EYE IN THIS DIRECTION, IT WILL BEND.

LOAD

LOAD

RESULT

NEVER INSERT THE POINT OF A HOOK IN AN EYEBOLT.

INCORRECT

CORRECT
USE A SHACKLE.

SWIVELS TO ANY ANGLE AND DOES NOT REQUIRE LOAD DERATING.

SWIVEL EYEBOLT

301F07.EPS

Figure 7 ◆ Eyebolt and ringbolt installation and lifting criteria.

When installed, the shoulder of the eyebolt must be at right angles to the axis of the hole and must be in full contact with the working surface when the nuts are properly fastened. Washers or other suitable spacers may be used to ensure that the shoulders are in firm contact with the working surface. Tapped holes used with screwed-in eyebolts should have a minimum depth of 1½ times the bolt diameter. Swivel eyebolts may be installed instead of fixed eyebolts. These devices swivel to the desired lift position and do not require any load rating reduction.

2.4.0 Lifting Lugs

Lifting lugs (*Figure 8*) are typically welded, bolted, or pinned by a manufacturer to the object to be lifted. They are designed by engineers and located to balance the load and support it safely. Lifting lugs should be used for straight, vertical lifting only, unless specifically designed for angular loads. This is because they can bend if loaded at an angle.

Lifting lugs should be inspected before each use for deformation, cracks, corrosion, and defective welds. A lifting lug must be removed from service if such conditions are found.

2.5.0 Turnbuckles

Turnbuckles are available in a variety of sizes. They are used to adjust the length of rigging connections. Three common types of turnbuckles are the eye, jaw, and hook ends (*Figure 9*). They can be used in any combination. The safe working load for turnbuckles is based on the diameter of the threaded

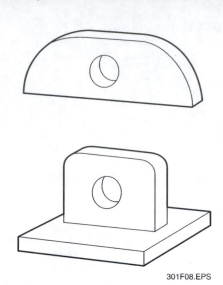

Figure 8 ◆ Two types of lifting lugs.

rods. The safe working load can be found in the manufacturer's catalog. The safe working load of turnbuckles with hook ends is less than that of the same size turnbuckles with other types of ends. The ratings of turnbuckles must be known before they are used.

Consider the following when selecting turnbuckles for rigging:

- Turnbuckles should be made of alloy steel and should not be welded.
- When using turnbuckles with multi-leg slings, do not use more than one turnbuckle per leg.
- Do not use jamb nuts on turnbuckles that do not come equipped with them.

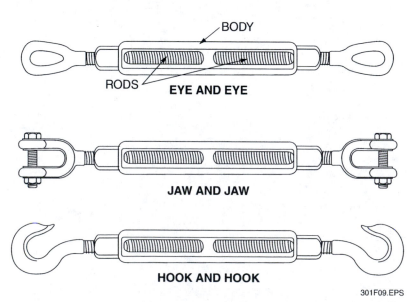

Figure 9 ◆ Turnbuckles.

- Turnbuckles should not be overtightened. Do not use a cheater to tighten them. Perform tightening with a wrench of the proper size, using as much force as a person can achieve by hand.
- Turnbuckles must not be used for angular loading unless permitted by the manufacturer, or if the angular loading has been calculated.

Turnbuckles should be inspected as shown in *Figure 10*. If a turnbuckle is damaged, remove it from service.

2.6.0 Beam Clamps

Beam clamps (*Figure 11*) are used to connect hoisting devices to beams so the beams can be lifted and positioned in place. The following are guidelines for using beam clamps:

- Do not use homemade clamps unless they are designed, load tested, and stamped by an engineer.
- Ensure that the clamp fits the beam and is of the correct capacity.
- Make sure beam clamps are securely fastened to the beam.
- Be careful when using beam clamps where angle lifts are to be made. Most are designed for essentially straight vertical lifts only. However, some manufacturers allow two clamps to be used with a long bridle sling if the angle of lift does not exceed 25 degrees from the vertical.
- Be certain the capacity appears on the beam clamp.

Figure 11 ◆ Typical beam clamp application.

- Do not place a hoist hook directly in the beam clamp lifting eye.
- Never overload a beam clamp beyond its rating capacity.

Beam clamps must be inspected before each use, and removed from service if you observe any of the following problems:

- The jaws of the beam clamp have been opened more than 15 percent of their normal opening.
- The lifting eye is bent or elongated.
- The lifting eye is worn.
- The capacity and beam size are unreadable.

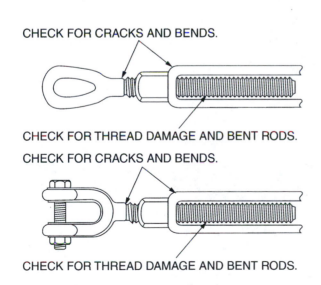

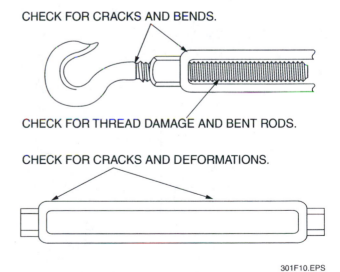

Figure 10 ◆ Turnbuckle inspection point.

2.7.0 Plate Clamps

Plate clamps (*Figure 12*) attach to structural steel plates to allow for easier rigging attachment and handling of the plate. There are two basic types of plate clamps: the serrated jaw type and the screw type.

Serrated clamps are designed to grip a single plate for hoisting and are available with a locking device. Screw clamps are considered the safest. They rely on a clamping action of a screw against the plate to secure it. Serrated clamps are used for vertical lifting, whereas screw clamps can be used from a horizontal position through 180 degrees. Plate clamps are designed to lift only one plate at a time. Always follow the manufacturer's recommendations for use and safe working load.

Remove plate clamps from service if any of the following are present during inspection:

- Changes in the opening at the jaw plate or wear of cam teeth
- Cracks in body
- Loose or damaged rivets
- Worn, bent, or elongated lifting eye
- Excessive rust or corrosion
- Unreadable capacity size

2.8.0 Rigging Plates and Links

Rigging plates and links (*Figure 13*) are made for specific uses. The holes in the plates or links may be different sizes and may be placed in different locations in the plates for different weights and types of lifts. Plates with two holes are called rigging links. Plates with three or more holes are called equalizer plates. Equalizer plates can be

(A) SCREW-ADJUSTED
CAM CLAMP

(B) BASIC NONLOCKING
CAMMING CLAMP

301F12.EPS

Figure 12 ◆ Standard lifting clamps.

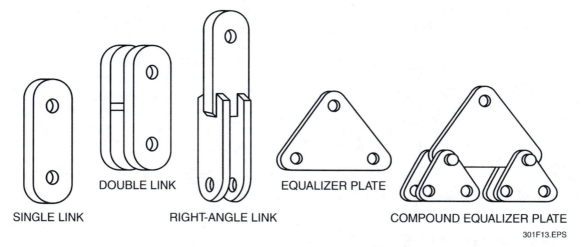

SINGLE LINK   DOUBLE LINK   RIGHT-ANGLE LINK   EQUALIZER PLATE   COMPOUND EQUALIZER PLATE

301F13.EPS

Figure 13 ◆ Rigging plates and links.

used to level loads when the legs of a sling are unequal. Plates are attached to the rigging with high-strength pins or bolts.

Inspect rigging plates and links, and remove them from service if any of the following are present:

- Cracks in body
- Worn or elongated lifting eye
- Excessive rust or corrosion

2.9.0 Spreader and Equalizer Beams

Spreader beams (*Figure 14*) are used to support long loads during lifting operations. If used properly, they help eliminate the hazard of the load tipping, sliding, or bending. They reduce low **sling angles** and the tendency of the slings to crush loads. Equalizer beams are used to balance the load on sling legs and to maintain equal loads on dual hoist lines when making tandem lifts.

Lifting beams, spreader beams, or spider frames (frames that allow the crane to lift circular or rectangular structures) all require engineering calculations. They must be used as they have been designed to be used.

Both types of beams are usually fabricated to suit a specific application. They are often made of heavy pipe, I-beams, or other suitable material. Custom-fabricated spreader or equalizer beams must be designed by an engineer and have their capacity clearly stamped on the side. They should be tested at 125 percent of rated capacity. Information on the beams should be kept on file. The capacity of beams designed for use with multiple attachment points depends upon the distance between attachment points.

Before use, a spreader or equalizer beam should be inspected for the following:

- Solid welds
- No cracks, nicks, gouges, or corrosion
- Condition of attachment points
- Capacity rating
- Sling angle tension at the points of attachment

3.0.0 ◆ SLINGS

The common types of slings include wire rope slings, synthetic slings, chain slings, and metal mesh slings. Wire rope and synthetic slings were covered in the *Core Curriculum* module, *Basic Rigging*. Metal mesh slings (*Figure 15*) are typically made of wire or chain mesh. They are similar in appearance to web slings and are suited for situations where the loads are abrasive, hot, or tend to cut other types of slings. Metal mesh slings resist abrasion and cutting, grip the load firmly without stretching, can withstand temperatures up to 550°F, are smooth, conform to irregular shapes, do not kink or tangle, and resist corrosion. These slings are available in several mesh sizes and can be coated with a variety of substances, such as rubber or plastic, to help protect the load they are handling.

3.1.0 Sling Capacity

Sling capacities depend on the sling material, construction, size of hitch configuration, quantity, and angle for the specific type of sling being used. The amount of tension on the sling is directly affected by the angle of the sling (*Figure 16*). For

301F14.EPS

Figure 14 ◆ Typical use of a spreader beam.

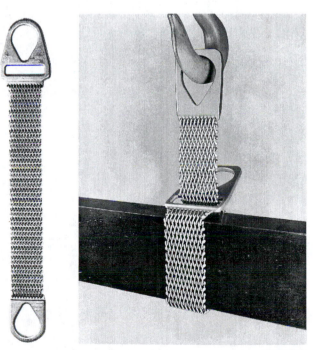

301F15.EPS

Figure 15 ◆ Metal mesh sling.

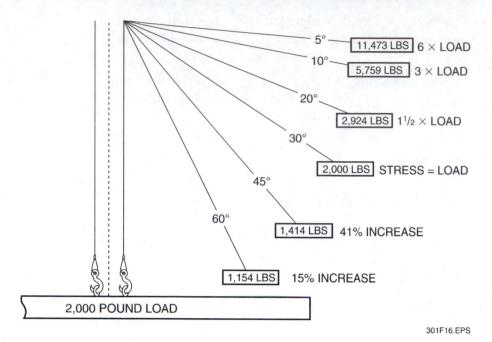

5° □11,473 LBS□ 6 × LOAD
10° □5,759 LBS□ 3 × LOAD
20°
□2,924 LBS□ 1½ × LOAD
30°
□2,000 LBS□ STRESS = LOAD
45°
60°
□1,414 LBS□ 41% INCREASE
□1,154 LBS□ 15% INCREASE
2,000 POUND LOAD

301F16.EPS

Figure 16 ◆ Sling angles.

this reason, proper sling angles are crucial to safe rigging. This information, along with other pertinent rigging information, is available from rigging equipment manufacturers and rigging trade organizations in the form of easy-to-use pocket guides like the ones shown in *Figure 17*. Sling capacity information is also given in *OSHA Regulation 29 CFR, Section 1926.251*, titled *Rigging Equipment for Material Handling*. *Table 1* shows an example of a typical capacity table used with wire rope slings.

The vertical columns in *Table 1* give, from left to right: the International Wire Rope Class; the size of the wire rope; the allowable load for a vertical lift over the eyebolt; the allowable load for a choker on the eye; allowable loads for the given angle of basket hitch; and the eye dimensions to be used.

Figure 17 ◆ Rigging pocket guides.

The D/d ratio applies to wire rope slings. It refers to the relationship between the diameter of the surface (D) the sling is around and the diameter of the sling (d). This relationship, specifically the severity of the bends in the sling, has an impact on the capacity of the sling. Efficiency charts have been developed to help determine the actual sling capacity. *Figure 18* shows load and hook diameter examples and a D/d ratio chart.

For example, consider a 1-inch-diameter rope sling in a basket hitch around a 2-inch shackle pin. The capacity of the sling in the basket hitch is listed as 17 tons. Look across the bottom of the chart for the 2/1 ratio and find the line that meets the ratio curve. It is opposite 65 percent efficiency. Multiply the rated load of 17 tons by 65 percent and you will get an actual capacity of 11 tons. Sling eyes with pin or body sizes that are at least equal to the rope diameter should be used with shackles and hooks. Always refer to the tag when selecting a wire rope sling.

3.2.0 Sling Care and Storage

You learned about inspection criteria for slings in the *Core Curriculum* module, *Basic Rigging*. The following are some important reminders for sling inspection, care, and storage:

• Store slings in a rack to keep them off the ground. The rack should be in an area free of moisture and away from acid or acid fumes and extreme heat.

Table 1 Example of a Wire Rope Sling Capacity Table

CLASS	SIZE (IN)	RATED CAPACITY - LBS*						EYE DIMENSIONS (APPROXIMATE)	
		VERTICAL	CHOKER**	BASKET HITCH				WIDTH (IN)	LENGTH (IN)
					30°	60°	90°		
6 x 19 IWRC	¼	1,120	820	2,200	2,200	1,940	1,580	2	4
	5/16	1,740	1,280	3,400	3,400	3,000	2,400	2½	5
	3/8	2,400	1,840	4,800	4,600	4,200	3,400	3	6
	7/16	3,400	2,400	6,800	6,600	5,800	4,800	3½	7
	½	4,400	3,200	8,800	8,600	7,600	6,200	4	8
	9/16	5,600	4,000	11,200	10,800	9,600	8,000	4½	9
	5/8	6,800	5,000	13,600	13,200	11,800	9,600	5	10
	¾	9,800	7,200	19,600	19,000	17,000	13,800	6	12
	7/8	13,200	9,600	26,000	26,000	22,000	18,600	7	14
	1	17,000	12,600	34,000	32,000	30,000	24,000	8	16
	1⅛	20,000	15,800	40,000	38,000	34,000	28,000	9	18
6 x 37 IWRC	1¼	26,000	19,400	52,000	50,000	46,000	36,000	10	20
	1⅜	30,000	24,000	60,000	58,000	52,000	42,000	11	22
	1½	36,000	28,000	72,000	70,000	62,000	50,000	12	24
	1⅝	42,000	32,000	84,000	82,000	72,000	60,000	13	26
	1¾	50,000	38,000	100,000	96,000	86,000	70,000	14	28
	2	64,000	48,000	128,000	124,000	110,000	90,000	16	32
	2¼	78,000	60,000	156,000	150,000	136,000	110,000	18	36
	2½	94,000	74,000	188,000	182,000	162,000	132,000	20	40

*RATED CAPACITIES FOR UNPROTECTED EYES APPLY ONLY WHEN ATTACHMENT IS MADE OVER AN OBJECT NARROWER THAN THE NATURAL WIDTH OF THE EYE AND APPLY FOR BASKET HITCHES ONLY WHEN THE D/d RATIO IS 25 OR GREATER, WHERE D=DIAMETER OF CURVATURE AROUND WHICH THE BODY OF THE SLING IS BENT, AND d=NOMINAL DIAMETER OF THE ROPE.
**SEE CHOKER HITCH RATED CAPACITY ADJUSTMENT CHART.

301T01.EPS

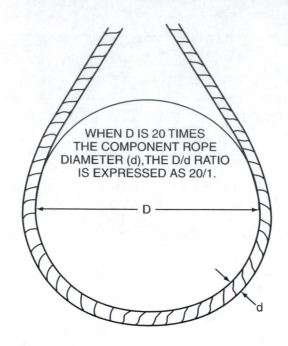

WHEN D IS 20 TIMES
THE COMPONENT ROPE
DIAMETER (d), THE D/d RATIO
IS EXPRESSED AS 20/1.

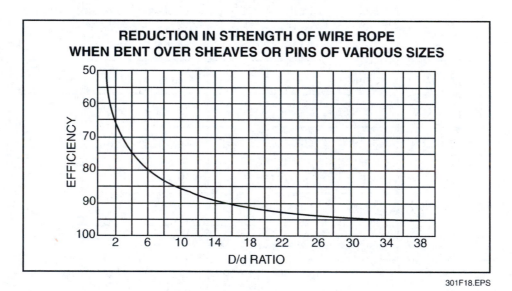

301F18.EPS

Figure 18 ◆ Load and hook diameter and D/d ratio chart for wire rope.

- Never let slings lie on the ground in areas where heavy machinery may run over them.
- Slings should be inspected at each use for broken wires, kinks, rust, or damaged fittings. Any slings found to be defective should be destroyed.

3.2.1 Wire Rope Slings

Wire rope slings should be destroyed if you discover any of the following conditions:

- *Broken wires* – Destroy a wire rope sling if you find ten randomly distributed broken wires in one rope lay, five broken wires in one strand in one rope lay, or any broken wire at any fitting.

- *Diameter reduction* – Some wear is normal; however, if the wear exceeds one-third of the original rope diameter, the sling must be destroyed. This type of wear is often caused by excessive abrasion of the outside wires, loss of core support, internal or external corrosion, inner wire failure, loosening of the rope lay, or stretching due to overloading.

- *Heat damage* – This is usually indicated by the discoloration and pitting associated with high temperature sources.

- *Corrosion* – Any noticeable rusting, pitting, or discoloration indicates corrosion in the rope or at the end fittings. Corrosion around end fittings is typically associated with broken wires near the end fittings. Corrosion can be difficult to detect when it develops internally in areas not visible from outside inspection.
- *Damaged end fittings* – This type of damage is easily detected and includes cracks, gouges, nicks, severe corrosion, and evidence that the end fitting is creasing into the rope.
- *Rope distortion* – This includes **kinking**, crushing, and **bird caging**.
- *Core protrusion* – This damage involves the core sticking out from between the strands or the strands separating, exposing the core.
- *Unlaying of a splice* – This is the unraveling of a splice.

3.2.2 Synthetic Web and Round Slings

When inspecting synthetic and round slings, look for the following conditions. If any of these appear, discard the slings immediately:

- Acid or caustic burns
- Melting or charring of any part of the sling
- Holes, cuts, tears, or snags
- Broken or worn stitching in loadbearing splices
- Excessive abrasive wear (to the point where the colored yarns are showing)
- Knots in any part of the sling
- Excessive pitting or corrosion, or cracked, distorted, or broken fittings
- Other visible damage that causes doubt as to the strength of the sling
- Missing or illegible tag

3.2.3 Metal Mesh Slings

Although metal mesh slings withstand abrasive or hot loads, they are still vulnerable to damage. Metal mesh slings should be discarded if any of the following conditions appear during inspection:

- A broken welded or brazed joint along the sling edge
- A broken wire in any part of the mesh
- A reduction of wire diameter of 25 percent due to abrasion or 15 percent due to corrosion
- Lack of flexibility due to distortion of the mesh
- Distortion of the choker fitting so that the depth of the slot is increased by 10 percent
- Distortion of either end fitting so that the width of the eye opening is decreased by more than 10 percent

- A 15 percent reduction in the original cross-sectional area of metal at any point around

3.3.0 Chain Slings

Some jobs are better suited for chain slings than for wire rope or web slings. Use of chain slings is recommended when lifting rough castings that would quickly destroy wire slings by bending the wires over rough edges. They are also used in high-heat applications or where wire chokers are not suitable, and for dredging and other marine work, because they withstand abrasion and corrosion better than cable. Information pertaining to sling angles and sling capacities described earlier for rope and web slings also apply to chain slings. *Figure 19* shows some common configurations of chain slings and hooks.

A chain link consists of two sides. The failure of either side would cause the link to open and drop the load. Wire rope is frequently composed of as many as 114 individual wires, all of which must fail before it breaks. Chains have less reserve strength and should be more carefully inspected. They must never be used for routine lifting operations.

Chains will stretch under excessive loading. This causes elongating and narrowing of the links until they bind on each other, giving visible warning. If overloading is severe, the chain will fail with less warning than a wire rope. If a weld should break, there is little, if any, warning.

WARNING!
Chains must be carefully inspected for weak or damaged links. The failure of one link can cause the entire chain to fail.

Typically, iron hoisting chains should be **annealed** every two years to relieve work hardening. Chains used as slings should be annealed every year. After being annealed six times, the chain should be destroyed. Steel chains should not be heat-treated after leaving the factory.

Note that only Grade 8 or higher chain is permitted for overhead lifting.

3.3.1 Chain Sling Storage

To store chains properly, hang them inside a building or vehicle on racks to reduce deterioration due to rust or corrosion from the weather. Never let chain slings lie on the ground in areas where heavy machinery can run over them. Be aware that some manufacturers suggest lubrication of alloy chains while in use; however,

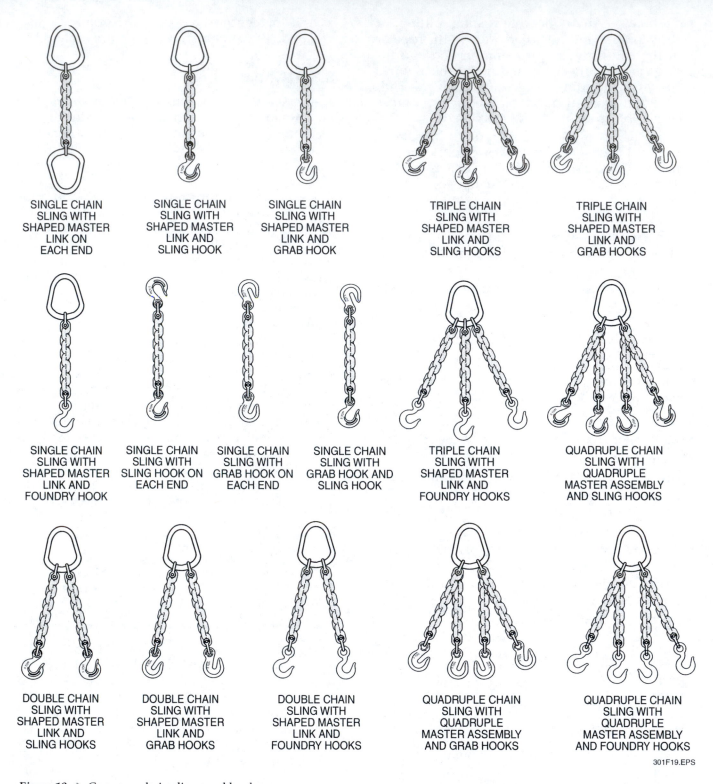

Figure 19 ◆ Common chain slings and hooks.

SINGLE CHAIN SLING WITH SHAPED MASTER LINK ON EACH END

SINGLE CHAIN SLING WITH SHAPED MASTER LINK AND SLING HOOK

SINGLE CHAIN SLING WITH SHAPED MASTER LINK AND GRAB HOOK

TRIPLE CHAIN SLING WITH SHAPED MASTER LINK AND SLING HOOKS

TRIPLE CHAIN SLING WITH SHAPED MASTER LINK AND GRAB HOOKS

SINGLE CHAIN SLING WITH SHAPED MASTER LINK AND FOUNDRY HOOK

SINGLE CHAIN SLING WITH SLING HOOK ON EACH END

SINGLE CHAIN SLING WITH GRAB HOOK ON EACH END

SINGLE CHAIN SLING WITH GRAB HOOK AND SLING HOOK

TRIPLE CHAIN SLING WITH SHAPED MASTER LINK AND FOUNDRY HOOKS

QUADRUPLE CHAIN SLING WITH QUADRUPLE MASTER ASSEMBLY AND SLING HOOKS

DOUBLE CHAIN SLING WITH SHAPED MASTER LINK AND SLING HOOKS

DOUBLE CHAIN SLING WITH SHAPED MASTER LINK AND GRAB HOOKS

DOUBLE CHAIN SLING WITH SHAPED MASTER LINK AND FOUNDRY HOOKS

QUADRUPLE CHAIN SLING WITH QUADRUPLE MASTER ASSEMBLY AND GRAB HOOKS

QUADRUPLE CHAIN SLING WITH QUADRUPLE MASTER ASSEMBLY AND FOUNDRY HOOKS

301F19.EPS

slippery chains increase handling hazards. Chains coated with oil or grease accumulate dirt and grit, which may cause abrasive wear. Coat chains to be stored in exposed areas with a film of oil or grease for rust and corrosion protection.

3.3.2 Chain Sling Care and Inspection

Chain slings should be visually inspected before every lift. Annual inspections are also required and records of inspections must be maintained. Discard chain slings if any of the following conditions are found during inspection:

- *Wear* – Any portion of the chain worn by 15 percent or more should be removed from service immediately. Wear will usually occur at the points where chain links bear on each other or on the outside of the link barrels.

- *Stretch* – Compare the chain with its rated length or with a new length of the same type chain. Any length increase means wear or stretch has occurred. If the length has increased 3 percent, further careful inspection is needed. If it is stretched more than 5 percent, it should be removed from service immediately. Significantly stretched links have an hourglass shape, and links tend to bind on each other. Be sure to check for localized stretching since a chain link can be overlooked easily.

- *Link condition* – Look for twisted, bent, cut, gouged, or nicked links.

- *Cracks* – Discard the chain if any cracks are found in any part.

- *Link welds* – Lifted fins at the weld edges signify overloading.

- *End fittings* – Check for signs of stretching, wear, twisting, bending, opening up, and corrosion.

- *Capacity tag* – Check for missing or illegible tags.

4.0.0 ◆ TAG LINES

Tag lines typically are natural fiber or synthetic rope lines used to control the swinging of the load in hoisting activities (*Figure 20*). Improper use of tag lines can turn the simplest hoisting operation into a dangerous situation. Tag lines should be used to control swinging of the load when a crane is traveling. Tag lines should also be used when

301F20.EPS

Figure 20 ◆ Use of a tag line to control swinging of a suspended load.

the crane is rotated if rotation of the load is hazardous. A non-conductive tag line is required during operation of a mobile crane within the vicinity of power lines.

When using a tag line to control swinging in a large lift, don't try to stop the swing in one movement. Pull hard when the object is swinging away from you, wait while it swings back toward you and maintain a rhythmic pull and relax until you have damped the swing. If the load starts to spin, keep a strain on the line steadily, and be careful not to let your rope get fouled or let yourself be jerked off balance.

When selecting a tag line, take certain factors into consideration. Natural fiber rope is inexpensive when compared to synthetic rope and is the most common type of rope used for tag lines. However, it is notably weaker than synthetic fiber (nylon, polyester, polypropylene, or polyethylene) ropes of the same size, and it is subject to ultraviolet deterioration and damage from heat and chemicals. Manila rope can also become a conductor of electricity when wet. Synthetic ropes are light and strong for their size, and most are resistant to chemicals. When dry, they are poor conductors of electricity, although some synthetic ropes will readily absorb water and conduct electricity. Dry polypropylene line is preferred for use as a tag line.

The diameter of a tag line should be large enough so that it can be gripped well even when wearing gloves. Rope with a diameter of ½ inch is common, but ¾-inch- and 1-inch-diameter rope are sometimes used on heavy loads or where the tag line must be extremely long. Tag lines should be of sufficient length to allow control of the load from its original lift location until it is safely placed or control is taken over by co-workers. Special consideration should be given to situations where a long tag line would interfere with safe handling of loads, such as steel erection or catalyst pours.

Tag lines should be attached to loads at a location that gives the best mechanical advantage in controlling the load. Long loads should have tag lines attached as close to the ends as possible. The tag line should be located in a place that allows personnel to remove it easily after the load is placed. Knots used to attach tag lines should be tied properly to prevent slipping or accidental loosening, but they need to be easily untied after the load is placed. Some recommended knots are the clove hitch with overhand safety and bowline. Riggers use different knots for different purposes. *Figure 21* shows several knots commonly used in rigging. As much as possible, tag lines should be of one continuous length, free of knots and splices, and **seized** on both ends. If joining two tag lines together is necessary, it should be done by splicing the lines. Knots tied in the middle or with free ends of tag lines can create difficulties.

To properly handle a tag line, determine the mechanical advantage intended by the tag line, and stand away from the load in order to have a clear

BOWLINE

ROUND TURN AND
TWO HALF-HITCHES

CLOVE HITCH

SQUARE KNOT

TIMBER HITCH

RUNNING BOWLINE

HALF-HITCH SAFETY

OVERHAND SAFETY

301F21.EPS

Figure 21 ◆ Common rigging knots.

area. When possible, stand where you can see the crane being used for lifting. Keep yourself and the tag line in view of the crane operator. Stay alert. Do not become complacent during the lift. Be aware of the location of any excess rope, and do not allow it to become fouled or entangled on anything.

Large loads often require the use of more than one tag line. In such cases, tag line personnel must work as a team and coordinate their actions.

5.0.0 ◆ BLOCK AND TACKLE

The block and tackle is the most basic lifting device. While you should know what a block and tackle is as it may be necessary for you to use one, it is a very old technology and is not often used. It is used to lift or pull light loads. A block consists of one or more sheaves or pulleys fitted into a wood or metal frame with a hook attached to the top. The tackle is the line that runs through the block and is used for lifting and pulling. Some block and tackle rigs have a brake that holds the load once it is lifted; others do not. The types that do not have a brake require continuous pull on the **hauling line**, or the hauling line must be tied off to hold the load. There are two types of block and tackle rigs: simple and compound.

A simple block and tackle consists of one sheave and a single line. It is used to lift or pull very light loads. The line hook is attached to the load, and the line is pulled by hand to lift the load. The load capacity of this type of block and tackle is equal to the capacity of the load line. The block must be attached to a building structure or other support by a method that provides adequate load capacity to support the load and the tackle. *Figure 22* shows a simple block and tackle.

A compound block and tackle uses more than one block. It has an upper, **fixed block** that is attached to the building structure or other support and a lower, movable block that is attached to the load. Each block may have one or more sheaves. The more sheaves the blocks have, the more **parts of line** the block and tackle has, and the higher the lifting capacity. The compound block and tackle is capable of handling heavier loads than the simple block and tackle and requires less effort to raise the load. *Figure 23* shows a compound block and tackle.

6.0.0 ◆ CHAIN HOISTS

A chain hoist, also called a chain fall, is another very useful and commonly used lifting device. It is used for much the same purposes as the block and tackle. Chain hoists should be used for straight, vertical lifts only. They are designed for straight lifts and may be damaged if used for angled lifts,

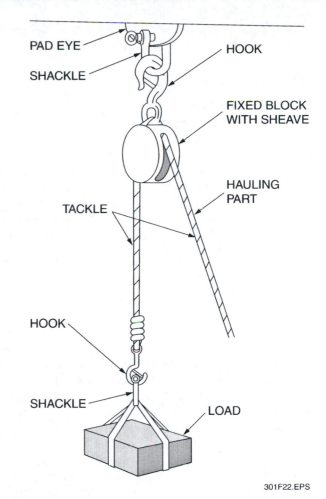

301F22.EPS

Figure 22 ◆ Simple block and tackle.

301F23.EPS

Figure 23 ◆ Compound block and tackle.

unless the load has been calculated beforehand. Always use chokers with chain hoists. The load chain of a chain hoist should never be wrapped around an object and used as a choker.

The load capacity of any chain hoist should be marked on the chain hoist. The capacity of a chain hoist must never be exceeded. Chain hoists are standard equipment in most shops and rigging departments because they are dependable, portable, and easy to use. Some common types of chain hoists are spur-geared and electric.

6.1.0 Spur-Geared Chain Hoists

The spur-geared chain hoist (*Figure 24*) has two chains. An endless chain, the hand chain, drives a single pocketed sheave, which in turn drives a gear-reduction unit. The load chain must have a mechanical brake, is fitted to the gear-reduction unit, and has a hook that attaches to the load. Roller chain is sometimes used as the load chain instead of link-type chain. The spur-geared chain hoist is the most efficient type of manual chain hoist and is most commonly used for heavier loads.

6.2.0 Electric Chain Hoists

Electric chain hoists (*Figure 25*) work much like manual chain hoists, except that they have an electric motor instead of a pull chain to raise and lower the load. Electric chain hoists are the fastest and most efficient type of chain hoist. Common electric chain hoists are available in capacities of ¼ to 5 tons, but special-application electric chain hoists are available in much higher capacities. They are controlled from a handheld push button control that is suspended on a wire from the chain hoist. Special care should be taken when raising a load with an electric chain hoist, since it is hard to tell how much force is being exerted by the electric motor. If the load gets caught on something, the chain hoist could be overloaded and damaged. The electrically driven pneumatic hoist is operated by a rotary air pump, and the motor of the hoist will stall before the chain or hoist mechanism can break.

6.3.0 Care of Chain Hoists

Like other lifting devices, chain hoists must be carefully inspected before each use. In particular, the chain must be checked to ensure that it has no defects. Follow these steps to select, inspect, use, and maintain a chain hoist:

Step 1   Select a chain hoist of adequate capacity to handle the load.

Step 2   Inspect the load chain and hook to ensure that they are not excessively worn, bent, or deformed in any way.

Step 3   Inspect the sheaves to ensure that they are not bent or excessively worn.

Step 4   Ensure that the chain hoist has proper lubrication.

301F24.EPS

Figure 24 ◆ Spur-geared chain hoist.

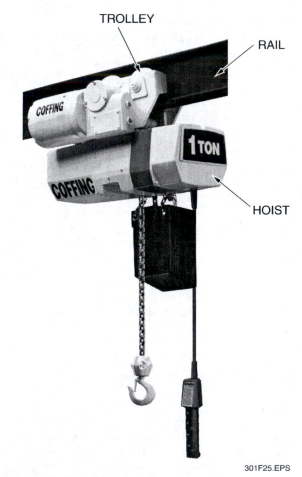

301F25.EPS

Figure 25 ◆ Electric chain hoist.

Step 5   Hang the chain hoist on a suitable support using the proper rigging.

Step 6   Lower the load hook, and connect it to the load using the proper rigging.

Step 7   Raise the load just barely off the ground, and stop to check whether the brake will hold the load. This is a much better place to find out that it won't than when the load is 20 feet up and over the site.

NOTE

To raise the load using a manual chain hoist, pull the hand chain. To raise the load using an electric chain hoist, press and hold the UP push button on the handheld control.

Step 8   Place the load.

Step 9   Disconnect the load hook from the load.

Step 9   Remove the chain hoist from its support.

Step 10   Coil the chain so that it will not get tangled.

Step 11   Store the chain hoist in its proper place.

7.0.0 ◆ RATCHET-LEVER HOISTS AND COME-ALONGS

Ratchet-lever hoists and come-alongs (*Figure 26*) are used for short pulls on heavy loads. The term come-along is widely used to identify both types, but they are not the same thing. A come-along uses a cable, whereas a ratchet-lever hoist uses a chain. Ratchet-lever hoists can be used for vertical lifts, but cable-type come-alongs can only be used for horizontal pulls. Come-alongs and ratchet-lever hoists are portable, easy to store, and easy to use. They are available in capacities ranging from ½ to 6 tons.

WARNING!

Never use a come-along for vertical lifts. They do not have the same safety braking mechanisms as the ratchet-lever hoist to prevent the load from slipping.

A ratchet-lever hoist consists of the following parts:

• *Suspension hook* – A steel hook used to hang the hoist.

COME-ALONG   RATCHET-LEVER HOIST

301F26.EPS

Figure 26 ◆ Come-along and ratchet-lever hoist.

• *Load hook* – A steel hook that attaches to the load.
• *Load chain* – The chain that attaches to the load hook and lifts the load.
• *Ratchet handle* – A handle that operates the ratchet that takes up the chain and lifts or lowers the load.
• *Fast wind handle* – A handle that takes up or lets out the chain without using the ratchet handle. It cannot be used to raise or lower the load.
• *Ratchet release* – A device that releases the ratchet so the chain can be pulled out. It also switches the device to the up or down position.

Follow these steps to select, inspect, use, and maintain a ratchet-lever hoist:

Step 1   Select a device of adequate capacity to handle the load.

Step 2   Inspect the chain and hooks to ensure that they are not excessively worn, bent, or deformed in any way.

Step 3   Inspect the device to ensure that it is not damaged in any way.

Step 4   Hang the device on a suitable support.

Step 5   Turn the ratchet release to the middle position.

Step 6   Pull the chain out enough to attach it to the load.

Step 7 Attach the load hook to the load, using the proper rigging.

Step 8 Turn the fast wind handle to take the slack out of the chain.

Step 9 Turn the ratchet release to the up position.

Step 10 Pump the ratchet handle to raise and position the load. Stop and check the brake as soon as the load clears the ground, then continue the lift.

Step 11 Turn the ratchet release to the down position.

Step 12 Pump the ratchet handle to lower the load into position and until there is slack in the chain.

Step 13 Disconnect the load hook from the load.

Step 14 Store the device in its proper place.

8.0.0 ◆ JACKS

A jack is a device used to raise or lower equipment. Jacks are also used to move heavy loads a short distance, with good control over the movement. The following are the three basic types of jacks:

- Ratchet
- Screw
- Hydraulic

8.1.0 Ratchet Jacks

The ratchet jack (*Figure 27*), also called a railroad jack, is used only to raise loads under 25 tons. It uses the lever-and-fulcrum principle. The downward stroke of the lever raises the rack bar one notch at a time. A latching mechanism, called a

301F27.EPS

Figure 27 ◆ Ratchet jack.

pawl, automatically springs into position, holding the load and releasing the lever for the next lifting stroke.

Ratchet jacks permit safe lifting, lowering, and leveling. They can lift full jack capacity on the toe or on the cap.

8.2.0 Screw Jacks

Screw jacks (*Figure 28*) are used to lift heavier loads. There are two general types of screw jacks: upright and inverted. The screw jack uses the

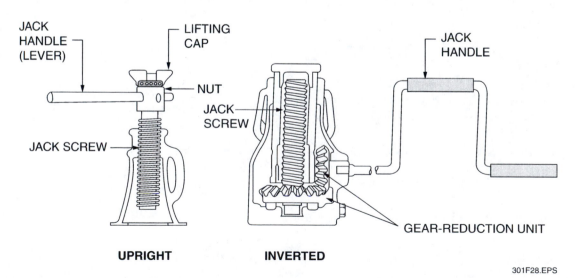

Figure 28 ◆ Upright and inverted screw jacks.

301F28.EPS

screw-and-nut principle. For lighter loads, a simple lever will apply enough power to turn the screw. For heavier loads, gear-reduction units and ratchet devices are used to increase the operator's strength. In the heaviest jacks, the screw jack is operated by an air motor for faster lifting and lowering.

8.3.0 Hydraulic Jacks

Hydraulic jacks (*Figure 29*), the most useful jacks for general purposes, are operated by the pressure of an enclosed liquid. There are two types of hydraulic jacks. One type has the pump built into the jack. The other uses a separate pump. Both jacks can lift heavy loads with little effort. The hydraulic jack with a separate pump has the following advantages:

- The craftworker can operate the pump at a safe distance from the load.
- The ram provides the greatest amount of travel for a given closed height.
- Pumps can be powered by hand, electric motor, air motor, or gasoline engine.

The hydraulic jack is not as fast as the ratchet jack, but it is excellent for lifting heavy loads.

8.4.0 Inspecting and Using Jacks

In time, a jack can become damaged or weakened and can fail under a load. To avoid such failures, all jacks should be carefully inspected before each use. Apply the following guidelines when using jacks:

- Inspect jacks before using them to ensure that they are not damaged in any way.

- Use wood softeners when jacking against metal.
- Use the proper jack handle, and remove it from the jack when it is not in use.
- Never step on a jack handle to create additional force.
- Use blocking or cribbing under the load when jacking. Never leave a jack under a load without having the load blocked up.
- Position jacks properly, and raise the load evenly to prevent the load from shifting or falling.
- Never place jacks directly on the ground when lifting. Use a solid footing.
- Position jacks so the direction of force is perpendicular to the base and the surface of the load.
- Never exceed the load capacity of the jack.
- Do not use extensions to the jack handles.
- Brace loads to prevent the jacks from tipping.
- Lash or block jacks when using them in a horizontal position to move an object.
- Never jack against rollers.
- Match ratchet jacks for uniform lifting.
- Keep fingers away from ratchet parts when raising or lowering an object.
- Thoroughly clean all hydraulic hose connectors before connecting them.

9.0.0 ◆ TUGGERS

Tuggers are actually pneumatically operated winches. They can be used for lifting and lowering or for pulling. The tugger must be securely

100 TON CAPACITY

2 TO 10 TON CAPACITY

301F29.EPS

Figure 29 ◆ Hydraulic jacks.

anchored to the floor or a steel beam by a method that will support any load that may be put on the tugger. When securing a tugger to a steel beam, use a softener on the choker. Tuggers can be located in almost any convenient location in a building, and the load line can be run through a series of blocks to the area where the lift is to be made. The blocks and all hardware must be of adequate load capacity to handle the load being lifted. The controls may be located on the tugger, or the tugger may have remote controls. Many times, the tugger is in a location where the operator cannot see the load being lifted. In this case, a flagman is needed to direct the operator, using the same hand signals that are used when flagging a crane. The tugger has a band-type brake to hold the load. *Figure 30* shows a common tugger.

Follow these steps to select, inspect, use, and maintain a tugger:

Step 1 Ensure that the tugger is of adequate capacity to handle the load.

Step 2 Ensure that an adequate air supply is connected to the tugger and is in service.

301F30.EPS

Figure 30 ◆ Tugger.

Step 3 Push the control handle toward the down position and reel off enough cable to attach to the load.

Step 4 Attach the cable to the load using the proper rigging.

 NOTE
Many times, the load is far removed from the tugger. In this case, the tugger operator will need the help of a rigger to attach the cable to the load and direct the movement of the object.

Step 5 Ensure that the cable is running properly through all blocks.

Step 6 Push the control handle toward the up position to raise the load.

Step 7 Push the control handle toward the down position to position the load at its destination.

 WARNING!
Always stand to the side of the tugger behind the safety shield while operating the tugger.

NOTE
Tuggers are variable-speed. Pushing the control handle slightly toward the up position raises the load slowly. Start slowly; then, push the handle further toward the up position to speed up. Slow down as the load approaches its destination.

Step 8 Push the control handle toward the down position to slacken the cable.

Step 9 Disconnect the load hook from the load.

Step 10 Operate the control handle to position the cable in its normal stored position.

Step 11 Disconnect air supply.

1. Mixing rigging hardware of different lifting capacities is okay, as long as the capacity of one of the hardware items exceeds the weight of the load.
   a. True
   b. False

2. The safe working load of a rigging hook is accurate only when the load is suspended from the _____ of the hook.
   a. body
   b. eye
   c. throat
   d. saddle

3. OSHA requires that a hook be replaced if the throat has opened _____ percent or more from its original size.
   a. 5
   b. 10
   c. 15
   d. 20

4. A hook that is point-loaded can carry about _____ percent of the rated load.
   a. 20
   b. 40
   c. 60
   d. 75

5. Slings are attached together using _____.
   a. spacers
   b. sling eyes
   c. shackles
   d. lifting lugs

6. All of the following are true *except* _____.
   a. shackles with pins worn more than 10 percent should be replaced
   b. the size of a shackle is determined by its pin size
   c. when using a shackle on a hook, the pin of the shackle should be hung on the hook
   d. if a shackle is pulled at an angle, its capacity is reduced

7. Tapped holes used with screwed-in eyebolts should have a minimum depth of _____ times the bolt diameter.
   a. 1¼
   b. 1½
   c. 1¾
   d. 2

8. Which of the following statements about lifting lugs is correct?
   a. They can only be used for vertical lifts.
   b. They can only be used for angled lifts.
   c. They are used primarily for vertical lifts, but some are also used for angled lifts if so designed.
   d. They are used primarily for angled lifts, but some are also designed for vertical lifts.

9. Screw-type plate clamps are used for vertical lifting of plates only.
   a. True
   b. False

10. A(n) _____ can be used to level loads when the legs of a sling are unequal.
    a. rigging link
    b. equalizer plate
    c. plate clamp
    d. beam clamp

11. Custom-fabricated spreader and equalizer beams should be tested at _____ percent of their rated capacity.
    a. 100
    b. 125
    c. 150
    d. 200

12. Long loads are supported during lifting operations with _____.
    a. spreader beams
    b. equalizer beams
    c. compound equalizer plates
    d. serrated clamps

13. One of the slings best suited for situations where the load is abrasive, hot, or tends to cut is a _____ sling.
    a. wire rope
    b. synthetic web
    c. endless grommet
    d. metal mesh

14. The term D/d ratio refers to the relationship between the sling and the _____.
    a. distance from the hook to the load
    b. diameter of the surface the sling is wrapped around
    c. distance from the load to the ground
    d. diameter of the load

15. Dredging and other marine work requires _____ slings because they withstand corrosion better.
    a. chain
    b. synthetic web
    c. wire rope
    d. mesh

16. A chain sling should be discarded if any portion of the chain is worn by _____ percent or more.
    a. 5
    b. 10
    c. 15
    d. 20

17. Which of the following statements about a compound block and tackle is correct?
    a. It has greater capacity, but requires more effort.
    b. The fixed block is attached to the load.
    c. The more sheaves the blocks have, the greater the load capacity.
    d. It is always necessary to tie off the hauling line to keep the load from falling.

18. The capacity range of come-alongs is _____ tons.
    a. ½ to 6
    b. 2 to 5
    c. 3 to 7
    d. 8 to 10

19. The ratchet jack can lift heavier loads than the screw jack or hydraulic jack.
    a. True
    b. False

20. The most useful general-purpose jack is the _____ jack.
    a. inverted screw
    b. hydraulic
    c. upright screw
    d. ratchet

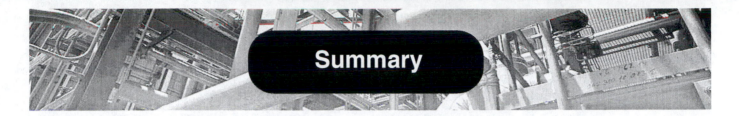

Summary

Selecting and setting up hoisting equipment, hooking cables to the load to be lifted or moved, and directing the load into position are all part of the process called rigging. Performing this process safely and efficiently requires the selection of the proper equipment for each hoisting job, an understanding of safety hazards, and knowledge of how to prevent accidents. Riggers must have the greatest respect for the hardware and tools of their trade because their lives may depend on them working correctly.

Notes

Trade Terms Introduced in This Module

Anneal: To soften a metal by heat treatment.

Bird caging: A deformation of wire rope that causes the strands or lays to separate and balloon outward like the vertical bars of a bird cage.

Equalizer beam: A beam used to distribute weight on multi-crane lifts.

Fixed block: The upper block of a block and tackle. The block that is attached to the support.

Hauling line: The line of a lifting device that is pulled by hand to raise the load.

Hydraulic: Operated by fluid pressure.

Kinking: Bending a rope so severely that the bend is permanent and individual wires or fibers are damaged.

Parts of line: The number of ropes between the upper and lower blocks of a block and tackle. These lines carry the load. Parts of line are also called falls.

Seizing: A binding to prevent rope from untwisting after it is cut.

Sling angle: The angle formed by the legs of a sling with respect to the horizontal when tension is put upon the load.

Resources & Acknowledgments

Additional Resources

This module is intended to be a thorough resource for task training. The following reference work is suggested for further study. This is optional material for continued education rather than for task training.

Occupational Safety and Health Standards for the Construction Industry, 29 CFR Part 1926. Washington, DC: OSHA Department of Labor, U.S. Government Printing Office.

Figure Credits

Topaz Publications, Inc., 301F01, 301F11, 301F14, 301F17

J.C. Renfroe & Sons, Inc., 301F12

Lift-All Company, Inc., 301F15

Cianbro Corporation, 301F20

Alan W. Grogono, M.D., 301F21 (bowline, round turn, clove hitch, square knot)

Dave Root, 301F21 (running bowline, timber hitch)

Lincoln Fire & Rescue, Lincoln, NE, 301F21 (half-hitch safety, overhand safety)

Lehman's Non-Electric Catalog, 301F23

Chester Hoist Company, 301F24

Coffing Hoists, 301F25, 301F26 (rachet-lever hoist

Lug-All, 301F26 (come-along)

Duff-Norton, 301F27, 301F29

Vernon Smith, 301F30

NCCER makes every effort to keep these textbooks up-to-date and free of technical errors. We appreciate your help in this process. If you have an idea for improving this textbook, or if you find an error, a typographical mistake, or an inaccuracy in NCCER's Contren® textbooks, please write us, using this form or a photocopy. Be sure to include the exact module number, page number, a detailed description, and the correction, if applicable. Your input will be brought to the attention of the Technical Review Committee. Thank you for your assistance.

Instructors – If you found that additional materials were necessary in order to teach this module effectively, please let us know so that we may include them in the Equipment/Materials list in the Annotated Instructor's Guide.

Write: Product Development and Revision
National Center for Construction Education and Research
3600 NW 43rd St, Bldg G, Gainesville, FL 32606

Fax: 352-334-0932

E-mail: curriculum@nccer.org

Craft _____   Module Name _____

Copyright Date _____   Module Number _____   Page Number(s) _____

Description _____
_____
_____
_____

(Optional) Correction _____
_____
_____

(Optional) Your Name and Address _____
_____
_____

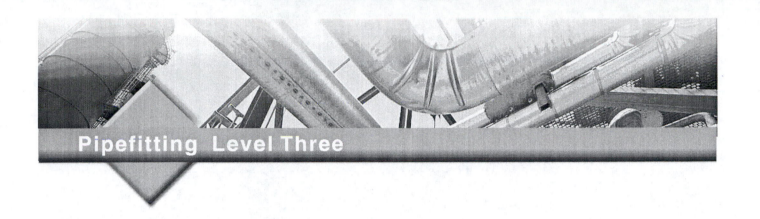

08302-07

Rigging Practices

08302-07
Rigging Practices

Topics to be presented in this module include:

Overview

Rigging practices are the ways that lifting equipment is properly used and controlled. This includes the signals used to communicate with the crane operator, the methods used to calculate equipment loads and capabilities, and the ways that different types of equipment can be used. You will learn in this module how to plan, calculate, and safely perform lifts.

Objectives

When you have completed this module, you will be able to do the following:

1. Identify and use the correct hand signals to guide a crane operator.
2. Identify basic rigging and crane safety procedures and determine the center of gravity of a load.
3. Identify the pinch points of a crane and explain how to avoid them.
4. Identify site and environmental hazards associated with rigging.
5. Properly attach rigging hardware for routine lifts and pipe lifts.
6. Identify the components of a lift plan.

Trade Terms

Anti-two-blocking devices

Blocking

Center of gravity

Cribbing

Required Trainee Materials

1. Pencil and paper
2. Appropriate personal protective equipment

Prerequisites

Before you begin this module, it is recommended that you successfully complete *Core Curriculum*; *Pipefitting Level One*, *Pipefitting Level Two*, and *Pipefitting Level Three*, Module 08301-07.

This course map shows all of the modules in the third level of the *Pipefitting* curriculum. The suggested training order begins at the bottom and proceeds up. Skill levels increase as you advance on the course map. The local Training Program Sponsor may adjust the training order.

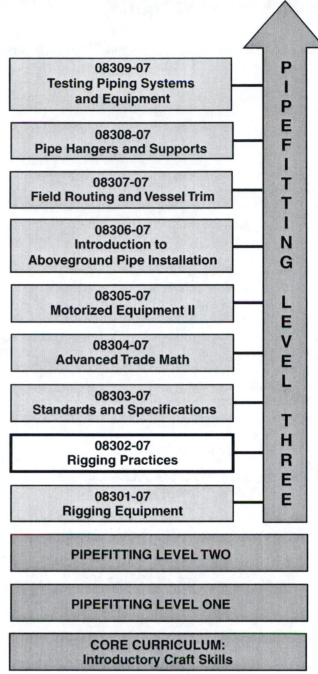

302CMAP.EPS

1.0.0 ◆ INTRODUCTION

Rigging involves the lifting and moving of heavy, and often bulky, objects. For that reason, safety must be the foremost consideration in any rigging job.

One important job that a rigger performs is guiding the crane operator. This is often done using a standard set of hand signals, which everyone involved in rigging must memorize.

Anyone doing rigging work must know how to select the right equipment, with the right capacity, for every rigging job. Failure to use the right equipment and correct rigging method can lead to serious consequences, including death or injury to workers and damage to equipment or materials.

Every crane and lifting device has capacity limits and conditions in which they are designed to operate safely. If you are not aware of these limits, you could make a fatal error. If you make the effort to learn the right and safe way to do the job, you will be successful. Always err on the side of caution.

This module is intended to instruct you on the basic safety precautions and rigging practices that will enable you to assist in rigging equipment and materials.

2.0.0 ◆ METHODS AND MODES OF COMMUNICATION

The methods and modes of communication vary widely in lifting operations. The method refers to whether the communication is verbal (spoken) or nonverbal. The mode is what is used to facilitate the communication. Modes can include, for example, a bullhorn, a radio, hand signals, or flags.

2.1.0 Verbal Modes of Communication

Verbal modes of communication vary depending on the requirements of the situation. One of the most common modes of verbal communication is a portable radio (walkie-talkie). Compact, low-power, inexpensive units enable the crane operator and signal person to communicate verbally. These units are rugged and dependable and are widely used on construction sites and in industrial plants.

There are some disadvantages, however, to using low-power and inexpensive equipment in an industrial setting. One disadvantage is interference. With low-power, inexpensive units, the frequency used to carry the signal may have many other users. This crowding of the frequency

could disrupt the signal from a more powerful unit. Another disadvantage is high background noise. In attempting to send a signal in a high-noise area, the person sending the signal may transmit unintended noise, resulting in a garbled, unintelligible signal for the receiver. On the receiving end, the individual may not be able to hear the transmission due to a high level of background noise in the cab of the crane.

There are several solutions to the problems associated with radio use. To overcome the shortcomings associated with low-power units, more expensive units with the ability to program specific frequencies and transmit at a higher power level may be needed. Some of these more expensive units may require licensing. To overcome the background noise problem, the use of an ear-mounted, noise-canceling microphone/headphone combination may be required (*Figure 1*).

Another solution is the use of an optional throat microphone. This device feeds the transmitted sound directly to the ear and picks up the voice communication from the jawbone at the ear junction. This prevents noise from entering the microphone and blocks out any background noise when listening. To avoid missed communication, the signal person's radio is usually locked in transmit so that the crane operator can tell if the unit is not transmitting. In any event, a feedback method should be established between the signal person and the crane operator so that the signal person knows the crane operator has received the signal.

Another mode of verbal communication is a hardwired system (*Figure 1*). These units overcome some of the disadvantages of radio use. When using this type of system, interference from another unit is unlikely because this system does not use a radio frequency to transmit information. Like a telephone system, occasional interference may be encountered if the wiring is not properly shielded from very strong radio transmissions. A hardwired unit is not very portable or practical when the crane is moved often. These units can also use an ear-mounted noise-canceling microphone/headphone combination to minimize the effects of background noise.

2.2.0 Nonverbal Modes of Communication

There is a wide range of nonverbal modes of communication. This is the most common type of communication used when performing crane operations. Several modes are available for use under this method. One mode is the use of signal flags, which may mean different colored flags or a specific positioning of the flags to communicate the desired message. Another mode is the use of sirens, buzzers, and whistles in which the number of repetitions and duration of the sound convey the message. The disadvantage of these two modes is that there is no established meaning to

any of the distinct signals unless they are prearranged between the sender and receiver. Also note that when sirens, buzzers, and whistles are used, background noise levels can be a problem.

The most common mode of communication reference used during crane operations is the *ASME B30.5 Consensus Standard* of hand signals (see *Figures 2* through *18*). In accordance with *ASME B30.5*, the operator must use standard hand signals unless voice communication equipment is utilized. It also requires that the hand signal chart be posted conspicuously at the job site. The advantage to using these hand signals is that they are well established and published in an industry-wide standard. This means that these hand signals are recognized by the industry as the standard hand signals to be used on all job sites. This helps ensure that there is a common core of knowledge and a universal meaning to the signals when lifting operations are conducted. Agreed-upon signals eliminate significant barriers to effective communication.

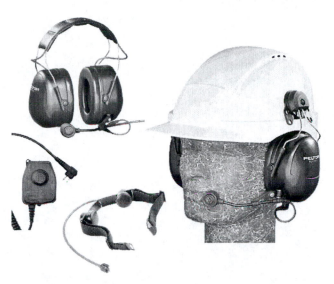

VOICE-OPERATED (VOX) OR PUSH-TO-TALK (PTT)
RADIO SYSTEM WITH THROAT MICROPHONE

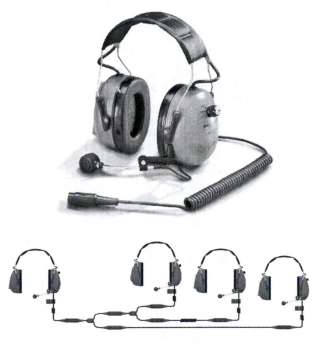

 NOTE
If it is apparent that the operator is not following the signal, immediately signal a stop.

Additions or modifications may be made for operations not covered by the illustrated hand signals, such as deployment of outriggers. The operator and signal person must agree upon these special signals before the crane is operated, and these signals should not be in conflict with any standard signal.

If you need to give instructions verbally to the operator instead of by hand signals, all crane motions must be stopped before doing so.

When a mobile crane is being moved without direction from the rigger, audible travel signals must be given using the crane's horn:

- *Stop* – One audible signal
- *Forward* – Two audible signals
- *Reverse* – Three audible signals

DUPLEX HARDWIRED SYSTEM WITH IN-LINE
AMPLIFIER AND EXTENSION CORDS FOR
SIMULTANEOUS TWO-WAY COMMUNICATION

302F01.EPS

Figure 1 ◆ Electronic communications systems.

HOIST: With forearm vertical, forefinger pointing up, move hand in small horizontal circle.

EXPECTED MACHINE MOVEMENT: The load attached to the block or ball rises vertically, accelerating and decelerating smoothly.

302F02.EPS

Figure 2 ◆ Hoist.

LOWER: With arm extended downward, forefinger pointing down, move hand in small horizontal circle.

EXPECTED MACHINE MOVEMENT: The load block or ball smoothly lowers vertically.

302F03.EPS

Figure 3 ◆ Lower.

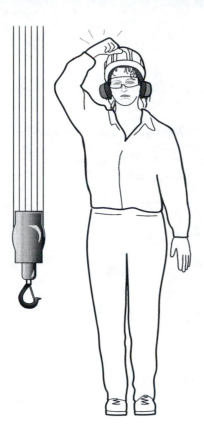

Figure 4 ◆ Use main hoist.

USE MAIN HOIST: Tap fist on head, then use regular hand signal.

EXPECTED MACHINE MOVEMENT: None. This signal is used only to inform the operator that the signal person has chosen the main hoist for the action to be performed as opposed to the auxiliary hoist.

302F04.EPS

Figure 5 ◆ Use whip line.

USE WHIP LINE (AUXILIARY HOIST): Tap elbow with open palm of one hand, then use regular hand signal.

EXPECTED MACHINE MOVEMENT: None. This signal is used only to inform the operator that the signal person has chosen the auxiliary hoist for the action to be performed as opposed to the main hoist.

302F05.EPS

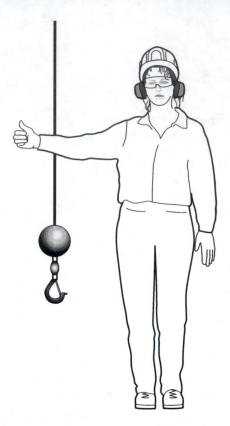

RAISE BOOM: Arm extended, fingers closed, thumb pointing upward.

EXPECTED MACHINE MOVEMENT: The boom rises, increasing the hook height and reducing the overall machine height clearance. The operating radius is slowly decreased, thus possibly increasing machine capacity and stability.

302F06.EPS

Figure 6 ◆ Raise the boom.

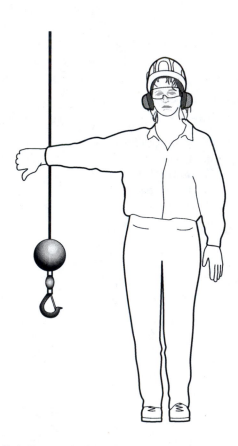

LOWER THE BOOM: Arm extended, fingers closed, thumb pointing downward.

EXPECTED MACHINE MOVEMENT: The boom will lower, decreasing the hook height and reducing the overall machine horizontal clearance. The operating radius is slowly increased, thus possibly decreasing machine capacity and stability.

302F07.EPS

Figure 7 ◆ Lower the boom.

MOVE SLOWLY: Use one hand to give any motion signal and place the other hand motionless over the hand giving the motion signal.

EXPECTED MACHINE MOVEMENT: Machine movement will vary depending on the signal being given.

302F08.EPS

Figure 8 ◆ Move slowly.

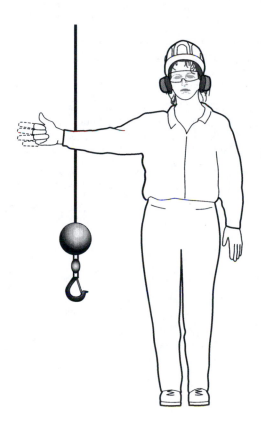

RAISE BOOM AND LOWER LOAD: Arm extended, fingers closed, thumb pointing up, flex fingers in and out as long as load movement is desired.

EXPECTED MACHINE MOVEMENT: The boom rises, reducing overall machine height clearance, as the load moves horizontally toward the crane. The operating radius is slowly decreased, thus possibly increasing machine capacity and stability.

302F09.EPS

Figure 9 ◆ Raise the boom and lower the load.

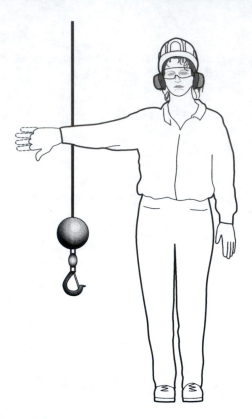

LOWER BOOM AND RAISE LOAD: Arm extended, fingers closed, thumb pointing down, flex fingers in and out as long as load movement is desired.

EXPECTED MACHINE MOVEMENT: The boom lowers, increasing overall machine height clearance, as the load moves horizontally away from the crane. The operating radius is slowly increased, thus possibly reducing both machine capacity and stability.

302F10.EPS

Figure 10 ◆ Lower the boom and raise the load.

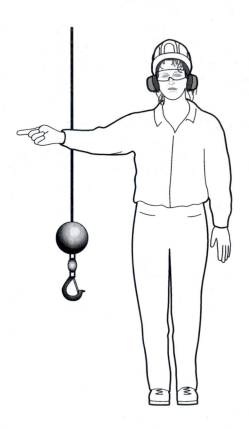

SWING: Arm extended, point with fingers in direction of boom swing. (Swing left is shown as viewed by the operator.) Use appropriate arm for desired direction.

EXPECTED MACHINE MOVEMENT: The boom moves about the center of rotation with the load (block or ball) swinging in an arc, either toward the right or left, while remaining approximately equidistant to a level plane.

302F11.EPS

Figure 11 ◆ Swing.

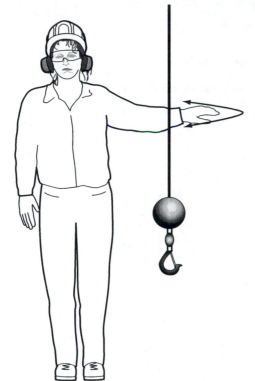

STOP: Arm extended, palm down, move arm back and forth horizontally.

EXPECTED MACHINE MOVEMENT: None. All movement of the machine ceases.

302F12.EPS

Figure 12 ◆ Stop.

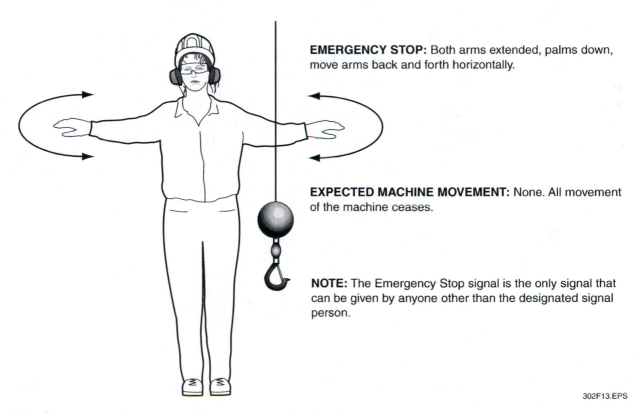

EMERGENCY STOP: Both arms extended, palms down, move arms back and forth horizontally.

EXPECTED MACHINE MOVEMENT: None. All movement of the machine ceases.

NOTE: The Emergency Stop signal is the only signal that can be given by anyone other than the designated signal person.

302F13.EPS

Figure 13 ◆ Emergency stop.

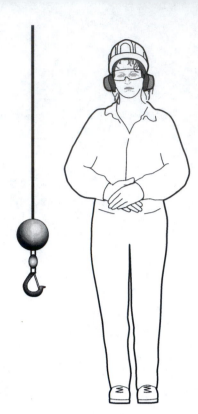

DOG EVERYTHING: Clasp hands in front of body.

EXPECTED MACHINE MOVEMENT: None.

302F14.EPS

Figure 14 ◆ Dog everything.

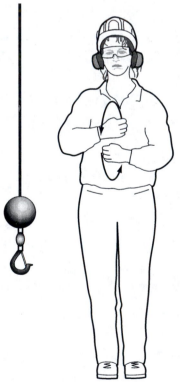

TRAVEL (BOTH TRACKS): Position both fists in front of body and move them in a circular motion, indicating the direction of travel (forward or backward).

EXPECTED MACHINE MOVEMENT: Machine travels in the direction chosen.

302F15.EPS

Figure 15 ◆ Travel (both tracks).

TRAVEL (ONE TRACK): Lock track on side of raised fist. Travel opposite track in direction indicated by circular motion of other fist, rotated vertically in front of body.

EXPECTED MACHINE MOVEMENT: Machine turns in the direction chosen.

302F16.EPS

Figure 16 ◆ Travel (one track).

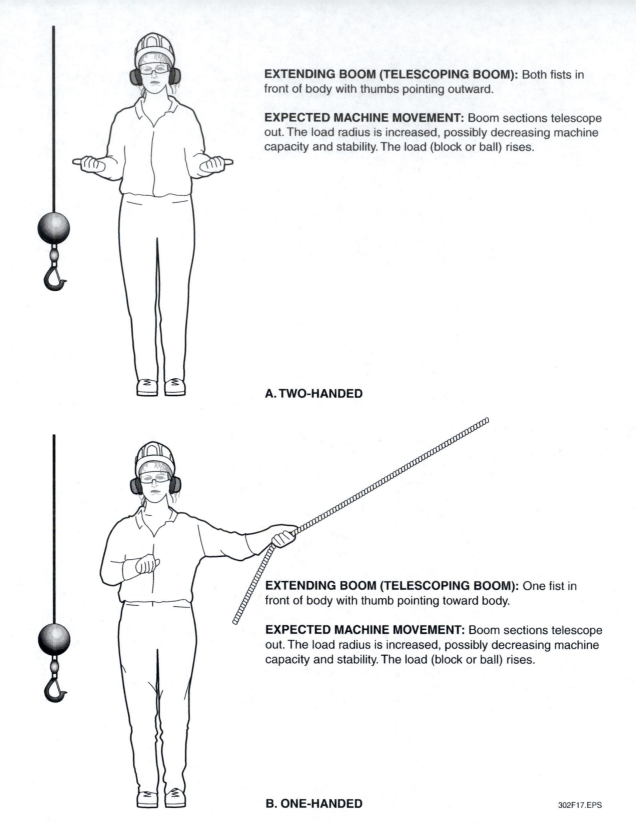

EXTENDING BOOM (TELESCOPING BOOM): Both fists in front of body with thumbs pointing outward.

EXPECTED MACHINE MOVEMENT: Boom sections telescope out. The load radius is increased, possibly decreasing machine capacity and stability. The load (block or ball) rises.

A. TWO-HANDED

EXTENDING BOOM (TELESCOPING BOOM): One fist in front of body with thumb pointing toward body.

EXPECTED MACHINE MOVEMENT: Boom sections telescope out. The load radius is increased, possibly decreasing machine capacity and stability. The load (block or ball) rises.

B. ONE-HANDED

302F17.EPS

Figure 17 ◆ Extend boom.

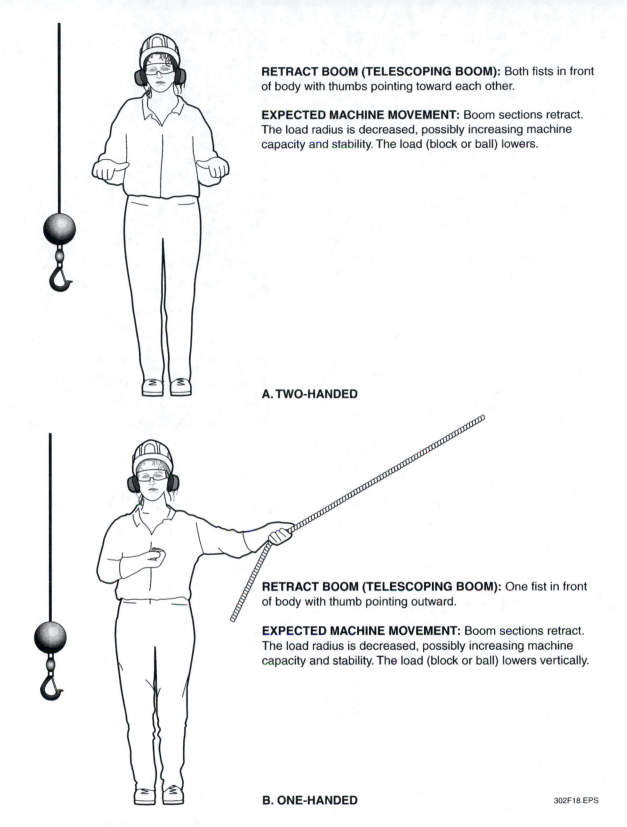

RETRACT BOOM (TELESCOPING BOOM): Both fists in front of body with thumbs pointing toward each other.

EXPECTED MACHINE MOVEMENT: Boom sections retract. The load radius is decreased, possibly increasing machine capacity and stability. The load (block or ball) lowers.

A. TWO-HANDED

RETRACT BOOM (TELESCOPING BOOM): One fist in front of body with thumb pointing outward.

EXPECTED MACHINE MOVEMENT: Boom sections retract. The load radius is decreased, possibly increasing machine capacity and stability. The load (block or ball) lowers vertically.

B. ONE-HANDED

302F18.EPS

Figure 18 ◆ Retract boom.

There are certain requirements that mandate the presence of a signal person. When the operator of the crane can't see the load, the landing area, or the path of motion, or can't judge distance, a signal person is required. A signal person is also required if the crane is operating near power lines or another crane is working in close proximity.

Not just anyone can be a signal person. Signal persons must be qualified by experience, be knowledgeable in all established communication methods, be stationed in full view of the operator, have a full view of the load path, and understand the load's intended path of travel in order to position themselves accordingly. In addition, they must wear high-visibility gloves and/or clothing, be responsible for keeping everyone out of the operating radius of the crane, and never direct the load over anyone.

NOTE

Do not give signals to the crane operator unless you have been designated as the signal person.

Although personnel involved in lifting operations are expected to understand these signals when they are given, it is acceptable for a signal person to give a verbal or nonverbal signal to an operator that is not part of the *ASME B30.5* standard. In cases where such non-standardized signals are given, it is important that both the operator and the signal person have a complete understanding of the message that is being sent.

3.0.0 ◆ GENERAL RIGGING SAFETY

When rigging, you must be aware of the unavoidable hazards associated with the work. You will be directing the movement of loads above and around other workers, where falling equipment and material can present a grave safety hazard (*Figure 19*). You may also work during extreme weather conditions where winds, slippery surfaces, and unguarded work areas exist. When working near cranes, always look up and be mindful of the hazards above and around you.

As a result of tighter regulations established by the Occupational Safety and Health Administration (OSHA), and more lost work days experienced by construction employees, individual construction trade contractors have put more stringent safety policies into place.

Safety consciousness is extremely important for the employee. The earning ability of injured employees may be reduced or eliminated for the rest of their lives. The number of employees

302F19.EPS

Figure 19 ◆ Overhead hazards.

injured can be lowered if each employee is committed to safety awareness. Full participation in the employer's safety program is your personal responsibility.

Safety consciousness is the key to reducing accidents, injuries, and deaths on the job site. Accidents can be avoided because most result from human error. Mistakes can be reduced by developing safe work habits derived from the principles of on-the-job safety.

3.1.0 Personal Protection

Workers on the job have responsibilities for their own safety and the safety of their fellow workers. Management has a responsibility to each worker to ensure that the workers who prepare and use the equipment, and who work with or around it, are well trained in operating procedures and safety practices.

Always be aware of your environment when working with cranes. Stay alert and know the location of equipment at all times when moving about and working within the job area.

Standard personal protective equipment, including hard hats, safety shoes, and barricaded work areas (*Figure 20*) are among the important safety requirements at any job site.

3.2.0 Equipment and Supervision

Your employer is responsible for ensuring that all hoisting equipment is operated by experienced, trained operators. Rigging workers must also be capable of selecting suitable rigging and lifting equipment, and directing the movement of the crane and the load to ensure the safety of all personnel. Rigging operations must be planned and supervised by competent personnel who ensure the following:

- Proper rigging equipment is available.
- Correct load ratings are available for the material and rigging equipment.
- Rigging material and equipment are well maintained and in good working condition.

A rigging supervisor is responsible for the following functions:

- Proper load rigging
- Crew supervision
- Ensuring that the rigged material and equipment meet the required capacity and are in safe condition
- Ensuring that the lifting bolts and other rigging materials and equipment are installed correctly
- Guaranteeing the safety of the rigging crew and other personnel

3.3.0 Basic Rigging Precautions

The most important rigging precaution is determining the weight of all loads before attempting to lift them. When the assessment of the weight load is difficult, safe-load indicators or weighing devices should be attached to the rigging equipment. It is equally important to rig the load so that it is stable and the **center of gravity** is below the hook.

The personal safety of riggers and hoisting operators depends on common sense. Always observe the following safety practices:

- Always read the manufacturer's literature for all equipment with which you work. This literature provides information on required start-up checks and periodic inspections, as well as inspection guidelines. Also, this literature provides information on configurations and capacities in addition to many safety precautions and restrictions of use.
- Determine the weight of loads, including rigging and hardware, before rigging. Site management must provide this information if it is not known.
- Know the safe working load of the equipment and rigging, and never exceed the limit.
- Examine all equipment and rigging before use. Discard all defective components.

302F20.EPS

Figure 20 ◆ Barricaded swing area.

- Immediately report defective equipment or hazardous conditions to the supervisor. Someone in authority must issue orders to proceed after safe conditions have been ensured.
- Stop hoisting or rigging operations when weather conditions present hazards to property, workers, or bystanders, such as when winds exceed manufacturer's specifications; when the visibility of the rigger or hoist crew is impaired by darkness, dust, fog, rain, or snow; or when the temperature is cold enough that hoist or crane steel structures could fracture upon shock or impact. Any time that temperatures are below 10°F, equipment should not be stressed heavily until it has warmed up for at least an hour, as the metal is relatively brittle.
- Recognize factors that can reduce equipment capacity. Safe working loads of all hoisting and rigging equipment are based on ideal conditions. These conditions are seldom achieved under working conditions.
- Remember that safe working loads of hoisting equipment are applicable only to freely suspended loads and plumb hoist lines. Side loads and hoist lines that are not plumb can stress equipment beyond design limits, and structural failure can occur without warning.
- Ensure that the safe working load of equipment is not exceeded if it is exposed to wind. Avoid sudden snatching, swinging, and stopping of suspended loads. Rapid acceleration and deceleration greatly increase the stress on equipment and rigging.
- Follow all manufacturer's guidelines, and consult applicable standards to stabilize mobile cranes properly (*Figure 21*).

Figure 21 ◆ Mobile crane with outriggers.

Figure 22 ◆ Using a tag line.

3.4.0 Load Path, Load Control, and Tag Lines

Physical control of the load beyond that of the ability of the crane may be required. Tag lines are used to limit the unwanted or inadvertent movement of the load as the load reacts to the motion of the crane. They are also used to allow the controlled rotation of the load for positioning (*Figure 22*). Tag lines are attached before the load is lifted. The rigger verifies that the load is balanced after it is lifted.

 WARNING!
Make sure that the tag line is not wrapped around any part of your body.

3.5.0 Barricades

Barricades should always be used to isolate the area of an overhead lift to prevent the possibility of injuring personnel who may walk into the area. Always follow the individual site requirements for proper barricade erection. If in doubt as to the proper procedure, ask your supervisor for guidance before proceeding with any overhead lifting operation. It is important to remember that accessible areas within the swing radius of the rear of the crane's rotating structure must be barricaded in such as manner as to prevent an employee or others from being struck or crushed by the crane (*Figure 23*).

SAFETY BARRIER

Figure 23 ◆ Use of a barrier to isolate the swing circle area of a crane.

3.6.0 Load-Handling Safety

The safe and effective control of the load involves the rigger's strict observance of load-handling safety requirements. This includes making sure that the swing path or load path is clear of personnel and obstructions (*Figure 24*). Keep the front and rear swing paths (*Figure 25*) of the crane clear for the duration of the lift. Most people watch the load when it is in motion, which prevents them from seeing the back end of the crane coming around.

Make sure the landing zone is clear of personnel, with the exception of the tag line tenders. Also make sure that the necessary **blocking** and **cribbing** for the load are in place before you position the load for landing. The practice of lowering the load just above the landing zone and then placing

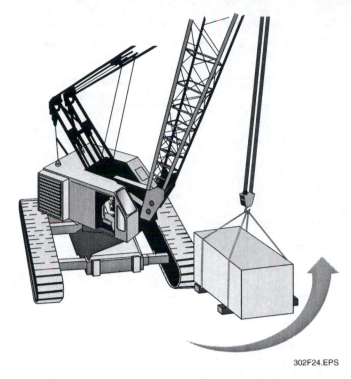

Figure 24 ◆ Front swing path.

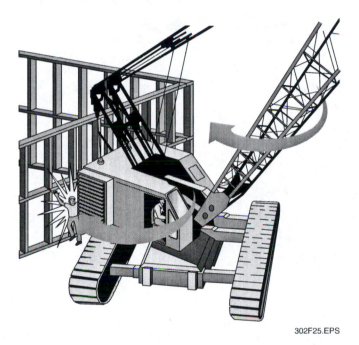

Figure 25 ◆ Rear swing path.

the cribbing and blocking can be dangerous. No one should work under the load. The layout of the cribbing can be completed in the landing zone before you set the load. Blocking of the load may have to be done after the load is set. In this case, do not take the load stress off the sling until the blocking is set and secured. Do not attempt to position the load onto the cribbing by manhandling it.

After the rigging has been set and whenever loads are to be handled, follow these procedures:

- Before lifting, make sure that all loads are securely slung and properly balanced to prevent shifting of any part.
- Use one or more tag lines to keep the load under control.
- Safely land and properly block all loads before removing the slings.
- Only use lifting beams for the purpose for which they were designed. Their weight and working load abilities must be visible on the beams.
- Never wrap hoist ropes around the load. Use only slings or other adequate lifting devices.
- Do not twist multiple-part lines around each other.
- Bring the load line over the center of gravity of the load before starting the lift.
- Make sure the rope is properly seated on the drum and in the sheaves if there has been a slack rope condition.
- Load and secure any materials and equipment being hoisted to prevent movement which, in turn, could create a hazard.
- Keep hands and feet away from pinch points as the slack is taken up.
- Wear gloves when handling wire rope.
- See that all personnel are standing clear while loads are being lifted and lowered or when slings are being drawn from beneath the load. The hooks may catch under the load and suddenly fly free. It is prohibited to pull a choker out from under a load that has been set on the choker.
- Never ride on a load that is being lifted.
- Never allow the load to be lifted above other personnel.
- Never work under a suspended load.
- Never leave a load suspended in the air when the hoisting equipment is unattended.
- Never make temporary repairs to a sling.
- Never lift loads with one or two legs of a multi-leg sling until the unused slings are secured.
- Ensure that all slings are made from the same material when using two or more slings on a load.
- Remove or secure all loose pieces from a load before it is moved.
- Lower loads onto adequate blocking to prevent damage to the slings.

4.0.0 ◆ WORKING AROUND POWER LINES

A competent signal person must be stationed at all times to warn the operator when any part of the machine or load is approaching the minimum safe distance from the power line. The signal person must be in full view at all times. *Figure 26* and *Table 1* show the minimum safe distance from power lines.

WARNING!

The most frequent cause of death of riggers and material handlers is electrocution caused by contact of the crane's boom, load lines, or load with electric power lines. To prevent personal injury or death, stay clear of electric power lines. Even though the boom guards, insulating links, or proximity warning devices may be used or required, these devices do not alter the precautions given in this section.

Table 1 High-Voltage Power Line Clearances

CRANE IN OPERATION (1)	
POWER LINE (kV)	BOOM OR MAST MINIMUM CLEARANCES (feet)
0 to 50	10
50 to 200	15
200 to 350	20
350 to 500	25
500 to 750	35
750 to 1000	45

CRANE IN TRANSIT (with no load and the boom or mast lowered) (2)	
POWER LINE (kV)	BOOM OR MAST MINIMUM CLEARANCES (feet)
0 to 0.75	4
0.75 to 50	6
50 to 345	10
345 to 7500	16
750 to 1000	20

Note 1: For voltages over 50kV, clearance increases 5 feet for every 150 kV.

Note 2: Environmental conditions such as fog, smoke, or precipitation may require increased clearances.

302T01.EPS

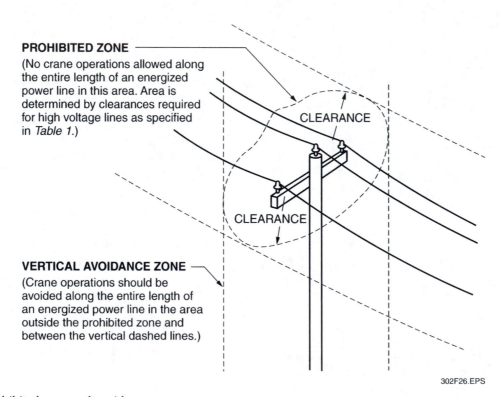

PROHIBITED ZONE
(No crane operations allowed along the entire length of an energized power line in this area. Area is determined by clearances required for high voltage lines as specified in *Table 1*.)

CLEARANCE

CLEARANCE

VERTICAL AVOIDANCE ZONE
(Crane operations should be avoided along the entire length of an energized power line in the area outside the prohibited zone and between the vertical dashed lines.)

302F26.EPS

Figure 26 ◆ Prohibited zone and avoidance zone.

The preferred working condition is to have the owner of the power lines de-energize and provide grounding of the lines that are visible to the crane operator. When that is not possible, observe the following procedures and precautions if any part of your boom can reach the power line:

- Make sure a power line awareness permit (see *Appendix*) or equivalent has been prepared.
- Erect non-conductive barricades to restrict access to the work area.
- Use tag lines of a non-conductive type if necessary for load control.
- The qualified signal person(s), whose sole responsibility is to verify that the proper clearances are established and maintained, shall be in constant contact with the crane operator.
- The person(s) responsible for the operation shall alert and warn the crane operator and all persons working around or near the crane about the hazards of electrocution or serious injury, and instruct them on how to avoid these hazards.
- All non-essential personnel shall be removed from the crane work area.
- No one shall be permitted to touch the crane or load unless the signal person indicates it is safe to do so.

If a crane or load comes in contact with or becomes entangled in power lines, assume that the power lines are energized unless the lines are visibly grounded. Any other assumption could be fatal.

The following guidelines should be followed if the crane comes in contact with an electrical power source:

- The operator should stay in the cab of the crane unless a fire occurs.
- Do not allow anyone to touch the crane or the load.
- If possible, the operator should reverse the movement of the crane to break contact with the energized power line.
- If the operator cannot stay in the cab due to fire or arcing, the operator should jump clear of the crane, landing with both feet together on the ground. Once out of the crane, the operator must take very short steps or hops with feet together until well clear of the crane.
- Call the local power authority or owner of the power line.
- Have the lines verified as secure and properly grounded within the operator's view before allowing anyone to approach the crane or the load.

5.0.0 ◆ SITE SAFETY

It takes a combined effort by everyone involved in crane operations to make sure no one is injured or killed. Site hazards and restrictions as well as crane manufacturer's requirements must be observed for safe crane operation.

WARNING!

Transmitters such as radio towers can also represent a hazard. These transmitters generate high-power radio frequency (RF) energy that can induce a hazardous voltage into the boom of a nearby crane. OSHA requires that the transmitter be de-energized in such situations, or that appropriate grounding methods be used to protect workers. Always consult your supervisor or site safety officer in such situations.

While induced RF voltages from transmitters are not an electrocution hazard like power line voltages, they can result in sparks to any object or person. The sparks can cause fires or explosions of combustible or flammable materials. They can also result in painful burns and shocks to personnel that come in close contact with the crane or the load being lifted by the crane.

5.1.0 Site Hazards and Restrictions

There are many site hazards and restrictions related to crane operations. These hazards include the following:

- Underground utilities such as gas, oil, electrical, and telephone lines; sewage and drainage piping; and underground tanks
- Electrical lines or high-frequency transmitters
- Structures such as buildings, excavations, bridges, and abutments

WARNING!

Power lines and environmental issues such as weather are common causes of injury and death during crane operations.

The operator and riggers should inspect the work area and identify hazards or restrictions that may affect the safe operation of the crane. This includes the following actions:

- Ensuring the ground can support the crane and the load

- Checking that there is a safe path to move the crane around on site
- Making sure that the crane can rotate in the required quadrants for the planned lift

The operator must follow the manufacturer's recommendations and any locally established restrictions placed on crane operations, such as traffic considerations or time restrictions for noise abatement.

6.0.0 ◆ EMERGENCY RESPONSE

Operators and riggers must react quickly and correctly to any crane malfunction or emergency situation that might arise. They must learn the proper responses to emergency situations. The first priority is to prevent injury and loss of life. The second priority is to prevent damage to equipment or surrounding structures.

6.1.0 Fire

Judgment is crucial in determining the correct response to fire. The first response is to cease crane operation and, if time permits, lower the load and secure the crane. In all cases of fire, evacuate the area even if the load cannot be lowered or the crane secured. After emergency services have been notified, a qualified individual may judge if the fire can be combated with a fire extinguisher. A fire extinguisher can be successful at fighting a small fire in its beginning stage, but a fire can get out of control very quickly. Do not be overconfident, and keep in mind that priority number one is preventing loss of life or injury to anyone. Even trained firefighters using the best equipment can be overwhelmed and injured by fires.

6.2.0 Malfunctions During Lifting Operations

Mechanical malfunctions during a lift can be very serious. If a failure causes the radius to increase unexpectedly, the crane can tip or the structure could collapse. Loads can also be dropped during a mechanical malfunction. A sudden loss of load on the crane can cause a whiplash effect that can tip the crane or cause the boom to fail.

The chance of these types of failures occurring in modern cranes is greatly reduced because of system redundancies and safety backups. However, failures do happen, so stay alert at all times.

If a mechanical problem occurs, the operator should lower the load immediately. Next, the operator should secure the crane, tag the controls out of service, and report the problem to a supervisor. The crane should not be operated until it is repaired by a qualified technician.

6.3.0 Hazardous Weather

Mobile crane operations generally take place outdoors. Under certain environmental conditions, such as extreme hot or cold weather or in high winds, work can become uncomfortable and possibly dangerous. For example, snow and rain can have a dramatic effect on the weight of the load and on ground compaction. During the winter, the tires, outriggers, and crawlers can freeze to the ground. This may lead the operator or rigger to the false conclusion that the crane is on stable ground. In fact, as weight is added during the lifting operation, it may cause an outrigger float, tire, or crawler to sink into the ground below the frozen surface. Severe rain can cause the ground under the crane to become unstable, resulting in instability of the crane due to erosion or loss of compaction of the soil. There are other specific things to be aware of when working under these adverse conditions.

6.3.1 High Winds and Lightning

High winds and lightning may cause severe problems on the job site (*Figure 27*). They are major weather hazards and must be taken seriously. Crane operators and riggers must be prepared to handle extreme weather in order to avoid accidents, injuries, and damages. It is very rare for high winds or lightning to arrive without some warning. This gives operators and riggers time to react appropriately.

High winds typically start out as less dramatic gusts. Operators and riggers must be aware of changing weather conditions, such as worsening winds, to determine when the weather becomes a hazardous situation. With high winds, the operator must secure crane operations as soon as it is practical. This involves placing the boom in the lowest possible position and securing the crane. Once this is done, all personnel should seek indoor shelter away from the crane.

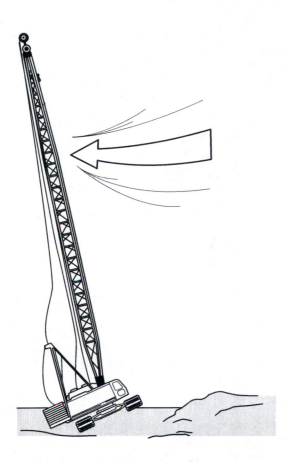

Figure 27 ◆ Wind and lightning hazards.

Because cranes' booms extend so high and are made of metal, they are likely targets for lightning. Operators and riggers must be constantly aware of this threat. Lightning can usually be detected when it is several miles away. As a general rule of thumb, when you hear thunder, the lightning associated with it is not more than 6 to 8 miles away. Be aware, however, that successive lightning strikes can touch down up to 8 miles apart. That means once you hear thunder or see lightning, it is close enough to present a hazard.

In some high-risk areas, proximity sensors provide warnings when lightning strikes within a 20-mile radius. Once a warning is given or lightning is spotted, crane operations must be secured as soon as practical, following the crane manufacturer's recommendations for doing so.

Once crane operations have been shut down, all personnel should seek indoor shelter away from the crane. Always wait a minimum of 30 minutes from the last observed instance of lightning or thunder before resuming work.

 WARNING!
The difference between the speed of light and the speed of sound means that when you see lightning, the sound wave is traveling toward you at a 1/5 mile per second. So if you see the lightning and start counting seconds, the lightning struck one mile away for every five seconds of interval from flash to thunder.

 WARNING!
Even with the boom in the lowest position, it may be taller than surrounding structures and could still be a target for lightning strikes.

6.3.2 Cold Weather

The amount of injury caused by exposure to abnormally cold temperatures depends on wind speed, length of exposure, temperature, and humidity. Freezing is increased by wind and humidity or a combination of the two factors.

Follow these guidelines to prevent injuries such as frostbite during extremely cold weather:

- Always wear the proper clothing.
- Limit your exposure as much as possible.
- Take frequent, short rest periods.
- Keep moving. Exercise fingers and toes if necessary, but do not overexert.
- Do not drink alcohol before exposure to cold. Alcohol can dull your sensitivity to cold and make you less aware of over-exposure.
- Do not expose yourself to extremely cold weather if any part of your clothing or body is wet.
- Do not smoke before exposure to cold. Smoking causes the smaller blood vessels in the arms, legs, feet, and hands to shrink, which increases the risk of frostbite.
- Learn how to recognize the symptoms of over-exposure and frostbite.
- Place cold hands under dry clothing against the body, such as in the armpits.

6.3.3 Cold Exposure Symptoms and Treatment

If you live in a place with cold weather, you will most likely be exposed to it when working. Spending long periods of time in the cold can be dangerous. It is important to know the following symptoms of cold weather exposure and how to treat them:

- Shivering
- Numbness
- Low body temperature
- Drowsiness
- Weak muscles

Follow these steps to treat cold exposure:

Step 1 Get to a warm inside area as quickly as possible.

Step 2 Remove wet or frozen clothing and anything that is binding, such as necklaces, watches, rings, and belts.

Step 3 Rewarm by adding clothing or wrapping in a blanket.

Step 4 Drink hot liquids, but do not drink alcohol.

Step 5 Check for frostbite. If frostbite is found, seek medical help immediately.

6.3.4 Symptoms and Treatment of Frostbite

Frostbite is an injury resulting from exposure to cold elements. It happens when crystals form in the fluids and underlying soft tissues of the skin. The nose, cheeks, ears, fingers, and toes are usually affected. Affected skin may be slightly flushed just before frostbite sets in. Symptoms of frostbite include the following:

- Skin becomes white, gray, or waxy yellow. The color indicates deep tissue damage. Victims are often not aware of frostbite until someone else recognizes the pale, glossy skin.
- Skin tingles and then becomes numb.
- Pain in the affected area starts and stops.
- Blisters show up on the area.
- The area of frostbite swells and feels hard.

Use the following steps to treat frostbite:

Step 1 Protect the frozen area from refreezing.

Step 2 Warm the frostbitten part as soon as possible.

Step 3 Get medical attention immediately.

6.3.5 Hot Weather

Hot weather can be as dangerous as cold weather. When someone is exposed to excessive amounts of heat, he or she runs the risk of overheating. The following conditions are associated with overheating:

- *Heat exhaustion* – Heat exhaustion is characterized by pale, clammy skin; heavy sweating with nausea and possible vomiting; a fast, weak pulse; and possible fainting.
- *Heat cramps* – Heat cramps can occur after an attack of heat exhaustion. Cramps are characterized by abdominal pain, nausea, and dizziness. The skin becomes pale with heavy sweating, muscular twitching, and severe muscle cramps.
- *Heat stroke* – Heat stroke is an immediate, life-threatening emergency that requires urgent medical attention. It is characterized by headache, nausea, and visual problems. Body temperature can reach as high as 106°F. This will be accompanied by hot, flushed, dry skin; slow, deep breathing; possible convulsions; and loss of consciousness.

Follow these guidelines when working in hot weather in order to prevent heat exhaustion, cramps, or heat stroke:

- Drink plenty of water.
- Do not overexert yourself.
- Wear lightweight clothing.

- Keep your head covered and face shaded.
- Take frequent, short work breaks.
- Rest in the shade whenever possible.

First aid for heat exhaustion involves these steps:

Step 1   Remove the victim from heat.

Step 2   Have the victim lie down and raise his or her legs six to eight inches.

Step 3   If the victim is nauseous, lay victim on his or her side.

Step 4   Loosen clothing, and remove any heavy clothing.

Step 5   Apply cool, wet cloths.

Step 6   Fan the victim, but stop if victim develops goosebumps or shivers.

Step 7   Give the victim one-half glassful of water to drink every 15 minutes if he or she is fully conscious and can tolerate it.

Step 8   If the victim's condition does not improve within a few minutes, call emergency medical services (911).

If you or anyone else experiences heat cramps, follow these guidelines:

- Sit or lie down in a cool area.
- Drink one-half glassful of water every 15 minutes.
- Gently stretch and massage cramped muscles.

First aid for heat stroke involves these steps:

Step 1   Call emergency medical services (911) immediately.

Step 2   Remove the victim from heat.

Step 3   Have the victim lie down on his or her back.

Step 4   If the victim is nauseous, lay victim on his or her side.

Step 5   Move all nearby objects, as heat stroke may cause convulsions or seizures.

Step 6   Cool the victim by fanning, spraying with cool water mist, covering with a wet sheet, or wiping with a wet cloth.

Step 7   If the victim is alert enough to do so, and not nauseous, give small amounts of cool water (a cup every 15 minutes).

Step 8   Place ice packs under the armpits and groin area.

7.0.0 ◆ USING CRANES TO LIFT PERSONNEL

Although using cranes to lift people was common in the past, OSHA regulations, as spelled out in *29 CFR 1926.550*, now discourage the practice. Using a crane to lift personnel is not specifically prohibited by OSHA, but the restrictions are such that it is only permitted in special situations where no other method is suitable. When it is allowed, certain controls must be in place, including the following:

- The rope design factor is doubled.
- No more than 50 percent of the crane's capacity, including rigging, may be used.
- Free-falling is prohibited.
- **Anti-two-blocking** devices are required on the crane boom.
- The platform must be specifically designed for lifting personnel.
- Before the personnel basket is used, it must be tested with appropriate weight, and then inspected.
- Every intended use must undergo a trial run with weights rather than people.

Figures 28 and *29* illustrate the requirements for personnel lifts.

7.1.0 Personnel Platform Loading

The personnel platform must not be loaded in excess of its rated load capacity. When a personnel platform does not have a rated load capacity, the personnel platform must not be loaded in excess of its maximum intended load.

The number of employees, along with material, occupying the personnel platform must not exceed the limit established for the platform and the rated load capacity or the maximum intended load.

Personnel platforms must be used only for employees, their tools, and the materials necessary to do their work, and must not be used to hoist only materials or tools when not hoisting personnel.

Materials and tools for use during a personnel lift must be secured to prevent displacement. These items must be evenly distributed within the confines of the platform while the platform is suspended.

7.2.0 Personnel Platform Rigging

When a wire-rope bridle sling is used to connect the personnel platform to the load line, each bridle leg shall be connected to a master link or shackle in such a manner as to ensure that the load is evenly divided among the bridle legs (*Figures 30* and *31*).

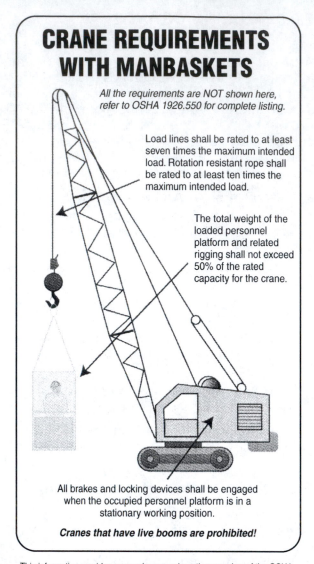

CRANE REQUIREMENTS WITH MANBASKETS

All the requirements are NOT shown here, refer to OSHA 1926.550 for complete listing.

Load lines shall be rated to at least seven times the maximum intended load. Rotation resistant rope shall be rated to at least ten times the maximum intended load.

The total weight of the loaded personnel platform and related rigging shall not exceed 50% of the rated capacity for the crane.

All brakes and locking devices shall be engaged when the occupied personnel platform is in a stationary working position.

Cranes that have live booms are prohibited!

This information provides a generic, non-exhaustive overview of the OSHA standard on suspended personnel platforms. Standards and interpretations change over time, you should always check current OSHA compliance requirements for your specific requirements.

29 CFR 1926.550 addresses the use of personnel hoisting in the construction industry, and *29 CFR 1910.180* addresses the use of personnel hoisting in general industry.

302F28.EPS

Figure 28 ◆ Crane requirements with manbaskets.

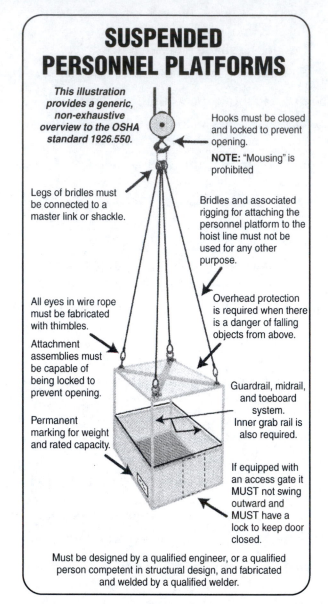

SUSPENDED PERSONNEL PLATFORMS

This illustration provides a generic, non-exhaustive overview to the OSHA standard 1926.550.

Hooks must be closed and locked to prevent opening.

NOTE: "Mousing" is prohibited

Legs of bridles must be connected to a master link or shackle.

Bridles and associated rigging for attaching the personnel platform to the hoist line must not be used for any other purpose.

All eyes in wire rope must be fabricated with thimbles.

Attachment assemblies must be capable of being locked to prevent opening.

Overhead protection is required when there is a danger of falling objects from above.

Permanent marking for weight and rated capacity.

Guardrail, midrail, and toeboard system. Inner grab rail is also required.

If equipped with an access gate it MUST not swing outward and MUST have a lock to keep door closed.

Must be designed by a qualified engineer, or a qualified person competent in structural design, and fabricated and welded by a qualified welder.

The OSHA rules on crane suspended personnel platforms contain many specifics that are not covered in this book. Refer to *29 CFR 1926.550* for the currrent OSHA compliance requirements.

302F29.EPS

Figure 29 ◆ Suspended personnel platforms.

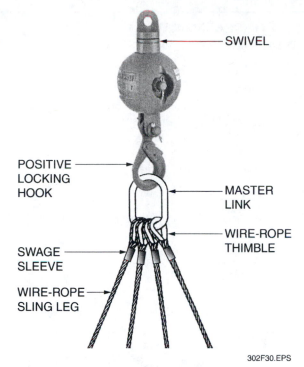

Figure 30 ◆ Bridle sling using a master link.

SWIVEL

POSITIVE LOCKING HOOK

MASTER LINK

WIRE-ROPE THIMBLE

SWAGE SLEEVE

WIRE-ROPE SLING LEG

302F30.EPS

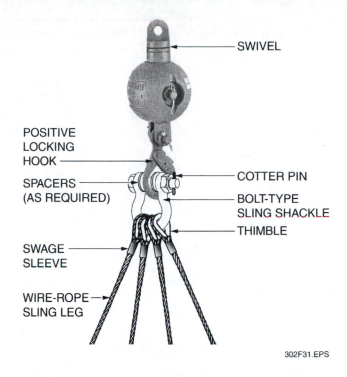

Figure 31 ◆ Bridle sling using a shackle.

SWIVEL

POSITIVE LOCKING HOOK

SPACERS (AS REQUIRED)

COTTER PIN

BOLT-TYPE SLING SHACKLE

THIMBLE

SWAGE SLEEVE

WIRE-ROPE SLING LEG

302F31.EPS

Hooks on headache ball assemblies, lower load blocks, or other attachment assemblies shall be of a type that can be closed and locked, eliminating the hook throat opening. Alternatively, an alloy anchor type shackle with a bolt, nut, and retaining pin may be used.

Wire rope, slings, shackles, rings, master links, and other rigging hardware must be capable of supporting, without failure, at least five times the maximum intended load applied or transmitted to that component. Where rotation-resistant rope is used, the slings shall be capable of supporting without failure at least ten times the maximum intended load. All eyes in wire rope slings shall be fabricated with thimbles.

Bridles and associated rigging for attaching the personnel platform to the hoist line shall be used only for the platform and the necessary employees, their tools, and the materials necessary to do their work, and they shall not be used for any other purpose when not hoisting personnel.

8.0.0 ◆ LIFT PLANNING

Before conducting a lift, most construction sites or companies require that a lift plan be completed and signed by competent personnel. Lift plans are mandatory for steel erection and multiple-crane lifts. A lift plan contains information relative to the crane, load, and rigging, and it lists any special instructions or restrictions for the lift. It is important to remember that any deviation from the original lift plan, no matter how small, requires a new lift plan.

8.1.0 Lift Plan Data

Critical lifts always require a lift plan. As a matter of policy, all personnel involved in a lift should require or prepare a plan for all lifts. The rigger is particularly concerned with crane placement, rope, and sizing of slings. A typical lift plan is shown in *Figure 32*.

PRE-LIFT CHECKLIST <u>USE FOR LIFTS EXCEEDING 75% OF CRANE'S RATED CHART CAPACITY</u> LAND BASED CRANES ONLY. CRANE APPROPRIATELY SUPPORTED AND LEVEL WITHIN 1%.

1.   Crane Type: _____

2.   Load Description:_____

3.   Operating Radius:_____ Boom Length:_____ Boom tip Elevation:_____

4.   Loaded Boom Angle:_____ Minimum Boom Length:_____

5.   Elevation of Lift:_____

6.   Crane Capacity at Working Radius: (6)_____
   What is the Maximum Radius with total load? _____

7.   Weight of Load: (7) – (_____)
                                                Subtotal _____

8.   Crane Deductions:
    a. Block Capacity   _____   Block WT. +_____
    b. Ball(s) Capacity   _____   Ball WT.   +_____
    c. Jib Effective WT   +_____
    d. Wire Rope # of parts Required: _____
    e. Other+_____   +_____
       WT/Ft. x # of ft. x parts being used ____x ____x____ = +_____

9.   Total Crane Deductions: (9) – (_____)
                                                  Subtotal _____

10.   Rigging Deductions:
    a. Slings: Qty.____   Type ____   Size ____   Cap. ____   WT. +____
                   Qty.____   Type ____   Size ____   Cap. ____   WT. +____
    b. Shackles: Qty.____   Type ____   Size ____   Cap. ____   WT. +____
                   Qty.____   Type ____   Size ____   Cap. ____   WT. +____
    c. Lifting Beam(s)   Qty.____   Cap. ____   WT. +____
    d. Softeners: Qty.____   Type ____   Size ____   WT. +____
    e. Misc. Rigging: (Snatch Blocks, Turn Buckles, etc.)   WT. +____
                                                    (11) – (_____)

11.   Total Rigging Deductions:

12.   For Total Load (add 7 + 9 + 11) = (12)_____
   To double check safety margin subtract total load # 12 from line # 6 _____
   This number and line # 13 should match.

13.   Safety Margin   (13) _____

Lifting Information

14.   # of Cranes in lift: _____ Use additional checklist for each crane.

15.   Ground conditions: _____ Use appropriate blocking mats required.   Y____ N____

16.   Weather conditions: _____ Wind speed _____ Direction _____

17.   Tag lines required: Y____ N____ Size____ Length _____ No. of lines____

18.   Designated Signal Person: Name _____ Spotter (s) Y____ N____ No. of _____

19.   Signal Method: Hand ____ Radio _____

20.   Has matching Activity Plan been done and reviewed by crew? Y____ N____

Sign – off

Crane Operator: _____   Competent Rigger: _____

Crew: _____   _____   _____

       _____   _____   _____

Superintendent: _____   Date: _____   Time: _____

302F32A.EPS

Figure 32 ◆ Lift plan (1 of 2).

Diameter (inches)	Weight (lbs/ft)	Vertical	Choker	Vertical Basket
1⅜	0.26	1.4	1.1	2.9
½	0.46	2.5	1.9	5.1
⅝	0.72	3.9	2.9	7.8
¾	1.04	5.6	4.1	11
⅞	1.42	7.6	5.6	15
1	1.85	9.8	7.2	20
1⅛	2.34	12	9.1	24
1¼	2.89	15	11	30
1⅜	3.50	18	13	36
1½	4.16	21	16	42
1⅝	4.88	24	18	49
1¾	5.67	28	21	57
1⅞	6.50	32	24	64
2	7.39	37	28	73
2⅛	8.35	40	31	80
2¼	9.36	44	35	89
2⅜	10.4	49	38	99
2½	11.6	54	42	109
2¾	14.0	65	51	130

CROSBY SCREW PIN ANCHOR SHACKLES
(G-209, S-209)

Size	Capacity (t)	Weight (lbs)
¾	4¾	3
⅞	6½	4
1	8½	5
1⅛	9½	8
1¼	12	10
1⅜	13½	14
1½	17	18
1¾	25	28
2	35	45
2½	55	86

302F32B.EPS

Figure 32 ◆ Lift plan (2 of 2).

9.0.0 ◆ CRANE LOAD CHARTS

One or more types of cranes may be used for a job site. These can include truck- or crawler-mounted lattice or hydraulic boom cranes, rough-terrain cranes, or tower cranes. In order to give or receive directions or calculate load capacities, the correct terms for the crane components must be understood.

9.1.0 Load Chart Requirements

Load chart requirements explain the critical measurements and cautions that must be considered when calculating lifting capacities. *Figure 33* is an example of a load chart. The following sections explain notes found on a typical lifting capacity chart.

MANITOWOC 4100 W-2

LIFTCRANE CAPACITIES

BOOM NO. 22C WITH OPEN THROAT TOP
146,400 LB. CRANE COUNTERWEIGHT
60,000 LB. CARBODY COUNTERWEIGHT
26'6" CRAWLERS EXTENDED

WARNING: This chart will apply only when two 12,000 lb. side ctwts. and two 30,000 lb. carbody ctwts. bear MEC registered Serial Numbers.

LIFTING CAPACITIES: Capacities for various boom lengths and operating radii may be based on percent of tipping, strength of structural components, operating speeds and other factors.

Capacities are for freely suspended loads and do not exceed 75% of a static tipping load. Capacities based on structural competence are shown by shaded areas.

Capacities are shown in pounds. Deduct 1200 pounds from capacities listed when single sheave upper boom point is attached and 1500 pounds when two sheave upper boom point is attached. To comply with B30.5 requirements, upper boompoint cannot be used on the 260 ft. boom. Weight of jib, (see chart A), all load blocks, hooks, weight ball, slings, hoist lines beneath boom and jib point sheaves, etc., is considered part of the main boom load. Boom is not to be lowered beyond radii where combined weights are greater than rated capacity. Where no capacity is shown, operation is not intended or approved.

OPERATING CONDITIONS: Machine to operate in a level position on a firm surface with crawlers fully extended and gantry in working position and be rigged in accordance with and under conditions referred to in rigging drawing No. 190693 and load line specification chard No. 6592-A.

Crane operator judgement must be used to allow for dynamic load effects of swinging, hoisting or lowering, travel, as well as adverse operating conditions & physical machine depreciation.

OPERATOR RADIUS: Operating is the horizontal distance from the axis of rotation to the center of vertical hoist line or load block with the load freely suspended. Add 14" to boom point radius for radius of sheave when using single part hoist line.

Boom angle is the angle between horizontal and centerline of boom butt and inserts and is an indication of operating radius. In all cases, operating radius shall govern capacity.

BOOM POINT ELEVATION: Boom point elevation, in feet, is the vertical distance from ground level to centerline of boom point shaft.

MACHINE EQUIPMENT: Machine equipped with 26'6" extendible crawlers, 48" treads, 17' retractable gantry, 12 part boom hoist reeving, four 1 3/8" boom pendants, 1st ctwt. 41,900 lbs., 2nd ctwt. 41,500 lbs., 3rd ctwt. 39,000 lbs., two 12,000 lbs. side ctwt's. and two 30,000 lbs. carbody ctwts.

LOAD AND WHIP LINE SPECIFICATIONS

LOAD LINE:	1-1/8" - 6 x 31 Warrington-Seale, Extra Improved Plow Steel, Regular Lay, IWRC. Minimum Breaking Strength 65 Ton. (Approx. Weight Per Ft. in Lbs. 2.34)
WHIP LINE:	1-1/8" - Warrington-Seale, Improved Plow Steel, Regular Lay, IWRC. Minimum Breaking Strength 56.5 Ton. Maximum Load - 28,300 Lbs. Per Line. (Approx. Weight Per Ft. in Lbs. 2.34)

MAXIMUM BOOM AND JIB LENGTH LIFTED UNASSISTED

OVER FRONT OF BLOCKED CRAWLERS		OVER SIDE OF EXTENDED CRAWLERS		DEDUCT FROM CAPACITIES WHEN JIB IS ATTACHED	
BOOM LENGTH	JIB NO. 123	BOOM LENGTH	JIB NO. 123	JIB LENGTH	JIB NO. 123
260'	--	260'	--	30'	3,000 lbs.
250'	--	250'	--	40'	3,600 lbs.
240'	40'	240'	40'	50'	4,200 lbs.
230'	60'	230'	60'	60'	4,900 lbs.

Load block, hook and weight ball on ground to start.

FOR JIB CAPACITIES, CONSULT JIB CHART.

HOIST REEVING FOR MAIN LOAD BLOCK

No. Parts Of Line	1	2	3	4	5	6
Max Load - Lbs.	32,500	65,000	97,500	130,000	162,500	195,000
No. Parts of Line	7	8	9	10	11	12
Max. Load - Lbs.	227,500	260,000	292,500	325,000	357,500	400,000
No. Parts of Line	13					
Max. Load - Lbs.	430,000					

BOOM LGTH FEET	OPER. RAD. FEET	BOOM ANG. DEG.	BOOM POINT ELEV.	CAPACITY: CRAWLERS EXTENDED
70	16.5	79.7	75.9	460,000
	17	79.3	75.8	400,000
	18	78.5	75.6	380,100
	19	77.6	75.4	363,000
	20	76.8	75.1	347,300
	22	75.1	74.6	319,600
	24	73.4	74.1	293,400
	26	71.7	73.5	266,100
	28	69.9	72.8	237,500
	30	68.2	72.0	214,300
	32	66.4	71.2	195,100
	34	64.6	70.2	178,900
	36	62.8	69.3	165,200
	38	60.9	68.2	153,300
	40	59.1	67.0	143,000
	45	54.1	63.7	122,100
	50	48.9	59.8	106,300
	55	43.2	54.9	93,900
	60	36.9	49.0	84,000
	65	29.4	41.3	75,800
	70	19.5	30.3	63,900

BOOM LGTH FEET	OPER. RAD. FEET	BOOM ANG. DEG.	BOOM POINT ELEV.	CAPACITY: CRAWLERS EXTENDED
80	17	80.6	85.9	392,800
	18	79.9	85.8	378,900
	19	79.2	85.6	361,800
	20	78.5	85.4	346,100
	22	77.0	84.9	318,400
	24	75.5	84.5	292,500
	26	74.0	83.9	265,600
	28	72.5	83.3	237,000
	30	71.0	82.7	213,800
	32	69.5	81.9	194,600
	34	68.0	81.2	178,500
	36	66.4	80.3	164,700
	38	64.8	79.4	152,800
	40	63.3	78.4	142,400
	45	59.2	75.7	121,500
	50	54.9	72.5	105,700
	55	50.4	68.6	93,300
	60	45.6	64.1	83,400
	65	40.3	58.8	75,200
	70	34.4	52.2	68,300
	75	27.4	43.9	52,500
	80	18.2	32.0	53,900

BOOM LGTH FEET	OPER. RAD. FEET	BOOM ANG. DEG.	BOOM POINT ELEV.	CAPACITY: CRAWLERS EXTENDED
90	18	81.1	95.9	355,400
	19	80.4	95.7	346,900
	20	79.8	95.6	336,900
	22	78.5	95.2	317,400
	24	77.2	94.7	291,700
	26	75.9	94.3	264,800
	28	74.5	93.7	236,600
	30	73.2	93.2	213,400
	32	71.9	92.5	194,200
	34	70.5	91.9	178,000
	36	69.2	91.1	164,200
	38	67.8	90.3	152,300
	40	66.4	89.5	142,000
	45	62.9	87.1	121,100
	50	59.3	84.4	105,200
	55	55.5	81.2	92,800
	60	51.5	77.5	82,900
	65	47.3	73.2	74,700
	70	42.8	68.2	67,800
	75	37.9	62.3	62,000
	80	32.4	55.2	57,000
	85	25.8	46.2	52,600
	90	17.1	33.5	45,900

BOOM LGTH FEET	OPER. RAD. FEET	BOOM ANG. DEG.	BOOM POINT ELEV.	CAPACITY: CRAWLERS EXTENDED
100	19	81.4	105.9	332,900
	20	80.8	105.7	327,100
	22	79.6	105.4	316,200
	24	78.5	105.0	290,800
	26	77.3	104.5	263,900
	28	76.1	104.1	236,200
	30	74.9	103.6	212,900
	32	73.7	103.0	193,700
	34	72.5	102.4	177,500
	36	71.3	101.7	163,700
	38	70.1	101.0	151,800
	40	68.9	100.3	141,400
	45	65.8	98.2	120,500
	50	62.6	95.8	104,700
	55	59.3	93.0	92,300
	60	55.9	89.8	82,300
	65	52.4	86.2	74,100
	70	48.7	82.1	67,200
	75	44.8	77.4	61,400
	80	40.5	72.0	56,400
	85	35.9	65.6	52,000
	90	30.7	58.0	48,200
	95	24.5	48.5	44,900
	100	16.3	35.0	39,300

CAUTION! CHECK AMOUNT OF COUNTERWEIGHT ON MACHINE BEFORE USE OF THIS CHART

© Manitowoc 1997

302F33.EPS

Figure 33 ◆ Load chart.

9.1.1 Operating Conditions

Operating conditions are the conditions that affect the machine lifting capacities. When the machine is operated in an unlevel condition, the boom is subjected to high side loads that can cause boom damage or tipping. The machine should be leveled to within one percent of grade. To check the level of the machine, place a spirit level on the roller path or use the cross level provided in or near the operator's cab. *Figure 34* shows a crane that is not level and a crane that is level.

Every attempt to verify soil conditions must be made before making a lift. If conditions cannot be verified, a support mat must be installed on the area under the crane. Crane operators must use judgment to allow for the effects of increased load caused by swinging, hoisting, lowering, traveling, adverse operating conditions, and machine wear.

9.1.2 Operating Radius

Load charts require the measurement of the operating radius. The operating radius is the horizontal distance, in feet, from the center line of rotation to the center of the vertical hoist line or load block. This measurement is taken with the load freely suspended.

9.1.3 Boom Angle and Length

The boom angle is the angle between the horizontal plane and the center line of the boom base, or boom butt, and inserts. Boom length is measured from the boom pivot point to the boom point sheave pin. These measurements are shown in *Figure 35*.

9.1.4 Boom Point Elevation

The boom point elevation is the vertical distance from ground level to the center line of the main boom point sheave (*Figure 35*). This distance is measured in feet and changes when the boom angle changes.

9.1.5 Hoist Reeving

The load chart instructions for hoist reeving the main load block must be followed to ensure that the proper parts of line are used. The maximum load of the reeving should be greater than or equal to the maximum capacity that will be used. Use the corresponding parts of line for the maximum load selected. Parts of line are based on the catalogue breaking strength of the rope, with a recommended safety factor. *Figure 36* shows a typical

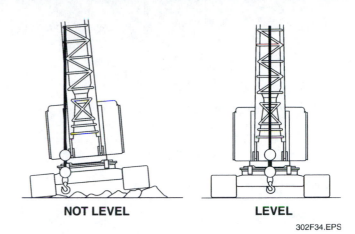

NOT LEVEL   **LEVEL**

302F34.EPS

Figure 34 ◆ Unlevel and level cranes.

chart for hoist reeving for the main load block taken from the example crane load chart. *Figure 37* shows an example of ten-part reeving.

9.1.6 Load and Whip Line Specifications

The load line and whip line are parts of the machine equipment. Their strength requirements must be met before the capacity chart can be used. The maximum whip line capacity for a given operating radius is limited by the maximum load per line noted in the specifications. The load cannot exceed the capacities shown for the radius on the main boom capacity chart. The wire rope must meet the recommended requirements.

9.1.7 Maximum Boom and Jib Lengths Lifted Unassisted

The crane load chart also lists the maximum boom and the boom and jib lengths that can be lifted unassisted. Do not attempt to raise a longer boom or jib than is allowed under the conditions specified on the chart.

9.1.8 Deductions from Capacities

The weight of certain crane components must be deducted from the load charts to determine the maximum load capacity of the crane. These are sometimes referred to as above-the-hook deductions. The deductions for a typical lattice boom crane are shown in *Figure 38*. These deductions are furnished as part of the load charts. Other deductions, sometimes called below-the-hook deductions, consist of the weight of all rigging equipment including spreader beams, cables, slings rings, and clamps, exclusive of the load.

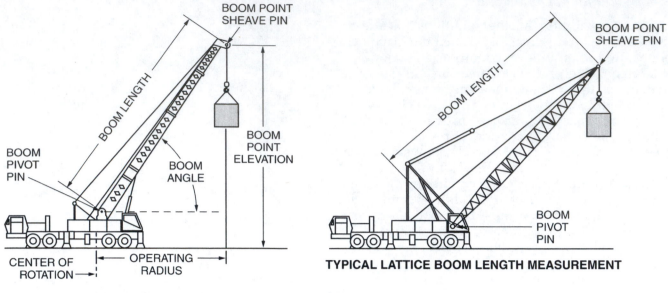

TYPICAL LOAD CHART DIMENSIONS

TYPICAL LATTICE BOOM LENGTH MEASUREMENT

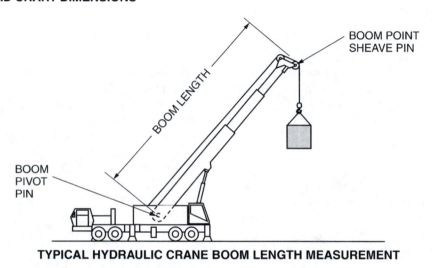

TYPICAL HYDRAULIC CRANE BOOM LENGTH MEASUREMENT

302F35.EPS

Figure 35 ◆ Boom length, boom angle, load operating radius, and boom point elevation measurements.

HOIST REEVING FOR MAIN LOAD BLOCK						
No. Parts of Line	1	2	3	4	5	6
Max Load - Lbs.	32,500	65,000	97,500	130,000	162,500	195,000
No. Parts of Line	7	8	9	10	11	12
Max. Load - Lbs.	227,500	260,000	292,500	325,000	357,500	400,000
No. Parts of Line	13					
Max. Load - Lbs.	430,000					

302F36.EPS

Figure 36 ◆ Hoist reeving chart.

10-PART REEVING

302F37.EPS

Figure 37 ◆ Heavy-lift crane with ten-part reeving.

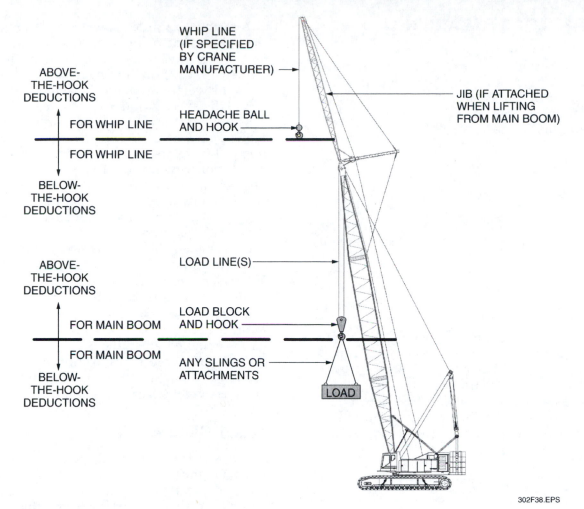

Figure 38 ◆ Typical deductions for a lattice boom crane.

The lifting capacity of the machine changes when using different styles of boom tips or when adding a jib. A jib deduction chart lists the number of pounds to be deducted from the load chart when the jib is attached and not being used. The deduction depends on the length of the jib.

The jib offset angle must be determined before using the jib load chart. The jib offset angle is the angle from the center line of the boom top section to the center line of the jib. Make sure the chart being used is for the desired jib offset angle. *Figure 39* shows the jib offset angle.

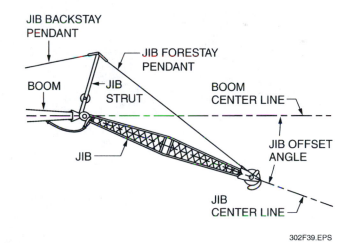

Figure 39 ◆ Jib offset angle.

10.0.0 ◆ LOAD BALANCING

Load balancing is equalizing the weight of the load at various lifting points. If the weight is concentrated at one point, the object being lifted may bend, break, or fall. Distributing the weight over several lifting points ensures that the load is balanced and lifted safely.

10.1.0 Center of Gravity

The crane's center of gravity is the point around which its weight is concentrated. The location of a crane's center of gravity in relation to its tipping fulcrum directly affects its leverage and stability.

To understand the relationship between a crane's center of gravity and stability, you must understand the principle of leverage. The principle of leverage can be illustrated with a simple teeter-totter (*Figure 40*). When the weight of the heavy load multiplied by the distance (X) from its center of gravity to the tipping fulcrum is equal to the weight of the lighter load multiplied by the distance (Y) from its center of gravity at its tipping fulcrum, a condition of balance has occurred.

This relationship is illustrated again with the long lever bent upwards in *Figure 41*.

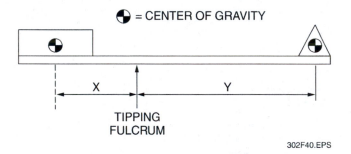

Figure 40 ◆ Teeter-totter demonstrating leverage.

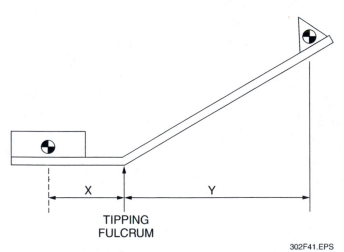

Figure 41 ◆ Bent lever.

A further example shows the relationship with the lighter load being suspended below the long lever in *Figure 42* like a crane. Note that in all three figures, the loads will remain balanced as long as the load weights remain the same and the horizontal distances X and Y remain the same.

The weight and center of gravity of various mobile crane members are determined by calculation or by weighing. This data then becomes the basis upon which stability ratings are calculated. The ratings and accuracy of the weight and center-of-gravity data are then confirmed by testing.

The center of gravity can change depending on the crane's quadrant of operation. When the crane is in over-the-side, over-the-front, or over-the-rear configurations, the center of gravity is shifted either closer or farther away from the tipping fulcrum. This, in turn, may increase or decrease the crane's leverage. For a rough-terrain crane, maximum true lifting capacity (greatest stability margin) is obtained when the crane boom is operated directly over the front. Maximum true lifting capacity of a truck crane is obtained when the boom is over the rear.

10.2.0 Sling Angles

The angle formed by the legs of a sling with respect to the horizontal when tension is put upon the load is called the sling angle. The amount of tension on the sling is directly affected by the angle of the sling. For this reason, proper sling angles are crucial to safe rigging. *Figure 43* shows the effect of sling angles on sling loading. To actually determine the load on a sling, a factor table is used. *Table 2* shows a sling factor table.

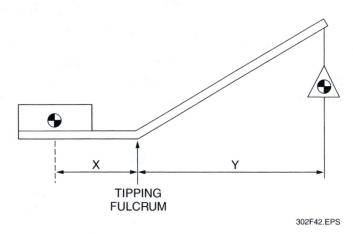

Figure 42 ◆ Crane-shaped lever.

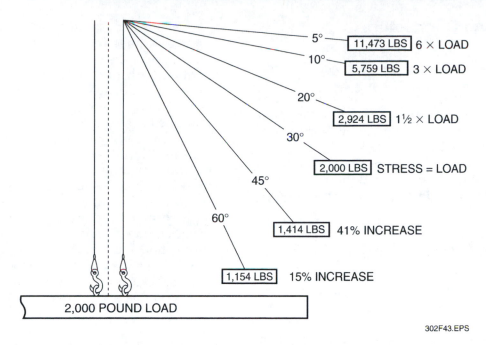

Figure 43 ◆ Sling angles.

Table 2 Sling Factor Table

SLING ANGLE	LOAD ANGLE FACTOR
5°	11.490
10°	5.747
15°	3.861
20°	2.924
25°	2.364
30°	2.000
35°	1.742
40°	1.555
45°	1.414
50°	1.305
55°	1.221
60°	1.155
65°	1.104
70°	1.064
75°	1.035
80°	1.015
85°	1.004
90°	1.000

302T02.EPS

In the example shown in *Figure 43*, two slings are being used to lift 2,000 pounds. When the slings are at a 45-degree angle when measured from the horizontal (90 degrees between the legs), there are 1,414 pounds of tension on each sling. This can be determined mathematically by taking the weight of the load (2,000 pounds) and dividing it by the number of slings (2); this equals 1,000 pounds. Then, multiply 1,000 pounds by the corresponding factor in *Table 2* for 45 degrees (1.414). The result is 1,414 pounds of tension on each sling based on the sling angle.

Optimum sling angles fall between 90 degrees and 60 degrees to horizontal (0 degrees and up to 60 degrees between sling legs). Further examination of *Figure 43* shows that the tension on the sling legs is higher when the legs are positioned at an angle of 30 degrees relative to the horizontal (120-degree angle between the legs) than when the legs are at an angle of 60 degrees relative to the horizontal (60-degree angle between the legs). Angles beyond this range are considered hazardous.

10.3.0 Lifting Connectors

Many pieces of equipment and machinery come with lifting connectors already attached at proper locations for balanced lifting. Lifting connectors may be manufactured or field-fabricated only if the lifting connectors have been engineered. Field-fabricated connectors must be fabricated of approved material and installed using approved methods to ensure that they are of adequate capacity for the load being lifted. Some common lifting connectors are lifting lugs and lifting eyebolts.

10.3.1 Lifting Lugs

Lifting lugs are welded, bolted, or pinned to the object to be lifted. They are designed and located to balance the load and support it safely. Unless angular lifts are specifically allowed by the manufacturer, lifting lugs should be used for straight, vertical lifting only, because they will bend if loaded at an angle. *Figure 44* shows two types of lifting lugs.

10.3.2 Lifting Eyebolts

Lifting eyebolts are also used to balance the weight of the load. There are three types of lifting eyebolts: the plain, or shoulderless, type; the forged alloy steel eyebolt with shoulders or collars; and the swivel eyebolt. The shoulderless eyebolt is used for straight, vertical lifts only, because it will bend when loaded at an angle.

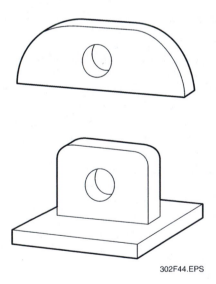

302F44.EPS

Figure 44 ◆ Lifting lugs.

Eyebolts with shoulders can be used for straight or angled lifts. The shoulder of the bolt must be tight against the surface of the load and make full contact with the surface. Washers can be used under shoulder eyebolts to ensure that tight contact is made. The eyebolts must be aligned so that force is applied to the plane of the eye to prevent the eyebolt from bending during an angled pull. One type of shouldered eyebolt, called a ringbolt, is equipped with an additional lifting ring. The safe working load of shouldered eyebolts and ringbolts is reduced with angular loading. Swivel eyebolts allow lifting from any angle and do not require load derating. *Figure 45* shows eyebolt and ringbolt installation and lifting criteria.

When lifting an object using eyebolts, always use a shackle for each eyebolt. Never use a hook in an eyebolt, because a hook may bend or disfigure the eye.

11.0.0 ◆ RIGGING PIPE

The following sections explain how to determine the weight of a load of pipe, and how to block, choke, lift, and balance the load. Guidelines are also given for landing a load of pipe.

11.1.0 Determining the Weight of the Pipe

The weight of the pipe must be determined before it is lifted. Every foot of pipe weighs a given amount, depending on the wall thickness and nominal size of the pipe. Therefore, if you know the weight per foot and the total number of feet, the total weight of the lift can be determined. *Table 3* lists the weights of carbon steel pipe.

The numbers in the far left column represent nominal pipe size in inches. The designations and numbers across the top represent wall thickness, or schedule. To use the table, find the point at which the nominal size and wall thickness of the pipe being used meet. The number in this block is the weight of 1 foot of pipe. To find the total weight, multiply the total number of feet by the weight per foot.

For example, to find the weight of a 20-inch schedule 40 carbon steel pipe that is 10 feet long, read across the table from the 20 in the nominal pipe size column until you reach the schedule 40 column. The number you find there is 123.1. This is the weight, in pounds, of 1 foot of pipe. Multiply this number by 10 to find the total weight of the pipe. The pipe weighs 1,231 pounds.

USE OF EYEBOLTS

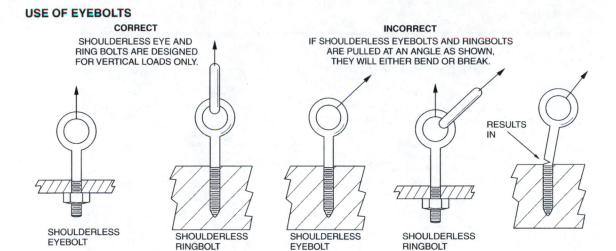

CORRECT
SHOULDERLESS EYE AND
RING BOLTS ARE DESIGNED
FOR VERTICAL LOADS ONLY.

INCORRECT
IF SHOULDERLESS EYEBOLTS AND RINGBOLTS
ARE PULLED AT AN ANGLE AS SHOWN,
THEY WILL EITHER BEND OR BREAK.

RESULTS
IN

SHOULDERLESS
EYEBOLT

SHOULDERLESS
RINGBOLT

SHOULDERLESS
EYEBOLT

SHOULDERLESS
RINGBOLT

USE OF SHOULDER TYPE EYEBOLTS AND RINGBOLTS

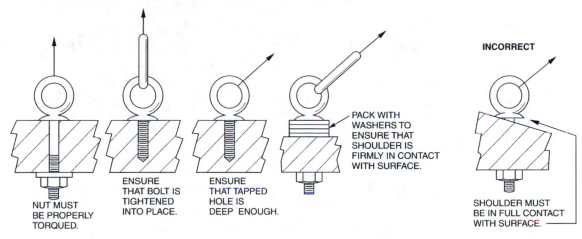

CORRECT
FOR SHOULDER TYPE EYEBOLTS AND RINGBOLTS
PROVIDING LOADS ARE REDUCED TO ACCOUNT FOR ANGULAR LOADING.

INCORRECT

PACK WITH
WASHERS TO
ENSURE THAT
SHOULDER IS
FIRMLY IN CONTACT
WITH SURFACE.

NUT MUST
BE PROPERLY
TORQUED.

ENSURE
THAT BOLT IS
TIGHTENED
INTO PLACE.

ENSURE
THAT TAPPED
HOLE IS
DEEP ENOUGH.

SHOULDER MUST
BE IN FULL CONTACT
WITH SURFACE.

ORIENTATION OF EYEBOLTS

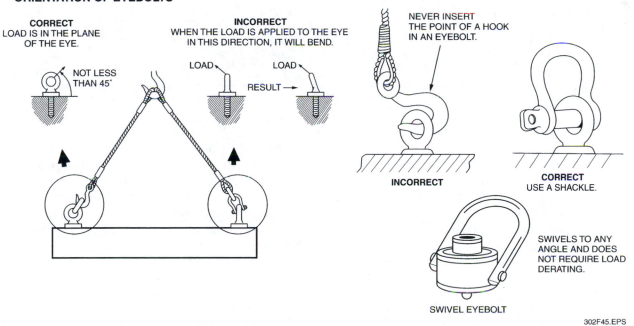

CORRECT
LOAD IS IN THE PLANE
OF THE EYE.

NOT LESS
THAN 45°

INCORRECT
WHEN THE LOAD IS APPLIED TO THE EYE
IN THIS DIRECTION, IT WILL BEND.

LOAD

LOAD

RESULT →

NEVER INSERT
THE POINT OF A HOOK
IN AN EYEBOLT.

INCORRECT

CORRECT
USE A SHACKLE.

SWIVELS TO ANY
ANGLE AND DOES
NOT REQUIRE LOAD
DERATING.

SWIVEL EYEBOLT

302F45.EPS

Figure 45 ◆ Eyebolt and ringbolt installation and lifting criteria.

Table 3 Carbon Steel Pipe Weights

Nominal Pipe Size (inches)	Wall Thickness								
	STD	XS	XXS	10	40	60	80	120	160
	Weight Per Foot in Pounds								
2.0	3.65	5.02	9.03	–	3.65	–	5.02	–	7.01
2.5	5.79	7.66	13.7	–	5.79	–	7.66	–	10.01
3.0	7.58	10.25	15.85	–	7.58	–	10.25	–	14.31
3.5	9.11	12.51	22.85	–	9.11	–	12.51	–	–
4.0	10.79	14.98	27.54	–	10.79	–	14.98	18.98	22.52
6.0	18.97	28.57	53.16	–	18.97	–	28.57	36.42	45.34
8.0	28.55	43.39	72.42	–	28.55	35.66	43.39	60.69	74.71
10.0	40.48	54.74	104.1	–	40.48	54.74	64.40	89.27	115.7
12.0	49.56	65.42	125.5	–	53.56	73.22	88.57	125.5	160.3
14.0	54.57	72.09	–	36.71	63.37	85.01	106.1	150.5	189.2
16.0	62.58	82.77	–	42.05	82.77	107.5	136.6	192.4	245.2
18.0	70.59	93.45	–	47.39	104.8	138.2	170.8	244.1	308.6
20.0	78.60	104.1	–	52.73	123.1	166.5	208.9	296.4	379.1
22.0	86.61	114.8	–	58.07	–	197.4	250.8	353.6	451.1
24.0	94.62	125.5	–	63.41	171.2	238.3	296.5	429.5	542.1
26.0	102.6	136.2	–	85.73	–	–	–	–	–
28.0	110.6	146.9	–	92.41	–	–	–	–	–
30.0	118.7	157.5	–	99.08	–	–	–	–	–
32.0	126.7	168.2	–	105.8	229.9	–	–	–	–
34.0	134.7	178.9	–	112.4	244.6	–	–	–	–
36.0	142.7	189.6	–	119.1	282.4	–	–	–	–
42.0	166.7	221.6	–	–	330.4	–	–	–	–

302T03.EPS

11.2.0 Blocking

Everything on the job site, including structural steel beams, machinery, and pipe, must be placed on blocking or dunnage. Blocks should be hardwood, preferably oak. They should be thick enough to protect the object from contact with the ground and to enable a rigger to slip a choker beneath the load.

Pipe placed on blocks requires wedges to keep the pipe from rolling off the blocking. Large diameter pipe requires chocks to keep it from rolling. The general rule regarding chocks is that there should be 1 inch of chock per 1 foot of diameter. Thus, a 42-inch pipe requires about a 3½-inch-thick chock.

11.3.0 Choking

When an item more than 12 feet long is being rigged, the general rule is to use two choker hitches spaced far enough apart to provide the stability required to transport the load (*Figure 46*).

To lift a bundle of loose items, or to maintain the load in a certain position during transport, a double-wrap choker hitch (*Figure 47*) may be useful. The double-wrap choker hitch is made by wrapping the sling completely around the load, and then wrapping the choke end around again and passing it through the eye like a conventional choker hitch. This enables the load weight to produce a constricting action that binds the load into the middle of the hitch, holding it firmly in place throughout the lift.

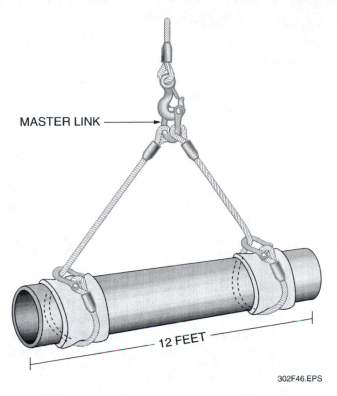

MASTER LINK

12 FEET

302F46.EPS

Figure 46 ◆ Double choker hitch.

Forcing the choke down will drastically increase the stress placed on the sling at the choke point. To gain gripping power, use a double-wrap choker hitch. The double-wrap choker uses the load weight to provide the constricting force, so there is no need to force the sling down into a tighter choke.

A double-wrap choker hitch is ideal for lifting bundles of items, such as pipes and structural steel. It will also keep the load in a certain position, which makes it ideal for equipment installation lifts. Lifting a load longer than 12 feet requires two of these hitches.

Use the following guidelines when choking a load of pipe:

- Pipe of the same length should be lifted together. If pipes of greatly different lengths must be lifted, lift each different length separately.
- Lay the pipe on blocks so the chokers can be wrapped around the load.
- Use two choker hitches to lift a load of pipe. This allows a level lift to be made and ensures that the pipe will not slip out of the sling. Make sure both chokers grab all pipes.

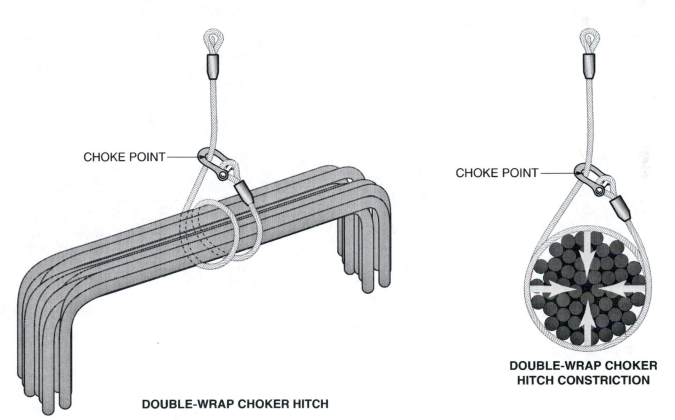

CHOKE POINT

CHOKE POINT

DOUBLE-WRAP CHOKER HITCH

DOUBLE-WRAP CHOKER HITCH CONSTRICTION

302F47.EPS

Figure 47 ◆ Double-wrap choker hitch and double-wrap choker hitch constriction.

- Ensure that the open part of the hook is facing away from center when using slings with fixed or sliding hooks.
- Do not use carbon steel chokers on painted or stainless steel pipe because of cross-contamination of the metals. Use nylon web slings on stainless steel pipe.

11.4.0 Lifting

Follow these guidelines when lifting a load of pipe:

- Use the hook and lifting line of the crane as the center line with which to line up the load.
- Be sure to check the boom angle of the crane to make sure that the load will not be moved forward or backward when tension is put on the line.
- Stand clear of the load and where the operator can clearly see you, and signal any necessary adjustments, using the proper hand signals. Never stand underneath the load.
- As soon as the load clears the ground, check its orientation. If it is not level, signal the operator to stop lifting the load, and guide the load back onto the blocks to correct the position of the chokers.
- Handle all loads with a tag line.
- Keep the load as low to the ground as possible.

11.5.0 Landing

Follow these guidelines when landing the load:

- Store all pipe on level ground.
- Land the load on blocks that are long enough and thick enough to support the load.
- Stand to one side of the load, and guide it onto the blocks. Move to the end of the load before the rigging is released.
- Ensure that the chokers can be removed once the load is placed on the blocks.
- Remember that a load of pipe will roll when tension is released. Keep away from pinch points, and do not stand in a position where the pipe may roll on you.
- Chock the sides of the pipe to keep it from rolling.
- Lay pipe side by side. If at all possible, do not stack pipe.

12.0.0 ◆ RIGGING VALVES

Rigging a valve correctly involves knowing where to place the sling, how to place the sling, what kind of sling to use, and what kind of valve is being lifted. First, determine what the valve is made of. Do not use a carbon steel choker to rig a stainless steel valve. Use synthetic slings for rigging stainless steel valves.

To rig a valve correctly, place a synthetic sling on each side of the valve body between the bonnet and the flanges. Bring the slings up through the handwheel so that the valve cannot tilt from front to back and so that the weight of the valve is evenly supported by the body of the valve. Do not place a sling around the handwheel or through the valve. The handwheel is not built to support the weight of the valve. A valve rigged around the handwheel is unsafe, even if the valve is going to be moved a short distance. Placing a sling through the valve will destroy the inner workings of the valve, even if soft synthetic slings are used.

13.0.0 ◆ GUIDELINES FOR UNLOADING AND YARDING MATERIALS

Reinforcing bars and other structural steel components are ordered from the fabricator. Delivery of these materials is usually scheduled to coincide with the construction schedule. Structural and reinforcing steel usually arrives at the job site on flatbed trucks or tractor trailers. If a spur track is available at the job site, shipments may be received by flat or gondola railroad cars. Where delivery of steel is scheduled to meet daily placement requirements, truckloads are delivered to the points of placement. Workers should be aware of safety factors that are necessary to achieve safe, efficient handling, storage, and hoisting of structural materials.

Charts are available for most types of materials so that riggers can calculate the weight of a load. *Table 4* is an example of such a chart. By knowing the number of reinforcing bars and their lengths, the rigger can easily calculate the weight of the load. *Table 3* provides similar information for pipe.

Table 4 ASTM Standard Metric and Inch-Pound Reinforcing Bars

| Bar Size | | Nominal Characteristics* | | | | | |
| | | Diameter | | Cross-Sectional Area | | Weight | |
Metric	[in-lb]	mm	[in]	mm	[in]	kg/m	[lbs/ft]
#10	[#3]	9.5	[0.375]	71	[0.11]	0.560	[0.376]
#13	[#4]	12.7	[0.500]	129	[0.20]	0.944	[0.668]
#16	[#5]	15.9	[0.625]	199	[0.31]	1.552	[1.043]
#19	[#6]	19.1	[0.750]	284	[0.44]	2.235	[1.502]
#22	[#7]	22.2	[0.875]	387	[0.60]	3.042	[2.044]
#25	[#8]	25.4	[1.000]	510	[0.79]	3.973	[2.670]
#29	[#9]	28.7	[1.128]	645	[1.00]	5.060	[3.400]
#32	[#10]	32.3	[1.270]	819	[1.27]	6.404	[4.303]
#36	[#11]	35.8	[1.410]	1006	[1.56]	7.907	[5.313]
#43	[#14]	43.0	[1.693]	1452	[2.25]	11.38	[7.65]
#57	[#18]	57.3	[2.257]	2581	[4.00]	20.24	[13.60]

*The equivalent nominal characteristics of inch-pound bars are the values enclosed within the brackets.

302T04.EPS

13.1.0 Unloading

Trucks should be promptly unloaded to prevent job-site clutter and potential safety hazards. Structural materials that are stored on the ground must be placed on timbers or other suitable blocking to keep them free from mud and to allow safe and easy rehandling.

Bars and other structural stock are normally stored by size and shape, and by length within each size. Tags are kept at the same end of each piece for easy identification. When a bundle is opened and part of the stock removed, the unit with the tag should remain with the bundle, and a new number indicating the remaining inventory should be inserted. Machine oil or grease that collects on the stock must be removed using a solvent. All mud must be washed off before placement.

A spreader bar may be necessary when unloading bundles of bars or similar stock. They are usually made from a fabricated truss, a piece of heavy-duty pipe, or an I-beam.

Tag lines are required when bundles are being hoisted. Tag lines guide the load as it is being hoisted. Fiber rope is normally used for this purpose. Safe hoisting requires a skilled crane operator and an authorized signal person. Operators take their directions from the signal person using hand signals, radios, or telephone headsets.

13.2.0 Using Slings

Never pick up a bundle by the wire wrappings that are used to tie the bundle together. Slings must be used to lift bundles. Choose a sling appropriate to the material being lifted. Check the weight of each bundle on the bill of lading, and ensure that the sizes of the wire rope slings, hooks, and shackles are safe. Chokers and slings of sufficient strength must be selected to lift the load.

The load on each sling depends on the number of slings or chokers, the angle of the choker, and the total load. Although the total weight lifted can be divided between the supporting slings, the force of the load is down. The greater the angle of the sling, the greater the tension on the sling. If the hoisting line and the slings are the same material and size, the tension on each sling must be less than or equal to the tension in the hoist line. This means that the strength of the hoist line determines the maximum load-lifting capability of the combination.

If two slings are used to lift bundles of structural iron, always thread the loops in the same direction. If one sling is looped from the opposite side, the bundle will twist when being lifted, making it less safe and causing difficulty when the bars are removed from the bundle. If possible, double wrap the slings to help prevent slipping of the slings and items in the bundle.

1. The main advantage of a hardwired communication system is _____.
   a. low cost
   b. portability
   c. less radio interference
   d. ease of use

2. What action is signaled by tapping your fist on your head, then using regular hand signals?
   a. Use main hoist.
   b. Lower.
   c. Use whip line.
   d. Raise boom and lower load.

3. Which of the following is a mode of nonverbal communication?
   a. Buzzer
   b. Portable radio
   c. Bullhorn
   d. Hardwired radio

4. When the signal person gives the *dog everything* signal, the operator is expected to _____.
   a. turn the machine
   b. travel the machine
   c. retract the boom sections
   d. make no machine movement

5. If it is apparent that the crane operator has misunderstood a hand signal, the signal person should _____.
   a. climb up on the cab and knock on the door
   b. find a supervisor
   c. signal the operator to stop
   d. get away from the crane as quickly as possible

6. A one-handed signal that is given with the fist in front of the body and the thumb pointing toward the signaler's body means _____.
   a. extend the boom
   b. move the crane toward me
   c. swing the load toward me
   d. extend the jib

7. When calculating the weight of the load, the rigging and related hardware should *not* be considered.
   a. True
   b. False

8. The most important rigging precaution is determining the weight of all loads before attempting to lift them.
   a. True
   b. False

9. The correct time to determine weight loads for rigging is after the load has been lifted.
   a. True
   b. False

10. When safe working loads are determined, they are based on _____ conditions.
    a. ideal
    b. average
    c. rare
    d. worst-case

11. A(n) _____ is used to allow controlled rotation of the load for position.
    a. tag line
    b. equalizer beam
    c. hoist line
    d. safe-load indicator

12. _____ lines are used when physical control of a load beyond the ability of the crane is required.
    a. Lockout
    b. Tag
    c. Lead
    d. Plumb

13. Tag lines are attached after the rigger verifies that the load is balanced.
    a. True
    b. False

14. When using two or more slings on a load, the rigger must ensure that _____.
    a. the sling angle is greater than 30 degrees relative to the horizontal
    b. all slings are made from the same material
    c. only chain slings are used
    d. only wire rope slings are used

15. The minimum safe distance from power lines of up to 50,000 volts (or 50kV) is _____ feet.
    a. 10
    b. 15
    c. 20
    d. 25

16. If a crane boom has come into contact with a power line, the operator should _____.
    a. immediately jump off the crane
    b. stop and turn off the crane
    c. try to back the crane away from the power line
    d. find out if the line is energized

17. High winds and lightning, which pose danger to cranes, often arrive without warning.
    a. True
    b. False

18. The amount of injury caused by exposure to abnormally cold temperatures depends on all of the following *except* _____.
    a. wind speed
    b. length of exposure
    c. temperature
    d. altitude

19. When cranes are used to lift people, the lift capacity of the crane is _____ percent as a safety factor.
    a. increased 10
    b. increased 20
    c. decreased 20
    d. decreased 50

20. Before a crane is used for a lift, it must be leveled to within _____ percent of grade.
    a. one
    b. two
    c. five
    d. ten

21. The operating radius of a crane is _____.
    a. the length of the boom plus the distance from the end of the boom to the ground
    b. the distance from the center line of rotation to the center of the load block
    c. always the same regardless of the boom position
    d. the distance from the center line of rotation to the top of the boom or jib

22. When two slings positioned at a sling angle of 60 degrees are being used to lift a 1,000-pound load, what is the tension on each sling? Assume a sling angle load factor of 1.155.
    a. 577.5 pounds
    b. 652.5 pounds
    c. 777.5 pounds
    d. 1,000 pounds

23. Shoulderless eyebolts cannot be used for angled lifts.
    a. True
    b. False

24. Using the rule of thumb for sizing chocks, a 24-inch diameter pipe would require a chock that is _____ inches thick.
    a. 1.5
    b. 2
    c. 2.5
    d. 3.5

25. The correct way to rig a valve is to _____.
    a. lift it by the handwheel
    b. run the choker through the valve body
    c. place a synthetic sling on each side of the valve body
    d. bolt lifting eyes to the valve flanges

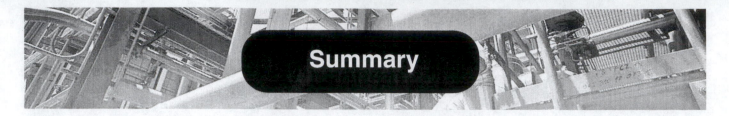

Summary

One of the important tasks a rigger performs is guiding the crane operator during a lift. In some instances, radio communications will be used. In others, it will be necessary to use the *ASME B30.5 Consensus Standard* hand signals. Both the rigger and operator must know these signals by heart.

The selecting and setting up of hoisting equipment, hooking cables to the load to be lifted or moved, and directing the load into position are all part of the process called rigging. Performing this process safely and efficiently requires selecting the proper equipment for each hoisting job, understanding safety hazards, and knowing how to prevent accidents. Riggers must have the greatest respect for the hardware and tools of their trade because their lives may depend on them working correctly.

The size of a crane and the huge loads it handles create a large potential for danger to anyone in the vicinity of the crane. The crane operator and rigging workers must be vigilant to ensure that the crane avoids power lines and that site personnel are not endangered by the movement of the crane or the load. Because of tipping hazards, it is important to make sure a crane is on level ground with outriggers extended, if applicable, before lifting a load. Although using cranes to lift personnel was once a common practice, it is discouraged by OSHA, and can only be done in special situations, and under special conditions.

Notes

Anti-two-blocking devices: Devices that provide warnings and prevent two-blocking from occurring. Two-blocking occurs when the lower load block or hook comes into contact with the upper load block, boom point, or boom point machinery. The likely result is failure of the rope or release of the load or hook block.

Blocking: Pieces of hardwood used to support and brace equipment.

Center of gravity: The point at which an object is perpendicular to and balanced in relation to the earth's gravitational field.

Cribbing: Timbers stacked in alternate tiers. Used to support heavy loads.

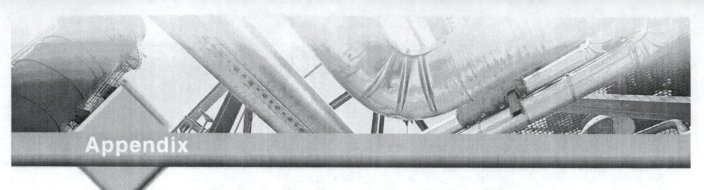

Powerline Awareness Permit

POWER LINE AWARENESS PERMIT

Today's Date _____ **Job Number** _____

Contractor Name	
Job Address	
Telephone Number	**Fax Number**

Emergency Contact Number	

Survey

Before beginning any project, you must first survey your work area to find power lines at the job site. (See job-site sketch on reverse side.)

Identify

After finding all of the power lines at your site, identify the activities you'll be doing that may put you or your workers at risk. Mark one or more of the following:

☐ Cranes (mobile or truck mounted)　　☐ Aerial lifts
☐ Drilling rigs　　　　　　　　　　　　☐ Dump trucks
☐ Backhoes/excavators　　　　　　　　☐ Ladders
☐ Long-handled tools　　　　　　　　　☐ Material handling & storage
☐ Other tools/high-reaching equipment　☐ Scaffolding
☐ Concrete pumper　　　　　　　　　　☐ Other_____

Eliminate or Control

After identifying the power line and high-risk activities on the job site, determine how to eliminate or control the risk of electrocution (a successful determination is often reached only after consultation with the utility).

Mark one or more of the following:

☐ Move the activity　　　　　　　　　　☐ Use barrier protection (insulated sleeves)
☐ Change the activity　　　　　　　　　☐ Use an observer
☐ Have the utility de-energize the power line　☐ Use warning lines with flags
☐ Have the utility move the power line　☐ Use non-conductive tools
　　　　　　　　　　　　　　　　　　　☐ Use a protective technology:
　　　　　　　　　　　　　　　　　　　　☐ Insulated link
　　　　　　　　　　　　　　　　　　　　☐ Boom cage guard
Always maintain your minimum safe clearance　☐ Proximity device
distance from the power line, except when the
utility has de-energized and visibly grounded the
power line.

Voltages	Distance from Power Line
Less than 50 kV	10 Feet
More than 50 kV	10'+(0.4")(# of kV over 50 kV)

WARNING!
It is unlawful to operate any piece of equipment within 10' of energized lines

CONSTRUCTION SAFETY COUNCIL

302A01.EPS

Job-Site Sketch

(Draw in location of power lines and their proximity to construction site, including such things as proposed excavations, location of heavy equipment, scaffolding, material storage areas, etc.)

Completed by_____Date_____

Title_____

Approved by_____Date_____

Title_____

CONSTRUCTION SAFETY COUNCIL

302A02.EPS

Resources & Acknowledgments

Additional Resources

This module is intended to be a thorough resource for task training. The following reference works are suggested for further study. These are optional materials for continued education rather than for task training.

Crane Safety on Construction Sites, 1998. Task Committee on Crane Safety on Construction Sites. Reston, VA: ASCE.

Rigging Handbook, 2003. Jerry A. Klinke. Stevensville, MI: ACRA Enterprises, Inc.

Figure Credits

Aearo Company, 302F01

Topaz Publications, Inc., 302F19, 302F20, 302F23

Link-Belt Construction Equipment Company, 302F21, 302F37, 302F38

Cianbro Corporation, 302F22

Rigging Handbook, 2003, by Jerry A. Klinke, **ACRA Enterprises, Inc.**, Stevensville, MI., 302F28, 302F29

The Crosby Group, Inc., 302F30 (photo), 302F31 (photo)

Wire Rope Sling Users Manual, Second Edition, 1997, **Wire Rope Technical Board**, 302F30 (art), 302F31 (art)

Manitowoc Crane Group, 302F33, 302F36

Construction Safety Council, Appendix

CONTREN® LEARNING SERIES — USER UPDATE

NCCER makes every effort to keep these textbooks up-to-date and free of technical errors. We appreciate your help in this process. If you have an idea for improving this textbook, or if you find an error, a typographical mistake, or an inaccuracy in NCCER's Contren® textbooks, please write us, using this form or a photocopy. Be sure to include the exact module number, page number, a detailed description, and the correction, if applicable. Your input will be brought to the attention of the Technical Review Committee. Thank you for your assistance.

Instructors – If you found that additional materials were necessary in order to teach this module effectively, please let us know so that we may include them in the Equipment/Materials list in the Annotated Instructor's Guide.

Write: Product Development and Revision
            National Center for Construction Education and Research
            3600 NW 43rd St, Bldg G, FL 32606

Fax: 352-334-0932

E-mail: curriculum@nccer.org

Craft   Module Name

Copyright Date   Module Number   Page Number(s)

Description

(Optional) Correction

(Optional) Your Name and Address

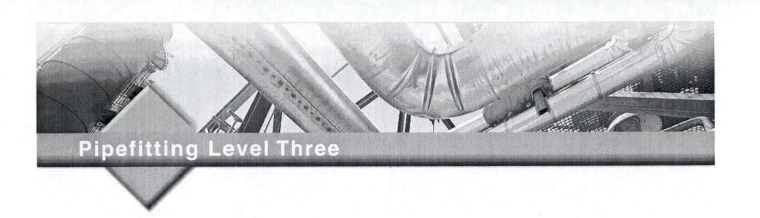

Pipefitting Level Three

08303-07

Standards and Specifications

08303-07
Standards and Specifications

Topics to be presented in this module include:

Overview

This module will prepare you to use the information provided by standards and codes. Standards and codes tell you what kind of materials to use, how those materials are identified, and what procedures to follow in using them to construct systems.

Objectives

When you have completed this module, you will be able to do the following:

1. Understand and interpret pipefitting standards and codes.
2. Read and interpret pipefitting specifications.
3. Identify pipe and components according to specifications.

Trade Terms

Addenda   Code

Required Trainee Materials

1. Pencil and paper
2. Appropriate personal protective equipment

Prerequisites

Before you begin this module, it is recommended that you successfully complete *Core Curriculum*; *Pipefitting Level One*, *Pipefitting Level Two*, and *Pipefitting Level Three*, Modules 08301-07 and 08302-07.

    This course map shows all of the modules in the third level of the *Pipefitting* curriculum. The suggested training order begins at the bottom and proceeds up. Skill levels increase as you advance on the course map. The local Training Program Sponsor may adjust the training order.

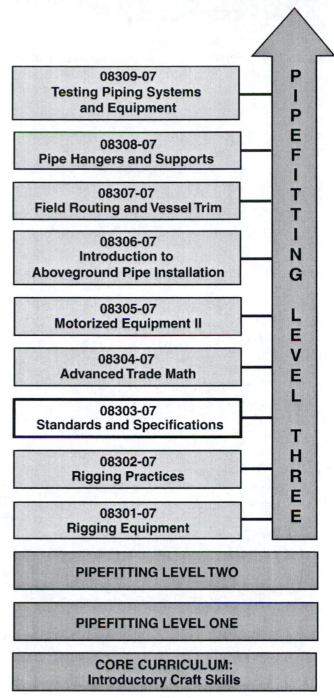

303CMAP.EPS

1.0.0 ◆ INTRODUCTION

Pipefitters use standards and specifications to obtain the information needed to install piping and associated items that meet both legal requirements and the specifications of the builder. Pipefitters must be able to read and understand standards and specifications to perform pipefitting duties. Any time there is a question about a standard or specification, consult your immediate supervisor, field engineer, quality control personnel, or other appropriate person at your job site before continuing.

2.0.0 ◆ STANDARDS AND CODES

The words standard and **code** are almost interchangeable. Standards and codes are documents that set up methods of assembly, manufacturing, testing, and good design practices, including those for safety and efficiency. They cover broad areas, are accepted by government agencies, and form a basis for legal obligations.

Standards and codes are based on proven engineering practices. They ensure the safeness of piping systems by providing the minimum requirements for materials, dimensions, design, erection, testing, and inspection. Revisions of the standards and codes are made periodically to reflect developments in industry.

Standards and codes are used for the following reasons:

- Hardware pieces made according to a standard are interchangeable and of known dimensions and characteristics.
- Following the applicable code or standard guarantees the performance, reliability, and quality of a piece of hardware and provides a basis for contract negotiations and for obtaining insurance.
- Standards and codes provide the basis for government safety regulations.

2.1.0 Standard and Code Sponsors

Standards and codes are developed and revised by national organizations called sponsors. Revisions are based on current industry practices and knowledge. *Table 1* lists the acronyms and names of organizations that issue standards relevant to pipefitters.

Table 1 Organizations Issuing Standards

Acronym	Organization
AASHTO	American Association of State Highway and Transportation Officials
ACPI	American Concrete Pipe Institute
ANSI	American National Standards Institute
API	American Petroleum Institute
ASME	American Society of Mechanical Engineers
ASTM International	American Society for Testing and Materials International
ASTM Committee 04	Committee on Clay Pipe
AWS	American Welding Society
AWWA	American Waterworks Association
BS	British Standards
CSA	Canadian Standards Association
DIPRA	Ductile Iron Pipe Research Association
ISO	International Standards Organization
MSS	Manufacturer's Standardization Society of the Valve and Fittings Industry
NRC	Nuclear Regulatory Commission
NFPA	National Fire Protection Association
PFI	Pipe Fabrication Institute
PPI	Plastic Pipe Institute

303T01.EPS

Of these, the American National Standards Institute (ANSI), the American Society for Testing and Materials International (ASTM International), and the American Society of Mechanical Engineers (ASME) are most often referenced in construction documents in America. Nonetheless, you may, and probably will see some of these other organizations referred to at some point in your career.

ANSI authorizes national standards from many sponsoring organizations. Each of these standards is identified by the ANSI acronym followed by the sponsor's acronym and a name and number code. For example, the ASME code for power piping is *ANSI/ASME B31.1.*

2.2.0 Code Changes

Periodically, codes are reviewed and changes are made. When a code is updated, either the entire code is reissued with an updated year prefix, or **addenda** sheets for the areas of the code affected by the changes are issued. Changes are typically noted to alert readers to the changes that have been made. For example, the American Welding Society (AWS) code indicates code changes with two vertical lines in the margin beside the changed area. Editorial changes to the AWS are indicated by a single vertical line in the margin.

ASME updates codes at intervals, usually every three years. An addendum is issued yearly. The yearly addendum is identified by an A in front of the year on the cover of the code. API updates the entire code every 5 years.

It is very important to recognize code changes. When a client specifies a code year, the code for only that year is used unless otherwise specified by the client. When referring to a code, be sure that the year on the code matches the specifications for the job.

2.3.0 Commonly Used Codes

Clients specify codes that must be followed when placing orders or awarding contracts. Severe penalties will be imposed for not conforming with the codes specified in the order or contract. Many of these codes are also mandated with the force of law by one or more government jurisdictions. Standards and codes are referenced in local, state, and federal specifications, usually by inclusion by reference. The specification will state that the standard is included in regulation by reference, meaning that the standard is seen as part of the regulation. The standard's requirements are not spelled out. It is then the responsibility of the contractor or engineer to meet the standard, and to be certain that the work

is correct. Always check the contract, order, or job specification for the specified codes. Some of the more common codes are as follows:

- *ASME Boiler and Pressure Vessel Code*
- *ANSI/ASME Code for Pressure Piping B31*
- *AWS Structural Welding Code D1.1*

2.3.1 ASME Boiler and Pressure Vessel Code

ASME is a scientific and technical organization of engineers, designers, producers, constructors, and individuals associated with industry who are experts in their fields and who share a technical interest in mechanical engineering and its application. The ASME engineers have developed a code that provides a collection of rules governing all of the activities involved in working with pressurized piping systems.

The *ASME Boiler and Pressure Vessel Code (ASME BPVC)* consists of a set of safety rules that govern design, fabrication, installation, and inspection during the construction of boilers, pressure vessels, and nuclear power plant components, such as piping, pumps, and valves that operate at high temperatures and pressures. Compliance with the code is required in the United States and in Canada in order to qualify for insurance. The ASME recognizes that there are many uses of pressure vessels, heating boilers, power boilers, and nuclear power plant vessels. Therefore, each of the sections within the code applies to a specific topic. The BPVC consists of the following sections:

I. *Power Boilers*
II. *Material Specifications*
    Part A – Ferrous Materials
    Part B – Nonferrous Metals
    Part C – Welding Rods, Electrodes and Filler Materials
    Part D – Properties
III. *Nuclear Power Plant Components*
    Subsection NCA General Regulations for Division 1 and Division 2
    i. Section III, Division 1
       a. Subsection NA: General Requirements
       b. Subsection NB: Class 1 Components
       c. Subsection NC: Class 2 Components
       d. Subsection ND: Class 3 Components
       e. Subsection NE: Class MC Components
       f. Subsection NF: Component Supports
       g. Subsection NG: Core Support Structures
       h. Appendices: Code Case N47 Class 1: Components in Elevated Temperature Service

ii. Section III Division 2: Codes for Concrete Reactor Vessel and Containment

IV. *Rules for Construction of Heating Boilers*

V. *Nondestructive Examinations*

VI. *Recommended Rules for Care and Operation of Heating Boilers*

VII. *Recommended Guidelines for the Care of Power Boilers*

VIII. *Pressure Vessels*

Division 1: Rules for the Construction of Pressure Vessels

Division 2: Alternative Rules for the Construction of Pressure Vessels

IX. *Welding and Brazing Qualifications*

X. *Fiber-Reinforced Plastic Pressure Vessels*

XI. *Rules for Inservice Inspection of Nuclear Power Plant Components*

Within Section III, *Nuclear Power Plant Components*, rules are established for classifying the plant systems into categories according to their importance and function. During the early stages of design and planning of a nuclear plant, the owners of the utility assign a safety class designation to each water, steam, and radioactive waste-containing component. As a guideline for assigning these safety classes, the Nuclear Regulatory Commission (NRC) has issued regulatory guides. These regulatory guides cover three quality groups known as Classes 1, 2, and 3, which are for work outside the reactor vessel itself, and Class MC, which is for work inside reactor containment vessels.

Class 1 systems require the strictest quality assurance and must be designed, fabricated, erected, and tested to the highest available national standards. Class 1 systems must be individually designed. Some components in a nuclear power plant that are designated as Class 1 include the reactor coolant system, reactor vessel, steam generator, safety valves, and relief valves. Class 2 systems require less strict quality assurance measures. Systems or components that are designated as Class 2 include those that are important to safety such as those that are designed for emergency reactor core cooling, post-accident heat removal, feedwater systems, and reactor shutdown systems. Class 3 systems require the least quality assurance of the three classes. Systems that are considered Class 3 include auxiliary feedwater systems, cooling water systems, and standby diesel generator systems. Class MC deals with the requirements for design and fabrication of metal containment vessels for nuclear reactors. Class MC work is designed and engineered to the very highest standards; welds and piping assemblies are X-rayed, and all work is held to a very high level. Class MC work is a specialty performed only by companies with a history of such work.

2.3.2 ASME Code for Pressure Piping B31

The *Code for Pressure Piping B31* provides engineering requirements for the safe design and construction of pressure piping. The code is not a design handbook but a guide that explains the basic design principles with specific requirements to ensure uniform piping practices and application of piping components. The code prohibits designs and practices that are known to be unsafe and also provides warnings where caution must be taken. The *Code for Pressure Piping B31* is divided into eight major sections specific to the industries and applications in which the piping is to be used. Each section of the code contains requirements that govern the following items:

- Material specifications and component standards, including dimensional requirements and pressure-temperature ratings
- Design of piping components and assemblies, including supports and hangers
- Information to help evaluate and control stresses, reactions, and movements caused by pressure, temperature changes, and other forces
- Recommendations and limitations on the selection of materials, components, and joining methods
- Fabrication, assembly, and erection of piping
- Examination, inspection, and testing of piping

The eight major sections of the *ASME Code for Pressure Piping B31* are as follows:

B31.1, *Power Piping*

B31.3, *Process Piping*

B31.4, *Pipeline Transportation Systems for Liquid Hydrocarbons and Other Liquids*

B31.5, *Refrigeration Piping and Heat Transfer Components*

B31.8, *Gas Transmission and Distribution Components*

B31.8S, *Managing System Integrity of Gas Pipelines*

B31.9, *Building Services Piping*

B31.11, *Slurry Transportation Piping Systems*

B31G, *1991 Manual for Determining the Remaining Strength of Corroded Pipelines*

The section numbers that are missing (*B31.2, B31.6, B31.7,* and *B31.10*) have either been consolidated with other sections because they overlapped too much, or have been transferred to the nuclear sections of ASME BPVC because the regulations were only relevant to the nuclear plant applications.

The most critical section of the code for pipefitters is B31.1, *Power Piping*. This section governs all pressure piping used with power boilers. The BPVC governs the actual boiler or vessel and the pipe on the vessel side of the first input/suction valve as well as the pipe between the vessel and the first output/discharge valve. *B31.1* governs the piping from these valves throughout the rest of the system. Some welds that fall under *B31.1* must pass X-ray testing by 100 percent according to specifications. All pipes and fittings must pass quality assurance testing and be marked with a heat number. When cutting pipe, you must always transfer this heat number to the pieces cut off that pipe and keep a good filing system for all required documentation. All plate or structural steel used on boiler code vessels or piping certification numbers must also be traced. Before making any alterations or installations on any boiler or pressurized piping system, refer to the applicable section of *B31.1* for specifications. *B31.1* is divided into the following chapters and parts:

I.   *Scope and Definitions*
II. *Design*
    Part 1 – Conditions and Criteria
    Part 2 – Pressure Design of Piping Components
    Part 3 – Fluid Service Requirements for Piping Components
    Part 4 – Fluid Service Requirements for Piping Joints
    Part 5 – Flexibility and Support
    Part 6 – Systems
III. *Materials*
IV. *Standards for Piping Components*
V.   *Fabrication, Assembly, and Erection*
VI. *Inspection, Examination, and Testing*

2.3.3 AWS Structural Welding Code D1.1

The *ANSI/AWS Structural Welding Code D1.1* provides standards and practices for the application of welding to the design and construction of structures. This code is not intended to apply to pressure vessels or pressure piping because these areas are covered by the ASME codes. The provisions of the code are mandatory when use of the code is specified in the engineering specifications. The code is divided into the following sections:

Section 1, *General Provisions*
Section 2, *Design of Welded Connections*
Section 3, *Workmanship*
Section 4, *Technique*
Section 5, *Qualification*
Section 6, *Inspection*
Section 7, *Stud Welding*
Section 8, *Statically Loaded Structures*
Section 9, *Dynamically Loaded Structures*
Section 10, *Tubular Structures*
Section 11, *Strengthening and Repairing Existing Structures*

3.0.0 ◆ SPECIFICATIONS

Information that cannot be shown on the working drawings is provided in written specifications. The working drawings explain only the shape, size, and location of each part of the building. The specifications are used to identify the quality and type of materials, colors, finishes, and workmanship required.

Specifications serve the following purposes:

- They are part of the contract, a legal document that gives instructions for bids, owner-contractor agreements, insurance, and bond forms.
- They provide information that can help prevent disputes between the builder and the owner.
- They specify the quality of materials to be used, eliminating conflicting opinions.
- They provide the information the contractor needs to estimate material and labor costs.

Specifications consist of listings for plant layout, piping, materials, supports, fabrication, insulation, welding, erection, painting, and testing. The pipefitter uses the plant layout and material specifications, which list the design requirements and materials for pipe, flanges, fittings, valves, and related items to be used to a particular project. The part of the specification dealing with a particular service can be identified from the piping drawing line number or P&ID number. All piping specifications must be strictly followed because they contain information supplied by the project design group.

The corporate engineering specifications for a given company usually have an index to the design criteria of the various services or processes. *Table 2* shows the design criteria section of a typical index to technical specifications.

Using the design criteria section of the index to technical specifications, determine what section of the specifications applies to your work as a pipefitter. In the above example, you would refer to Section H, *Piping*. At the beginning of each section is another index specific to that section. From the specification number in this index, you can locate the exact specifications needed for the job you are working on. Each section index is set up the same as the index at the beginning of the technical specifications. The specification title refers to the title of each specification that appears at the top of the first page of the specifications. The specification number refers to the number of the actual specification listed, and the revision number indicates the number of revisions. Always ensure that the specification revision number matches the revision number shown in the index. The title block at the bottom of each page in the specifications contains the specification number and other information. *Figure 1* shows a typical specification title block.

3.1.0 Information Required

Specifications provide information that applies to the area of work that is described in the title of the specification. This information is organized under separate headings within the specifications for easy reference. Examples of the headings found in a typical piping specification include the following:

- *Scope* – Explains the purpose of the specification and what the specification will cover.
- *Codes and standards* – Lists all of the applicable codes and standards that must be referred to within the specification.
- *Code conflicts* – Explains which code takes preference in case of a conflict. Many times, codes from two different sources or agencies will conflict.
- *Drawings* – Explains the various types of drawings that are provided with the specifications for construction.

Date:	Project Rev. No:	Master Rev. No:	Page / Total Pages	Specification No:
January 2006	P-2811, Rev. 1	0	1 of 5	H-001

303F01.EPS

Figure 1 ◆ Typical specification title block.

Table 2 Typical Index to Technical Specifications – Design Criteria Section

Spec. No.	Specification Title	Project Rev. No.
A-001	General	0
B-001	Civil Design Criteria	0
C-001	Concrete Design Criteria	1
D-001	Architectural Design Criteria	0
E-001	Structural Steel Design Criteria	1
F-001	Mechanical	0
G-001	HVAC General Design Criteria	0
G-002	HVAC Ventilation Rates Design Standards	0
G-003	HVAC Space Pressurization Design Criteria	0
G-004	HVAC General Design Criteria/Equipment	1
H-001	Piping Design Criteria	2
H-010	Piping Material Service Index	1
I-001	Plumbing Design Criteria	0
J-001	Fire Protection Design Criteria	0
K-001	Electrical Design Criteria	0
L-001	Instrumentation & Control System Design Criteria	1
M-001	Equipment Insulation Design Criteria	1
M-002	Piping Insulation Design Criteria	0
N-001	Painting	0

303T02.EPS

- *General piping practices* – Provides a list of general practices that must be complied with when designing and fabricating the piping systems. An example of these general practices includes providing drain connections and suitable valves on all pumps, tanks, and other related equipment.

- *Line sizing* – Provides guidelines explaining how to determine line size and the line sizes that must be avoided.

- *Fittings and flanges* – Gives the flange and fitting specifications, such as situations in which flat-faced or raised-face flanges or long- or short-radius elbows must be used.

- *Valves* – Specifies the minimum requirements for valves as well as the valve locations.

- *Piping* – Provides the line numbers in numerical order; the class or specification of each line and size; whether or not each line is insulated; what each line transports (liquid or vapor); starting and termination points; design pressure and temperature; and operating pressure and temperature.

- *Stress analysis* – Provides a guide to determine when stress analysis calculations should be made based on the code that governs the line, the operating temperature and pressure of the line, and the type and size of pipe.

- *Pipe hangers and supports* – Explains how to determine the locations of hangers and supports on the drawings and the minimum requirements for support materials.

- *Pipe sleeves* – Explains the minimum requirements for the length, width, and installation of pipe sleeves through floor penetrations and walls.

- *Insulation* – Specifies which piping should be insulated and the minimum requirements for pipe insulation.

- *Winterizing* – Specifies conditions that require winterizing and the methods used to winterize the piping systems and components.

- *Flushing and cleaning* – Specifies which lines are to be cleaned to remove foreign materials and provides information about any special cleaning procedures or chemicals that must be used.

- *Testing* – Explains the minimum testing requirements of all piping and components as determined by the governing code for each system.

3.2.0 Change Orders

Once the contract has been signed, the specifications cannot be changed unless the owner orders changes, such as additions, deletions, or other revisions. Any changes made in the specifications must be accompanied by a change order. This is a written order to the contractor authorizing a change in the work. It becomes part of the contract, and the changes listed must be performed. *Figure 2* shows a sample change order.

3.3.0 Welding Procedure Specifications

A welding procedure specification (WPS) is a written set of instructions for producing sound welds. Each WPS is written and tested in accordance with a particular welding code or specification and must be in accordance with industry practice. All welding requires that acceptable industry standards be followed, but not all welds require a WPS. If a weld does require a WPS, it must be followed. The consequences of not following a required WPS are severe. Consequences include producing an unsafe weldment, which could endanger life, as well as the rejection of the weldment and the possibility of a lawsuit. When a WPS is required, be sure to follow it. The requirement for the use of a WPS is often listed on job blueprints as a note or in the tail of the welding symbol. If you are unsure whether or not the welding being performed requires a WPS, do not proceed until you check with your supervisor.

Each WPS is written by the corporate quality control manager of a given company who knows welding codes, specifications, and acceptable industry practices. It then becomes the responsibility of each job site quality control manager to qualify the welders at that job site according to the WPS. The WPS is tested by welding test coupons and then testing the coupons according to the code. For SMAW of complete penetration groove welds, the testing required includes nondestructive testing (NDT), tensile strength and root, and face or side bends. The results of the testing are recorded on a Procedure Qualification Record (PQR). The WPS and PQR must be kept on file. *Figure 3* shows a typical WPS.

There are many different formats used for WPSs, but they all contain the same essential information. Information typically found on a WPS includes the following:

- *Scope* – Explains the welding process and base metals the WPS applies to as well as the governing code and/or specification to be used.

- *Base metal* – Provides the chemical composition or specification of the base metal the WPS applies to.

CHANGE ORDER

No. _____
Date _____

Project: County Power Co.
3250 North Oak
Allendale, Nebraska

To: XYZ Construction Co.
2900 Bluff Road
Canyon City, Oklahoma

Revised Contract Amount

Previous contract
Amount of this order
(decrease) (increase) _____

Revised contract

The contract time is hereby (increased) (decreased) (unchanged) by _____ days.

This order covers the contract revision described below:

The work covered by this order shall be performed under and be part of the original construction contract.

Changes Approved   Brown and Smith, Architects

_____   by _____
Owner

by _____
Contractor

303F02.EPS

Figure 2 ◆ Change order.

- *Welding process* – Identifies the welding process to use, such as uphill progression or downhill progression.
- *Filler metal* – Specifies the composition, identifying type, or classification of the filler metal to be used as well as the size electrodes for base metals of different thicknesses and various positions.
- *Type of current* – Specifies the type of current, either AC or DC. If DC is specified, the polarity must also be given.
- *Arc voltage and travel speed* – Specifies the arc voltage range. Ranges for travel speed for automatic processes are mandatory and are recommended for semiautomatic processes.
- *Joint design and tolerances* – Provides joint design details, tolerances, and welding sequences as a cross-sectional drawing or as a reference to drawings or specifications.

- *Joint and surface preparation* – Lists the methods that can be used to prepare joint faces and the degree of surface cleaning required.
- *Tack welding* - Provides details pertaining to tack welding. Tack welders must use the WPS.
- *Welding details* – Specifies the size of electrodes to use for different portions and positions of the joint, the arrangement of welding passes to fill the joint, and the pass width and weave limitations.
- *Positions of welding* – Lists the positions that welding can be done in.
- *Peening* – Provides the details for peening/slagging and the type of tool to be used.
- *Heat input* – Provides the details for controlling heat input.
- *Second side preparation* – Specifies the method used to prepare the second side when joints are welded from two sides.

WELDING PROCEDURE SPECIFICATION (WPS)   Yes ☐
PREQUALIFIED _____   QUALIFIED BY TESTING _____
or PROCEDURE QUALIFICATION RECORDS (PQR)   Yes ☒

Company Name **Red Inc.**
Welding Process(es) **FCAW**
Supporting PQR No.(s) **—**

Identification # **PQR 231**
Revision **1** Date **12-1-06** By **W. Lye**
Authorized by **J. Jones** Date **1-18-07**
Type—Manual ☐ Semi-Automatic ☐
Machine ☐ Automatic ☐

JOINT DESIGN USED
Type: **Butt**
Single ☒ Double Weld ☐
Backing: Yes ☒ No ☐
Backing Material: **ASTM A131A**
Root Opening **1/4"** Root Face Dimension **—**
Groove Angle: **52-1/2°** Radius (J–U) **—**
Back Gouging: Yes ☐ No ☒ Method **—**

POSITION
Position of Groove: **O.H.** Fillet: **—**
Vertical Progression: Up ☐ Down ☐

ELECTRICAL CHARACTERISTICS
Transfer Mode (GMAW) Short-Circuiting ☐
Globular ☒ Spray ☐
Current: AC ☐ DCEP ☒ DCEN ☐ Pulsed ☐
Other _____
Tungsten Electrode (GTAW)
Size: _____
Type: _____

BASE METALS
Material Spec. **ASTM A131**
Type or (Grade) **A**
Thickness: Groove **1"** Fillet **—**
Diameter (Pipe) **—**

FILLER METALS
AWS Specification **A5.20**
AWS Classification **E71T-1**

TECHNIQUE
Stringer or Weave Bead: **Stringer**
Multi-pass or Single Pass (per side) **Multipass**
Number of Electrodes **1**
Electrode Spacing Longitudinal **—**
Lateral **—**
Angle **—**

Contact Tube to Work Distance **3/4-1"**
Peening **None**
Interpass Cleaning: **Wire Brush**

SHIELDING
Flux **—** Gas **CO_2**
Composition _____
Electrode-Flux (Class)_____ Flow Rate _____
_____ Gas Cup Size _____

PREHEAT
Preheat Temp., Min **75° Ambient**
Interpass Temp., Min **75°F** Max **350°F**

POSTWELD HEAT TREATMENT
Temp. **N.A.**
Time _____

WELDING PROCEDURE

| Pass or Weld Layer(s) | Process | Filler Metals | | Current | | Volts | Travel Speed | Joint Details |
		Class	Diam.	Type & Polarity	Amps or Wire Feed Speed			
1	FCAW	E71T-1	.045"	DC+	180	26	8	
2-8	"	"	"	"	200	27	10	
9-11	"	"	"	"	200	27	11	
12-15	"	"	"	"	200	27	9	
16	"	"	"	"	200	27	11	

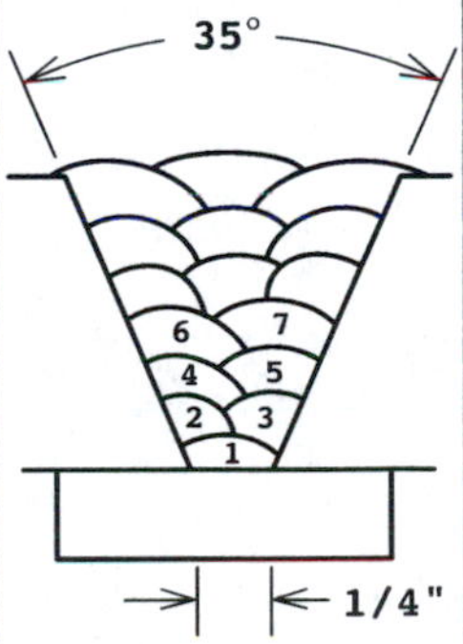

Form E-1 (Front)

303F03.EPS

Figure 3 ◆ WPS.

- *Postheat treatment* – Provides details about postheat treatment or a reference to a separate document.

4.0.0 ◆ IDENTIFICATION OF PIPE AND COMPONENTS

As shown in *Table 2*, specification number H-010 is the piping material service index. This specification lists all of the substances that flow through the plant, the pipe symbol used for each substance, the piping specification that identifies the type of pipe to use, and the valves that must be used. To use the piping material service index, you must be aware of the various types of pipe and components and be able to identify these items. *Table 3* shows a section from a typical piping material service index.

4.1.0 Pipe Identification

Pipes and fittings are available in many different alloys, sizes, and strengths. All piping is identified by the ASTM International identification system for weldable metals. In the ASTM International system, each metal is identified by a series of letters and numbers, called an identification number. This identification number is stamped or printed on the pipe and fittings. When the job drawings and specifications list a pipe by identification number, the pipefitter must use that kind of pipe for the job.

An example of an ASTM International identification number is *ASTM-A-120*. If the metal contains iron, it is considered a ferrous material, and the identification number starts with an A. If the metal does not contain iron, the number begins with a B. The three-digit identification number identifies the specific metal within the ferrous or nonferrous group.

The ASTM International identification system is only one of several systems used by engineers. The other systems are by ANSI and the American Petroleum Institute (API), each of which has its own numbering system. When selecting a pipe for any piping system, always check the drawings and specifications for the job, and then select a pipe with the correct identification number. *Table 4* shows the specifications for different types of pipe.

Table 3 Piping Material Service Index

Service	Pipe Sym.	Size	Spec No.	Shut-off	Valve Throttle	Check	Max Psi	Max F.
ACID NITRIC	AC	6" & Smaller	514	BA-152	GL-154	CK-160	150	140
ACID HYDROCHLORIC (Uninhibited)	AH	2" & Smaller	632	PL-156	PL-156	CK-166	150	100
		2-1/2" Thru 6"	632	PL-156	PL-156	CK-167		
ACID PHOSPHORIC	APH	2" & Smaller	514	PL-156	PL-156	CK-160	150	200
ACID SPENT (CLO2)	AS	1" & Thru 4"	633	PL-101	PL-101	CK-163	125	200
		6" Thru 8"	633	PL-156	PL-156	PL-156		
ACID SULFURIC (10% – 50%) CRSS SERV => 602	ASD	1" & Thru 4"	632	PL-156	PL-156	CK-166		
		6" Thru 8"	632	PL-156	PL-156	CK-167		
ACID SULFURIC (93% MIN.) *17 CRSS SERV => 600 2" & Smaller CRSS SERV => 601 2½" & Larger	ASC	2" & Smaller	409	PL-156	PL-156	CK-166	150	120
		2-1/2" Thru 6"	412	PL-156	PL-156	CK-167		
Date: January 2006	Project Rev. No: P-2811, Rev. 2	Master Rev. No: 0		Page/Total Pages 1 of 29			Specification No: H-010	

303T03.EPS

4.2.0 Component Identification

Valves, fittings, and flanges are identified according to the type of material, size, and pressure rating. The rating designation, also referred to as service designation, pressure designation, or pressure class, defines the pressure, fluid, and temperature limitations of the component. In most instances, the size and rating designation is stamped, forged, or cast onto the side of the component if the size of the component allows.

Valves have a pressure rating stamped on the side. If only a number is stamped on the valve, this number designates the working pressure of the valve when it is used with steam services. The number is often followed by a letter or several letters. These letters represent the type of fluid that the pressure rating is determined for. Examples of letters used to represent fluids are as follows:

- W – Water
- L – Liquid
- O – Oil
- G – Gas
- S – Steam

Sometimes, two pressure ratings are stamped on the side of the valve. This shows that the valve has a hot pressure rating and a cold pressure rating. If the valve must be installed so that flow travels through the valve in one direction only, the valve either has an arrow on the side indicating flow direction, or the ends are marked as the inlet and outlet. *Figure 4* shows valve markings.

Valves are identified on the blueprints according to the symbols and abbreviations noted in the job specifications. A typical valve number structure gives letters and a number designating the

Table 4 Specifications for Pipe (1 of 2)

ASTM or API Designation	ANSI Designation	Title
ASTM A53	B125.1	Welded and Seamless Steel Pipe
ASTM A106	B125.30	Seamless Carbon Steel Pipe for High-Temperature Service
ASTM A120	B125.2	Black and Hot-Dipped Zinc-Coated (Galvanized) Welded and Seamless Steel Pipe for Ordinary Uses
ASTM A134	B125.55	Electric-Fusion (Arc)-Welded Steel Plate Pipe (Sizes 16 in. and Over)
ASTM A135	B125.3	Electric-Resistance-Welded Steel Pipe
ASTM A139	B125.31	Electric-Fusion (Arc)-Welded Steel Plate Pipe (Sizes 4 in. and Over)
ASTM A155	B125.4	Electric-Fusion-Welded Steel Pipe for High-Pressure Service
ASTM A211	B122.56	Spiral-Welded Steel or Iron Pipe
ASTM A312	B125.16	Seamless and Welded Austenitic Stainless Steel Pipe
ASTM A333	B125.17	Seamless and Welded Steel Pipe for Low-Temperature Service
ASTM A335	B125.24	Seamless Ferritic Alloy Steel Pipe for High-Temperature Service
ASTM A358	B125.57	Electric-Fusion-Welded Austenitic Chromium-Nickel Alloy Steel Pipe for High-Temperature Service
ASTM A369	B125.27	Carbon and Ferritic Alloy Steel Forged and Bored Pipe for High-Temperature Service
ASTM A376	B125.25	Seamless Austenitic Steel Pipe for High-Temperature Central-Station Service
ASTM A381	B125.35	Metal-Arc-Welded Steel Pipe for High-Pressure Transmission Systems
ASTM A405	B125.26	Seamless Ferritic Alloy Steel Pipe Specially Heat Treated for High-Temperature Service
ASTM A523	G62.5	Plain End Seamless and Electric-Resistance-Welded Steel Pipe for High Pressure Pipe-Type Cable Circuits
ASTM A524	B125.37	Seamless Carbon Steel Pipe for Process Piping
ASTM A530	B125.20	General Requirements for Specialized Carbon and Alloy Steel Pipe
API 5L		Line Pipe
API 5LX		High-Test Line Pipe
API 5LS		Spiral Weld Line Pipe

303T04A.EPS

Table 4 Specifications for Pipe (2 of 2)

ASTM or AWWA Designation	Title
PVC Pipe Standards	
ASTM D3034	Type PSM PVC Sewer Pipe and Fittings
ASTM F679	PVC Large-Diameter Plastic Gravity Sewer Pipe and Fittings
ASTM F789	Type PS-46 and Type PS-115 PVC Plastic Gravity Sewer Pipe and Fittings
ASTM F794	PVC Profile Gravity Sewer Pipe and Fittings Based on Controlled I D
ASTM F949	PVC Corrugated Sewer Pipe with Smooth Interior and Fittings
ASTM F1803	PVC Closed Profile Gravity Sewer Pipe and Fittings Based on Controlled I D
ASTM D2241	PVC Pressure-Rated Pipe (SDR Series)
AWWA C900	PVC Pressure Pipe and Fabricated Fittings, 4 in. Through 12 in. For Water
AWWA C905	PVC Pressure Pipe and Fabricated Fittings, 4 in. Through 48 in. For Water
AWWA C909	Molecularly Oriented PVCO Pressure Pipe, 4 in. Through 24 in. For Water
ASTM F1483	Oriented PVCO Pressure Pipe, 4 in. Through 16 in.
Polyethylene Pipe Standards	
AWWA C901	PE Pipe, ½ in. Through 3 in. For Water Service
AWWA C906	PE Pipe, 4 in. Through 63 in. For Water Distribution and Transmission
ASTM F714	PE Pipe, Based on OD
ASTM D3035	PE Pipe, Based on Controlled OD
ASTM D2657	Standard Practice for Heat-Joining Polyolefin Pipe and Fittings
ASTM D2683	Socket-Type PE fittings for OD Controlled PE Pipe and Tubing
ASTM D3261	Butt Heat Fusion PE Plastic Fusion For PE Pipe and Tubing
ASTM D3350	PE Plastic Pipe and Fittings Materials
ASTM F1055	Electrofusion Type PE Fittings for OD Controlled PE Pipe and Tubing

303T04B.EPS

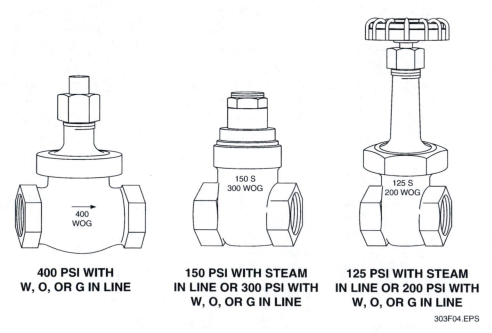

| 400 PSI WITH W, O, OR G IN LINE | 150 PSI WITH STEAM IN LINE OR 300 PSI WITH W, O, OR G IN LINE | 125 PSI WITH STEAM IN LINE OR 200 PSI WITH W, O, OR G IN LINE |

303F04.EPS

Figure 4 ◆ Valve markings.

valve to use in the system. This number must be looked up in the valve index in the specifications to determine exactly what type of valve to use. The valve index lists the valve type, pressure rating, end connection, construction, materials, and recommended manufacturer. *Table 5* shows a page from a sample valve index.

Flanges have their manufacturer's stamp, size, ASTM International specification, and pressure rating stamped on the side. The job specifications that govern the system specify the type and pressure rating that must be used.

5.0.0 ◆ REQUEST FOR INFORMATION (RFI)

It sometimes happens that the pipefitter in the field finds a problem that cannot be solved within the scope of the documents, or finds a problem that may involve a conflict with specifications or standards. In such a case, the pipefitter would file a request for information (RFI) with the owner or engineer to resolve the conflict. *Figure 5* is an example of an RFI form. The form may be filed with the field representative of the client for whom the work is being done, or it may be filed

Table 5 Sample Valve Index

Gate Valves (GA) Index								
	CONSTRUCTION			METALLURGY				MGRF
Valve	Bonnet	Disk	Seat	Body	Disk	Seat	Stem	Fig. No/Size
GA-100 125 PSIG SCR	Scr'd	Solid Wedge	Integ	Bronze	Bronze	Bronze	Bronze	Crane 428 Stockham B-105 ¼" thru 3"
GA-101 125 PSIG FLG	Bolted OS&Y	Solid Wedge	SCR	Cast Iron	Bronze	Bronze	Bronze	Crane 465-½, 2" thru 48" Stockham G-623 2" thru 36"
GA-102 125 PSIG SCR	Clamp OS&Y	Solid Wedge	Integ	Mall. Iron	Mall. Iron	Mall. Iron	Nickle Plated	Crane 484-½ ½" thru 3"
GA-103 125 PSIG FLG	Clamp OS&Y	Solid Wedge	Integ	Mall. Iron	Mall. Iron	Mall. Iron	Nickle Plated	Crane 485-½ 1½" thru 4"
GA-104 125 PSIG FLG	Bolted OS&Y	Solid Wedge	SCR	3% NI Iron	No Resist	18-8SMO	18-8SMO	Crane 14477 2" thru 16"
GA-105 125 PSIG FLG	Bolted OS&Y	Solid Wedge	Integ	Cast Iron	Cast Iron	Cast Iron	Nickle Plated	Crane 475-1/2 2" thru 48"
GA-106 125 PSIG FLG	Bolted OS&Y	Solid Wedge	SCR	NI-Resist	SS316	SS316	SS316	Crane 1671 2" thru 8"
GA-150 150 PSIG FLG	Bolted OS&Y	Solid Wedge	Weld	Cast Crbon	13% CHR	13% CHR	13% CHR Steel	Crane 47 XU 2" thru 30"
GA-151 150 PSIG SCR	Bolted OS&Y	Split Wedge	Integ	SS316	SS316	SS316	SS316	Crane 122 ½" thru 2"
Date: January 2006	Project Rev. No: P-2811, Rev. 2		Master Rev. No: 0		Page/Total Pages 1 of 29		Specification No: H-301	

303T05.EPS

by the contractor, and passed on to the field representative or the client's engineer. This constitutes a paper trail for questions that arise, and show diligence on the part of the contractor.

Date:	**Request for Information**	RFI number:

Information Request

Project _____   Contractor _____   Project Type _____

Project Name _____   Purchase Order No. _____

Requested By _____

Information Requested:

_____   _____   _____
Representative Signature   Contact Number   Date

Response Needed Date _____

RESPONSE

Action: Per conversation with _____ Dept. _____

No change

Repair

Reject

Other

          _____   _____   _____
          Engineer   Contact Number   Date
Resolved
          _____   _____
          Representative Signature   Date

303F05.EPS

Figure 5 ◆ Example RFI form.

Review Questions

1. Which of the following is an acronym for a piping standards organization?
   a. ACLE
   b. ACME
   c. ASME
   d. ISOP

2. The ASME power piping code is _____.
   a. *ACLU/ASME A44.1*
   b. *ANSI/ASME B31.1*
   c. *IROC/ISO B45.1*
   d. *ACME B35.5*

3. When the code is updated, either the entire code is reissued with an updated year prefix, or _____ are issued.
   a. addendum sheets
   b. correlation sheets
   c. change orders
   d. prerequisite orders

4. The set of safety rules that govern design and fabrication of boilers, pressure vessels, and nuclear power plant components is called the _____.
   a. *ASME BRM*
   b. *ASME BPVC*
   c. *ASME B32.1*
   d. *ASME BK4*

5. The highest quality level of nuclear power plant components is _____.
   a. Class A
   b. Class 3
   c. Class 1
   d. Class Y

6. The standby diesel generator system for a nuclear power plant would be a _____ component.
   a. Class A
   b. Class B
   c. Class 1
   d. Class 3

7. The *Code for Pressure Piping B31* is a design handbook for pipe.
   a. True
   b. False

8. The standards and practices for the application of welding to design and construction of structures are provided by _____.
   a. *ASME B31.1*
   b. *ANSI B16.5*
   c. *ANSI/AWS D1.1*
   d. *ACME D31.1*

9. Specifications prevent conflicts by expressing the _____ of materials required.
   a. quality
   b. quantity
   c. price
   d. shape

10. If there are two pressure ratings stamped on a valve, the two ratings are for _____.
    a. different directions of flow
    b. open or closed
    c. hot and cold ratings
    d. gas or water

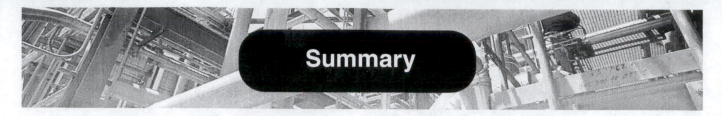

Summary

Standards establish methods for manufacturing and testing; codes establish good design practices, including the factors of safety and efficiency. Specifications identify the quality and type of materials, colors, finishes, and workmanship required for a particular job. Always ensure that the specified standards and codes that are found in the job specifications are followed, and if you ever have any question concerning the correct standard to use, check with your supervisor.

Notes

Trade Terms
Introduced in This Module

Addenda: Supplementary information typically used to describe corrections or revisions to documents.

Code: A set of regulations covering permissible materials, service limitations, fabrication, inspection, and testing procedures for piping systems.

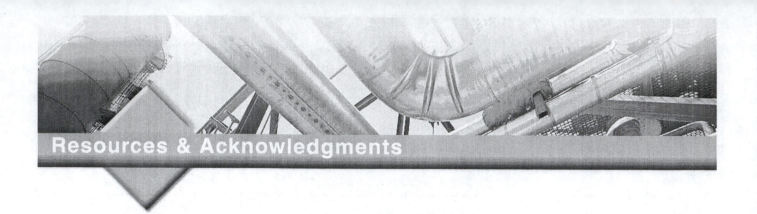

Additional Resources

This module is intended to be a thorough resource for task training. The following reference works are suggested for further study. These are optional materials for continued education rather than for task training.

American National Standards Institute (ANSI), 1430 Broadway, New York, NY 10018, (212) 642-4900.

The American Society of Mechanical Engineers (ASME), 345 East 47th Street, New York, NY 10017, (212) 705-7000.

Manufacturers Standardization Society of the Valve and Fittings Industry, Inc. (MSS), 127 Park Street N.E., Vienna, VA 22180, (703) 281-6613.

National Institute of Standards and Technology (NIST), US Department of Commerce, Gaithersburg, MD, (301) 975-2000.

Spring Manufacturers Institute, Inc. (SMI), 380 West Palatine Road, Wheeling, IL 60090, (847) 520-3290.

Underwriters Laboratories (UL), 333 Pfingsten Road, Northbrook, IL 60062, (847) 272-8800.

Figure Credits

American Welding Society, 303F03

08304-07

Advanced Trade Math

08304-07
Advanced Trade Math

Topics to be presented in this module include:

Overview

Pipefitters use many kinds of mathematics to calculate lengths, volumes, and weights. The mathematics involved includes geometry, ratios, trigonometry, and algebra. You will learn to calculate takeouts, lay out special angle cuts, and determine the different sides of a figure or an offset.

Objectives

When you have completed this module, you will be able to do the following:

1. Use tables of equivalents.
2. Perform right angle trigonometry.
3. Calculate takeouts using trigonometry.

Trade Terms

Adjacent side
Cosine
Hypotenuse
Opposite side
Ratio

Reference angle
Sine
Tangent

Required Trainee Materials

1. Pencil and paper
2. Appropriate personal protective equipment

Prerequisites

Before you begin this module, it is recommended that you successfully complete *Core Curriculum*; *Pipefitting Level One*; *Pipefitting Level Two*; and *Pipefitting Level Three*, Modules 08301-07 through 08303-07.

This course map shows all of the modules in the third level of the *Pipefitting* curriculum. The suggested training order begins at the bottom and proceeds up. Skill levels increase as you advance on the course map. The local Training Program Sponsor may adjust the training order.

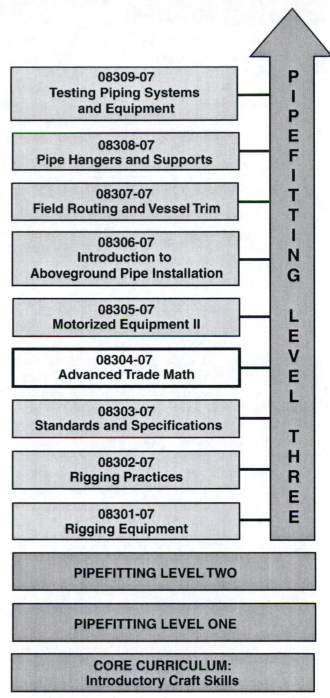

304CMAP.EPS

1.0.0 ◆ INTRODUCTION

In addition to basic math skills, a pipefitter needs advanced mathematical skills. This module explains tables of equivalents and unit conversions, how to perform right angle trigonometry, and how to calculate takeouts. All these skills can be applied to your everyday duties as a pipefitter, and the use of mathematics can become a valuable tool to make your job easier.

2.0.0 ◆ TABLES OF EQUIVALENTS

A measurement that is expressed in one type of unit can be quickly read as another type of unit by using a table of equivalents. These tables can be used to convert different types of numeric values and measures. Tables of equivalents are usually written so that the equivalent numbers or measurements are listed across from each other. To use a table of equivalents, locate the number that you wish to convert in the table. Follow that line across the table until you locate the equivalent measurement. *Table 1* lists decimal equivalents of common fractions. *Table 2* lists equivalent units of measure. You will find that some equivalencies will be so useful that you will memorize them over time. An example is the fraction equivalents in decimal form, especially quarters, halves, eighths, and sixteenths, because those are the fractions shown on the common tape measure. You can help yourself now to move toward that goal, by remembering that one sixteenth is 0.0625 and one eighth is 0.125. You have probably already memorized one quarter (0.25), one half (0.50), and three quarters (0.75), because of the coinage system, where a 25-cent coin is called a quarter.

3.0.0 ◆ UNIT CONVERSION TABLES

A number expressed in one unit of measurement can also be converted to another unit of measurement using multiplication or division. When a number is converted, its value is not changed, it is only expressed in different units. *Table 3* lists standard conversions obtained by multiplication.

Using *Table 3*, follow these steps to perform conversions:

Step 1   Locate the unit of measurement to be converted in the To Change column.

Step 2   Locate the desired unit of measurement in the second, or To, column.

Step 3   Multiply the corresponding number in the Multiply By column by the original quantity.

NOTE

If you are converting a quantity to smaller units, your answer will contain more of the smaller units. If you are converting a quantity to larger units, your answer will contain fewer of the larger units.

4.0.0 ◆ TRIGONOMETRY

Trigonometry is a branch of mathematics that deals with the study of angles, triangles, and distances. Every triangle has six components, which include three sides and three angles. If the length of one side is known and any two of the other components are known, the remaining angles and sides can be calculated using trigonometry. These calculations can be useful for the following:

- Laying out areas
- Determining machine and vessel placement
- Determining stress loads
- Laying out piping offsets

Right angle trigonometry is about the relationships between angles and sides of a right triangle. If the triangle from which you are obtaining dimensions and angles is not a right triangle, you can make two right triangles out of it (*Figure 1*), determine the angles and sides that you need for the original triangle, and then use that information to obtain the other angles and sides.

4.1.0 Pythagorean Theorem

The simplest triangles are right triangles. A right triangle has one 90-degree, or right, angle. The angles of any triangle add up to 180 degrees, so if you know any two angles, subtract the sum of the known angles from 180, and you will know the third angle. A right angle is usually indicated with a small box drawn in the angle. The right triangle is very important to pipefitters because it is used to determine the components of a piping offset.

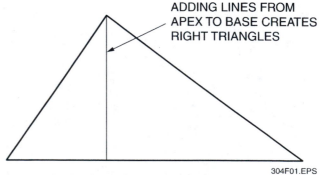

ADDING LINES FROM APEX TO BASE CREATES RIGHT TRIANGLES

304F01.EPS

Figure 1 ◆ Making two right triangles.

Table 1 Decimal Equivalents of Common Fractions

Fraction of an Inch	Decimal Equivalent		Fraction of an Inch	Decimal Equivalent	
	English (in)	Metric (mm)		English (in)	Metric (mm)
1/64	0.015625	0.3969	33/64	0.515625	13.0969
1/32	0.03125	0.7938	17/32	0.53125	13.4938
3/64	0.046875	1.1906	35/64	0.546875	13.8906
1/16	0.0625	1.5875	9/16	0.5625	14.2875
5/64	0.078125	1.9844	37/64	0.578125	14.6844
3/32	0.09375	2.3813	19/32	0.59375	15.0813
7/64	0.109375	2.7781	39/64	0.609375	15.4781
1/8	0.1250	3.1750	5/8	0.6250	15.8750
9/64	0.140625	3.5719	41/64	0.640625	16.2719
5/32	0.15625	3.9688	21/32	0.65625	16.6688
11/64	0.171875	4.3656	43/64	0.671875	17.0656
3/16	0.1875	4.7625	11/16	0.6875	17.4625
13/64	0.203125	5.1594	45/64	0.703125	17.8594
7/32	0.21875	5.5563	23/32	0.71875	18.2563
15/64	0.234375	5.9531	47/64	0.734375	18.6531
1/4	0.250	6.3500	3/4	0.750	19.0500
17/64	0.265625	6.7469	49/64	0.765625	19.4469
9/32	0.28125	7.1438	25/32	0.78125	19.8438
19/64	0.296875	7.5406	51/64	0.796875	20.2406
5/16	0.3125	7.9375	13/16	0.8125	20.6375
21/64	0.328125	8.3384	53/64	0.828125	21.0344
11/32	0.34375	8.7313	27/32	0.84375	21.4313
23/64	0.359375	9.1281	55/64	0.859375	21.8281
3/8	0.3750	9.5250	7/8	0.8750	22.2250
25/64	0.390625	9.9219	57/64	0.890625	22.6219
13/32	0.40625	10.3188	29/32	0.90625	23.0188
27/64	0.421875	10.7156	59/64	0.921875	23.4156
7/16	0.4375	11.1125	15/16	0.9375	23.8125
29/64	0.453125	11.5094	61/64	0.953125	24.2094
15/32	0.46875	11.9063	31/32	0.96875	24.6063
31/32	0.484375	12.3031	63/64	0.984375	25.0031
1/2	0.500	12.7000	1	1.000	25.4000

304T01.EPS

Table 2 Equivalent Units of Measure

Linear Measure

12 inches = 1 foot   3 feet = 1 yard   5½ yards = 1 rod
40 rods = 1 furlong   8 furlongs = 1 mile

10 millimeters   = 1 centimeter
100 centimeters = 1 meter
1,000 millimeters = 1 meter
1,000 meters   = 1 kilometer

Equivalent Values

Inches		Feet		Yards		Rods		Furlongs		Miles
36	=	3	=	1						
198	=	16.5	=	5.5	=	1				
7,920	=	660	=	220	=	40	=	1		
63,360	=	5,280	=	1,760	=	320	=	8	=	1

Square Measure

144 square inches   = 1 square foot
9 square feet   = 1 square yard
30¼ square yards   = 1 square rod
160 square rods   = 1 acre
640 acres   = 1 square mile

1 square mile = 640 acres = 102,400 square rods = 3,097,600 square yards

1 square mile = 27,878,400 square feet = 4,014,489,600 square inches

Cubic Measure

1,728 cubic inches   = 1 cubic foot
27 cubic feet   = 1 cubic yard
128 cubic feet   = 1 cord
24¾ cubic feet   = 1 perch
1 cubic yard   = 27 cubic feet = 46,656 cubic inches

Liquid Measure

1 gallon occupies 231 cubic inches
7.48 gallons occupy 1 cubic foot
1 liter occupies 1,000 cubic centimeters
1,000 liters occupy 1 cubic meter

304T02.EPS

The sides of a right triangle have been named for reference. The side opposite the right angle is always called the **hypotenuse**, and the two sides adjacent to, or connected to, the right angle are called the legs. If one of the other angles is labeled angle A, the leg of the triangle that is not connected to angle A is called its **opposite side**. The remaining leg that is connected to angle A is called the **adjacent side**. The reference angle is the angle to which a given set of trigonometric relationships refer.

The sides of the piping offset have also been named for reference. These sides are called the set, run, and travel. The set is the distance, measured center to center, that the pipeline is to be offset from the previous line of pipe. The run is the total linear distance required for the offset. The travel is the center-to-center measurement of the offset piping. The angle of the fittings is the number of degrees the piping changes direction. *Figure 2* shows a right triangle and a piping offset.

Table 3 Standard Conversions

To Change	To	Multiply By
Inches	Feet	0.0833
Inches	Millimeters	25.4
Feet	Inches	12.0
Feet	Yards	0.333
Yards	Feet	3.0
Square inches	Square feet	0.00694
Square feet	Square inches	144.0
Square feet	Square yards	0.11111
Square yards	Square feet	9.0
Cubic inches	Cubic feet	0.00058
Cubic feet	Cubic inches	1,728.0
Cubic feet	Cubic yards	0.03703
Cubic yards	Cubic feet	27.0
Cubic inches	Gallons	0.00433
Cubic feet	Gallons	7.48
Gallons	Cubic inches	231.0
Gallons	Cubic feet	0.1337
Gallons	Pounds of water	8.33
Pounds of water	Gallons	0.12004
Ounces	Pounds	0.0625
Pounds	Ounces	16.0
Inches of water	Pounds per square inch	0.0361
Inches of water	Inches of mercury	0.0735
Inches of water	Ounces per square inch	0.578
Inches of water	Pounds per square foot	5.2
Inches of mercury	Inches of water	13.6
Inches of mercury	Feet of water	1.1333
Inches of mercury	Pounds per square inch	0.4914
Ounces per square inch	Inches of mercury	0.127
Ounces per square inch	Inches of water	1.733
Pounds per square inch	Inches of water	27.72
Pounds per square inch	Feet of water	2.310
Pounds per square inch	Inches of mercury	2.04
Pounds per square inch	Atmospheres	0.0681
Feet of water	Pounds per square inch	0.434
Feet of water	Pounds per square foot	62.5
Feet of water	Inches of mercury	0.8824
Atmospheres	Pounds per square inch	14.696
Atmospheres	Inches of mercury	29.92
Atmospheres	Feet of water	34.0
Long tons	Pounds	2,240.0
Short tons	Pounds	2,000.0
Short tons	Long tons	0.89285

304T03.EPS

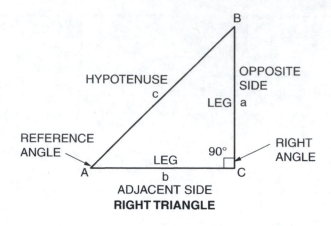

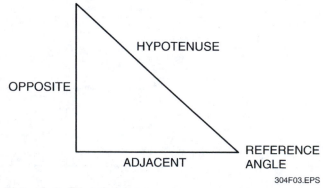

PIPING OFFSET

304F02.EPS

Figure 2 ◆ Right triangle and piping offset.

The Pythagorean theorem states that the square of the hypotenuse is equal to the sums of the squares of the other two sides. For example, in triangle abc with c being the hypotenuse and a and b being the two legs, the Pythagorean theorem states that $a^2 + b^2 = c^2$.

4.2.0 Trigonometric Functions

Trigonometry calculations are made by using **ratios** of two of the sides of a right triangle. The ratio of the two sides is directly related to the size of one of the angles in the triangle, known as the reference angle. These ratios are known as the trigonometric, or trig, functions of the reference angle. The major trig functions are the following:

- Sine
- Cosine
- Tangent

The ratios for sine, cosine, and tangent can easily be remembered by the following memory device: SOH, CAH, TOA. The first acronym means: Sine equals Opposite side divided by the Hypotenuse. The second says: Cosine equals the Adjacent side divided by the Hypotenuse. The third acronym stands for: Tangent equals the Opposite side divided by the Adjacent side. Another mnemonic (memory trick) for the three ratios is: Some Old Horse Caught Another Horse Taking Oats Away. *Figure 3* shows the sides and reference angle.

These trig functions are also used to find unknown sides and angles of piping offsets. *Figure 4* shows the relationship between trig functions and piping offsets.

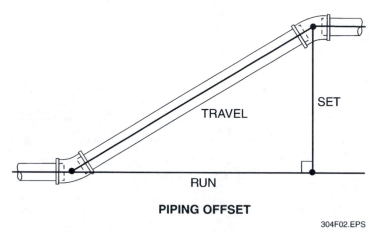

304F03.EPS

Figure 3 ◆ Triangle terms.

Finding another side when you have one side and the reference angle:

- Travel × sine = set
- Travel × cosine = run
- Run × tangent = set
- Run × cosine = travel
- Set/sine = travel
- Set/tangent = run

On the calculator (*Figure 5*), once you have set up the formula as 30 (the travel) × sin 30 degrees, pressing the = button will get the answer. If you are using a table, looking up sin 30 will give you 0.5000 as the value of the function. Insert that value in the formula, and the set will again be 15 inches.

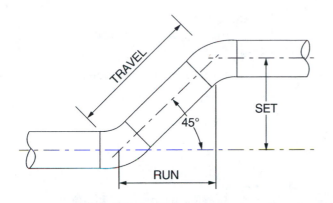

TO DETERMINE ANGLE OF OFFSET WHEN THE LENGTHS OF TWO SIDES ARE KNOWN

SET DIVIDED BY TRAVEL = SINE

RUN DIVIDED BY TRAVEL = COSINE

SET DIVIDED BY RUN= TANGENT

Length of sides when angle and one side are known	Angle of Offset				
	60°	45°	30°	22.5°	15°
Set = Travel × Sine	0.866	0.707	0.5	0.383	0.259
Run = Travel × Cosine	0.5	0.707	0.866	0.924	0.966
Set = Run × Tangent	1.732	1.0	0.577	0.414	0.268
Travel = Run/Cosine	0.5	0.707	0.866	0.924	0.966
Travel = Set/Sine	0.866	0.707	0.5	0.383	0.259
Run = Set/Tangent	1.732	1.0	0.577	0.414	0.268

304F04.EPS

Figure 4 ◆ Relationship between trig functions and piping offsets.

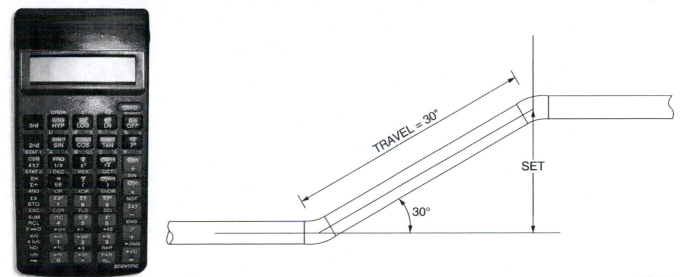

304F05.EPS

Figure 5 ◆ Scientific calculator with sin, cos, and tan keys.

4.2.1 Sine

The sine of the reference angle is the ratio formed by dividing the length of the side opposite the angle by the length of the hypotenuse. In triangle ABC (*Figure 6*), if the side opposite angle A is divided by the hypotenuse, the ratio is called the sine of angle A. It is usually written as sin A and read as the sine of A. The formula for sin A is written as follows:

sin A = side opposite A/hypotenuse

Follow these steps to find the sine of an angle:

NOTE

The following steps refer to triangle ABC in which the side opposite angle A is 3 inches long and the hypotenuse is 5 inches long.

Step 1   Find the length of the side opposite the angle.

   Example: 3 inches

Step 2   Find the length of the hypotenuse.

   Example: 5 inches

Step 3   Form a ratio, using the formula for sine.

   Example: sin A = 3/5

Step 4   Divide the numerator of the ratio by the denominator.

   Example: 3/5 = 0.6

In this example, sin A is equal to 0.600. The size of sin A can be found by using a table of trig functions. This type of table (*Table 4*) can be used to find the size of an angle from its decimal value or the decimal value from the size of the angle.

The size of angle A can be found by locating the closest value to 0.6000 under the sine column. This number is 0.6018, which corresponds to 37 degrees listed in the angle column. Therefore, angle A is approximately 37 degrees.

If a 16-foot ladder is set against a wall at a 61-degree angle, the height at which the ladder touches the wall can be found by using the formula for sine. *Figure 7* shows a representation of the ladder angle.

The height at which the ladder touches the wall is found as follows:

sin angle = side opposite/hypotenuse

sin 61 degrees = X/16 feet

sin 61 degrees = 0.8746

0.8746 × 16 = 13.9936

The height at which the ladder touches the wall is 14 feet.

Using the formulas from *Figure 4*, sine can be used to determine the set in a piping offset when the angle of the elbows and the travel are known. *Figure 8* shows piping offset 1.

In piping offset 1, the travel is given as 30 inches and the angle of the elbows is 30 degrees. The length of the set can be found as follows:

Set = travel × sin 30 degrees

Set = 30 inches × 0.500

Set = 15 inches

![Triangle ABC with hypotenuse 5", side 3", vertices A, B, C]

304F06.EPS

Figure 6 ◆ Triangle ABC.

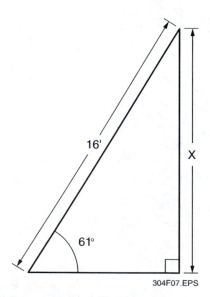

304F07.EPS

Figure 7 ◆ Representation of ladder angle.

Table 4 Trig Functions

Table of SIN (Angle)				Table of COS (Angle)				Table of TAN (Angle)			
Angle	SIN (A)	Angle	SIN (A)	Angle	COS (A)	Angle	COS (A)	Angle	TAN (A)	Angle	TAN (A)
0.0	0.0	46.0	.7193	0.0	1.00	46.0	.6947	0.0	0.0	46.0	1.0355
1.0	0.0174	47.0	.7314	1.0	0.9998	47.0	.6820	1.0	0.0175	47.0	1.0724
2.0	0.0349	48.0	.7431	2.0	0.9994	48.0	.6691	2.0	0.0349	48.0	1.1106
3.0	0.0523	49.0	.7547	3.0	0.9986	49.0	.6561	3.0	0.0524	49.0	1.1504
4.0	0.0698	50.0	.7660	4.0	0.9976	50.0	.6428	4.0	0.0699	50.0	1.1918
5.0	0.0872	51.0	.7772	5.0	0.9962	51.0	.6293	5.0	0.0875	51.0	1.2349
6.0	0.1045	52.0	.7880	6.0	0.9945	52.0	.6157	6.0	0.1051	52.0	1.2799
7.0	0.1219	53.0	.7986	7.0	0.9926	53.0	.6018	7.0	0.1228	53.0	1.3270
8.0	0.1392	54.0	.8090	8.0	0.9903	54.0	.5878	8.0	0.1405	54.0	1.3764
9.0	0.1564	55.0	.8191	9.0	0.9877	55.0	.5736	9.0	0.1584	55.0	1.4281
10.0	0.1736	56.0	.8290	10.0	0.9848	56.0	.5592	10.0	0.1763	56.0	1.4826
11.0	0.1908	57.0	.8387	11.0	0.9816	57.0	.5446	11.0	0.1944	57.0	1.5399
12.0	0.2079	58.0	.8480	12.0	0.9781	58.0	.5299	12.0	0.2126	58.0	1.6003
13.0	0.2249	59.0	.8571	13.0	0.9744	59.0	.5150	13.0	0.2309	59.0	1.6643
14.0	0.2419	60.0	.8660	14.0	0.9703	60.0	.5000	14.0	0.2493	60.0	1.7321
15.0	0.2588	61.0	.8746	15.0	0.9659	61.0	.4848	15.0	0.2679	61.0	1.8040
16.0	0.2756	62.0	.8829	16.0	0.9613	62.0	.4695	16.0	0.2867	62.0	1.8907
17.0	0.2924	63.0	.8910	17.0	0.9563	63.0	.4540	17.0	0.3057	63.0	1.9626
18.0	0.3090	64.0	.8988	18.0	0.9511	64.0	.4384	18.0	0.3249	64.0	2.0503
19.0	0.3256	65.0	.9063	19.0	0.9455	65.0	.4226	19.0	0.3443	65.0	2.1445
20.0	0.3420	66.0	.9135	20.0	0.9397	66.0	.4067	20.0	0.3640	66.0	2.2460
21.0	0.3584	67.0	.9205	21.0	0.9336	67.0	.3907	21.0	0.3839	67.0	2.3559
22.0	0.3746	68.0	.9272	22.0	0.9272	68.0	.3746	22.0	0.4040	68.0	2.4751
23.0	0.3907	69.0	.9336	23.0	0.9205	69.0	.3584	23.0	0.4245	69.0	2.6051
24.0	0.4067	70.0	.9397	24.0	0.9135	70.0	.3420	24.0	0.4452	70.0	2.7475
25.0	0.4226	71.0	.9455	25.0	0.9063	71.0	.3256	25.0	0.4663	71.0	2.9042
26.0	0.4384	72.0	.9511	26.0	0.8988	72.0	.3090	26.0	0.4877	72.0	3.0777
27.0	0.4540	73.0	.9563	27.0	0.8910	73.0	.2924	27.0	0.5095	73.0	3.2709
28.0	0.4695	74.0	.9613	28.0	0.8829	74.0	.2756	28.0	0.5317	74.0	3.4874
29.0	0.4848	75.0	.9659	29.0	0.8746	75.0	.2588	29.0	0.5543	75.0	3.7321
30.0	0.5000	76.0	.9703	30.0	0.8660	76.0	.2419	30.0	0.5773	76.0	4.0108
31.0	0.5150	77.0	.9744	31.0	0.8571	77.0	.2249	31.0	0.6009	77.0	4.3315
32.0	0.5299	78.0	.9781	32.0	0.8480	78.0	.2079	32.0	0.6249	78.0	4.7046
33.0	0.5446	79.0	.9816	33.0	0.8387	79.0	.1908	33.0	0.6494	79.0	5.1446
34.0	0.5592	80.0	.9848	34.0	0.8290	80.0	.1736	34.0	0.6745	80.0	5.6713
35.0	0.5736	81.0	.9877	35.0	0.8191	81.0	.1564	35.0	0.7002	81.0	6.3138
36.0	0.5878	82.0	.9903	36.0	0.8090	82.0	.1392	36.0	0.7265	82.0	7.1154
37.0	0.6018	83.0	.9926	37.0	0.7986	83.0	.1219	37.0	0.7535	83.0	8.1443
38.0	0.6157	84.0	.9945	38.0	0.7880	84.0	.1405	38.0	0.7813	84.0	9.5144
39.0	0.6293	85.0	.9962	39.0	0.7772	85.0	.0872	39.0	0.8098	85.0	11.430
40.0	0.6428	86.0	.9976	40.0	0.7660	86.0	.0698	40.0	0.8391	86.0	14.301
41.0	0.6561	87.0	.9986	41.0	0.7547	87.0	.0523	41.0	0.8693	87.0	19.081
42.0	0.6691	88.0	.9994	42.0	0.7431	88.0	.0349	42.0	0.9004	88.0	28.636
43.0	0.6820	89.0	.9998	43.0	0.7314	89.0	.0174	43.0	0.9325	89.0	57.290
44.0	0.6947	90.0	1.000	44.0	0.7193	90.0	0.0	44.0	0.9657	90.0	INFINITE
45.0	0.7071			45.0	0.7071			45.0	1.000		

304T04.EPS

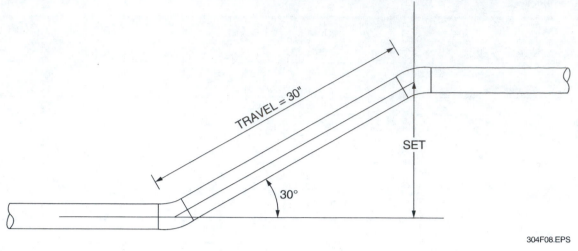

Figure 8 ◆ Piping offset 1.

304F08.EPS

If the set of 15 inches is known and the travel is unknown, you can use division with the sine function to determine the length of the travel. For example:

Set = travel × sin 30 degrees

15 inches = travel × 0.500

15 inches/0.500 = travel

30 inches = travel

4.2.2 Cosine

The ratio formed by dividing the length of the side adjacent to an angle by the length of the hypotenuse is called the cosine of that angle. If the side adjacent to angle A is divided by the hypotenuse, the ratio is called the cosine of angle A. It is usually written as cos A. The formula for cos A is written as follows:

cos A = side adjacent to A/hypotenuse

This formula can be used to find the cosine of angle A in triangle ABC when the side adjacent to A is 39 inches and the hypotenuse is 45 inches. *Figure 9* shows triangle ABC.

The cosine of angle A is found as follows:

cos A = side adjacent to A/hypotenuse

cos A = 39/45 = 13/15

cos A = 0.8667

Angle A is approximately 30 degrees (from *Table 4* or from calculator)

If the problem is to find the length of travel of a simple offset when the angles of the fittings are 45 degrees and the run is 12 inches, the length of the travel can be found using the formula for cosine. *Figure 10* shows piping offset 2.

The length of the travel is found as follows:

Run = travel × cos 45 degrees

12 inches = travel × 0.7071

12 inches/0.7071 = travel

16.97 inches = travel

The length of travel is 17 inches.

4.2.3 Tangent

The ratio formed by dividing the length of the side opposite an angle by the length of the side adjacent to the angle is called the tangent of that angle. If the side opposite angle A is divided by the side adjacent to angle A, the ratio is called the tangent of angle A. It is usually written as tan A. The formula for tan A is written as follows:

tan A = side opposite A/side adjacent to A

This formula can be used to find the tangent of angle A in triangle ABC when the side opposite

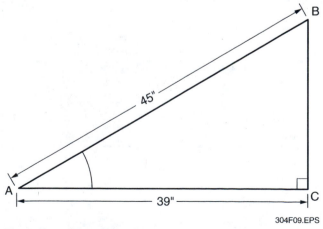

Figure 9 ◆ Triangle ABC.

304F09.EPS

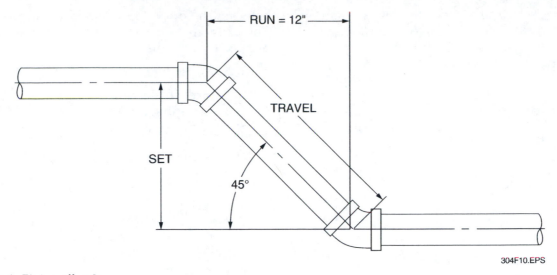

Figure 10 ◆ Piping offset 2.

angle A is 17 inches and the side adjacent to angle A is 21 inches. *Figure 11* shows triangle ABC.

The tangent of angle A is found as follows:

tan A = side opposite A/side adjacent to A

tan A = 17/21

tan A = 0.8095

Angle A is approximately 39 degrees (from *Table 4*).

The tangent trig function can be used to determine the set of a piping offset when the angles of the fittings and the length of the run are known. *Figure 12* shows piping offset 3.

In piping offset 3, the run is given as 14 inches and the angle of the elbows is 30 degrees. The length of the set is found as follows:

Set = run × tan 30 degrees

Set = 14 inches × 0.577

Set = 8.078 inches

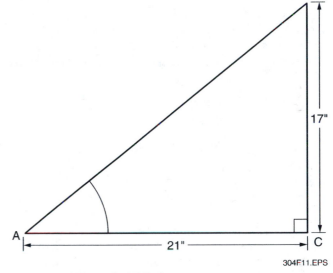

Figure 11 ◆ Triangle ABC.

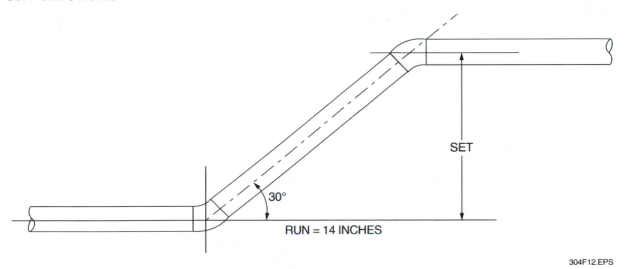

Figure 12 ◆ Piping offset 3.

Table 1, Decimal Equivalents of Common Fractions, shows that 8.078 converts to 8⁵⁄₆₄ inches, and rounded to the nearest ¹⁄₁₆-inch equals approximately 8¹⁄₁₆ inches.

Suppose that for piping offset 3, the set of 8¹⁄₁₆ inches and the run of 14 inches are known, but you need to determine the angle of the elbows. The tangent is the set divided by the run. The angle of the elbows is found as follows:

set/run = tan

8-1⁄6 inches/14 inches = tan

8.0625/14 = tan (from *Table 1*)

0.576 = tan

Angle = 30 degrees (from *Table 4*)

4.3.0 Triangle Calculation

To calculate the sine, cosine, and tangent, as shown in *Figure 13*, state the formula for each, substitute the line segment lengths, and use either a table or calculator to find the value of the function. For example:

tangent = opposite side over adjacent side

tangent angle A = ⁶⁄₈ = 0.75

Having determined the tangent of angle A, we can find the angle, to know what angle miter you need to use. Either the calculator or the table will tell you that the angle corresponding to tan = 0.75 is 36.87 degrees. Therefore, since there are 180 degrees in any triangle, angle B is 90 – 36.87 = 53.13 degrees. In this way you can know all the angles of any right triangle, given the sides.

Now let's try the same procedure with only one side and a reference angle. *Figure 14* gives us an example to work from. We have the angle, 60 degrees, and the side opposite, 48 inches long. To calculate the adjacent side, side b, we use the tangent ratio, because that states the relationship between the opposite side and the adjacent side. The tangent is given as equal to the opposite divided by the adjacent.

tan A = opposite/adjacent

Therefore, if we multiply both sides by the unknown length, b, we have:

adjacent × tan 60 = 48 inches

Now we divide both sides by tan 60, and get:

adjacent = 48/tan 60 = 27.7 inches

At this point we could use the Pythagorean theorem to find the length of the hypotenuse, or we can use the sine or cosine to find it. Since we started with side a, the opposite side, we can use the sine.

sin 60 = side a/hypotenuse

sin 60 × hypotenuse = side a = 48 inches

hypotenuse = 48/sin 60 = 55.43 inches

The angles are easy; we know that the right angle is 90 degrees, and the other angle we know is 60 degrees, so the third angle is 90 – 60 = 30 degrees.

4.3.1 Using a Scientific Calculator to Convert Sine, Cosine, or Tangent Values to Angles

To convert sine, cosine, or tangent values to angles using a scientific calculator, the calculator must have the SIN, COS, TAN, SIN⁻¹, COS⁻¹, and TAN⁻¹ keys as shown in *Figure 5*.

Before converting sine, cosine, or tangent values to angles, you must first decide which angle in the

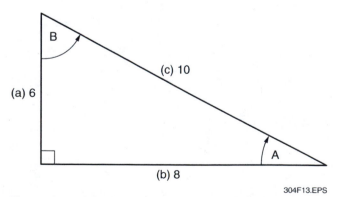

304F13.EPS

Figure 13 ◆ Calculating the sine, cosine, and tangent of a right triangle.

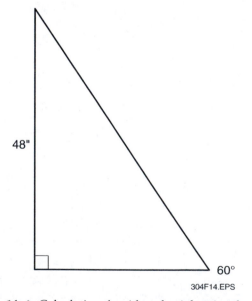

304F14.EPS

Figure 14 ◆ Calculating the sides of a right triangle.

right triangle you are trying to find. Then you must calculate the ratios of the line segments associated with that angle. Again, remember that the sine is the length of the line segment directly across from the angle divided by the hypotenuse; the cosine is the length of the line segment adjacent (connected) to the angle divided by the hypotenuse; and the tangent is the length of the line segment directly across from the angle divided by the length of the line segment adjacent to the angle.

As long as you know any two of the three lengths, you can calculate the angle with one of the three ratio methods. Begin by using the right triangle in *Figure 15*.

NOTE

You only need one of the three ratios to find the angle.

With the information given in *Figure 15*, you can find the length of line segment b using the Pythagorean theorem:

Step 1   Write the formula in the form that solves for line segment b:

$$b^2 = c^2 - a^2$$

Step 2   Replace the letters with the known numbers:

$$b^2 = 26.8^2 - 17.4^2$$

Step 3   Solve the formula using the calculator by
- Entering 26.8 on the calculator keyboard; the calculator displays 26.8.
- Pressing the x^2 key; the calculator displays 718.24.
- Pressing the minus (–) key; the calculator displays 718.24.
- Entering 17.4 on the calculator keyboard; the calculator displays 17.4.

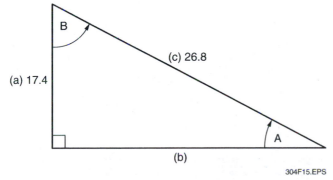

(b)

304F15.EPS

Figure 15 ◆ Example of solving ratio problems in right triangles.

- Pressing the x^2 key; the calculator displays 302.76.
- Pressing the equal (=) key; the calculator displays 415.48 ($b^2 = 415.48$).
- Pressing the $\sqrt{\ }$ key; the calculator displays 20.38332652 ($b = 20.4$ after rounding off to one decimal place like the other two line segments).

Find the degrees in angle A of *Figure 15* by applying the appropriate line segment ratio (sine, cosine, or tangent) and using the calculator. Because all three line segment lengths are now known, you can use any one of the three ratios. For this example, use sine:

sin A = side opposite divided by the hypotenuse

sin A = a/c

Step 1   Enter 17.4; the calculator displays 17.4.

Step 2   Press the divide key (÷); the calculator displays 17.4.

Step 3   Enter 26.8; the calculator displays 26.8.

Step 4   Press equal (=) key; the calculator displays 0.649253731.

sin A = 0.64925 when rounded off to five decimal places

This decimal answer is the ratio or sine between the two line segments. It is not the angle. In other words, the ratio between the two line segments is approximately $^{6}\!/_{10}$, 0.6, or $^{3}\!/_{5}$. To figure out the angle formed by the two line segments that have the calculated sine ratio, use the SIN^{-1} key on the scientific calculator. Use the SIN key to determine the sine ratio when the angle is known. When you know the sine but not the angle, you reverse the process. The small minus one ($^{-1}$) means that the process is reversed or inverted. The technical term for this reversal is **reciprocal**. SIN^{-1} is the reciprocal of SIN. The calculator does the math as long as you press the correct keys.

From the sine value of angle A in *Figure 15*, which is 0.64925, use your calculator's inverse or SIN^{-1} key to determine the degrees in angle A. The calculator illustrations are shown in *Figure 16*.

sin A = 0.64925

Step 1   Turn the calculator ON.

Step 2   Enter 0.64925; the calculator displays 0.64925 (*Figure 16A*).

Step 3   On this particular calculator, the SIN and the SIN^{-1} function are located on the same key. To switch from the SIN to the SIN^{-1} function, the key marked 2nd must

(A)   (B)   (C)

304F16.EPS

Figure 16 ◆ Steps for using a calculator to convert a sine value to an angle.

be pressed before pressing the SIN⁻¹ key (*Figure 16B*). By pressing the 2nd key first, the functions of all the calculator keys are changed to the functions shown on the upper part of the keys. If you wish to change back to the functions on the lower half of the keys, you press the key marked *3rd* first.

Step 4   Press the SIN⁻¹; the calculator displays 40.48507893, which is the number of degrees in angle A (*Figure 16C*).

Step 5   Round off answer to one decimal place— angle A = 40.5 degrees.

Use the same process to determine the angle if you know the cosine or tangent ratio, except use the COS⁻¹ key when you know the cosine ratio, and use the TAN⁻¹ key when you know the tangent ratio.

To determine the sine, cosine, or tangent ratio when the angle is known:

Step 1   Enter the value of the angle in degrees.

Step 2   Press the SIN, COS, or TAN key to display the ratio in decimal form.

NOTE

If your calculator has the SIN/SIN⁻¹, COS/COS⁻¹, or TAN/TAN⁻¹ combined with each other, as the calculator used in these illustrations does, you may have to press a key such as the 2nd or 3rd key to switch the calculator keys between the two functions.

4.3.2 Obtuse Triangle

An obtuse triangle (*Figure 17*) is not a right triangle. The largest angle of this triangle is more than 90 degrees. Since this is not a right triangle, we'll make two right triangles out of it to use our trig functions.

In this case, we could have done this with two sides and one angle, or, as we have here, with two angles. (We do know the third angle of the triangle, because the total of the three angles must be 180 degrees, so 30 + 45 = 75; 180 − 75 = 105.) First, take the length of the side we have. We take that as the hypotenuse of the triangle made by draw-

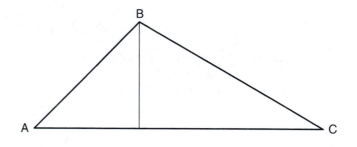

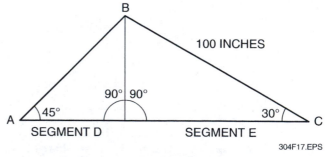

304F17.EPS

Figure 17 ◆ Solving an obtuse triangle.

ing a line from the apex of the triangle, at a right angle to the base of the triangle. Now, we want the length of our height line, so:

sin 30 = opposite divided by 100 inches

opposite = 100 × sin 30 = 50 inches

We have the height, and that gives the length of segment d, using the tangent of 45 degrees:

tan 45 = 50 inches / adjacent side (segment d)

segment d × tan 45 = 50 inches

segment d = 50 inches / tan 45 = 50 inches

Now we need the line segment from angle B to angle C. Using the Pythagorean theorem:

$(50^2 + 50^2) = BC^2$

$\sqrt{5,000} = BC = 70.71$ inches

We only need one more length, that of the base line. We already know the length of segment d. In order to get the length of the other segment, we can use either the Pythagorean theorem, as we just did, or use the cosine of angle A and the length of the line AC to find the answer.

cos 30 degrees = segment e / 100 inches

segment e = 100 inches × cos 30 = 86.6 inches

Our overall length for the base, then, is segment e + segment d, 86.6 + 50 = 136.6 inches.

4.4.0 Determining the Angles When Side Lengths Are Known

Determining the angles to be used in an offset bend is not difficult. Both angles must be the same number of degrees so that the piping run remains parallel. To determine the angle to use, use the ratios between the sides (*Figure 18*) to determine the value of the trig functions. Then use the function to determine the angle.

Using *Figure 19*, the sine of angle A is found as follows:

sin A = side opposite A/hypotenuse

sin A = 5/7.5

sin A = 0.667

Angle A is approximately 42 degrees.

If the length of a piping set is 17 inches and the travel is 32 inches, the offset angle can be determined using the formula for sine. *Figure 20* shows piping offset 4.

The angle of the offset is found as follows:

set/travel = sine

17/32 = sine

0.53125 = sine

Angle X is approximately 32 degrees.

The sine can also be used to calculate the travel of a rolling offset. The rolling offset is similar to the simple offset except there is one more factor to calculate. This is the roll of the offset. The roll is the distance that one end of the travel is displaced laterally. While a triangle best represents a simple offset, a rectangular box best represents the rolling

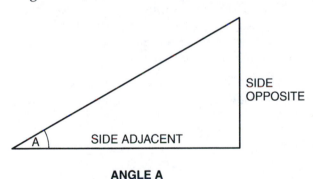

ANGLE A RELATIONSHIPS

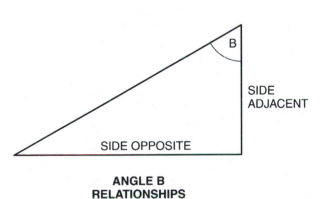

ANGLE B RELATIONSHIPS

304F18.EPS

Figure 18 ◆ Angle – side relationships.

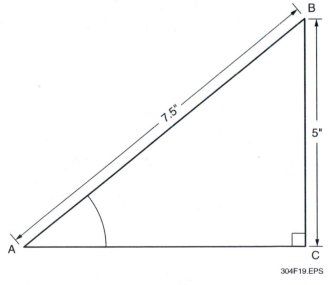

304F19.EPS

Figure 19 ◆ Triangle ABC.

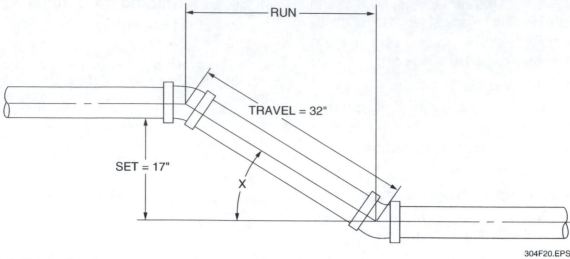

Figure 20 ◆ Piping offset 4.

offset with the travel moving diagonally across the box from corner to corner. *Figure 21* shows a rolling offset.

Assume that the roll of a 60-degree offset is 8 inches and the set is 15 inches.

Follow these steps to calculate the travel of a rolling offset, using a framing square.

Step 1 Lay out the roll on one side of the framing square and the height on the other side of the square. *Figure 22* shows laying out a rolling offset, using a framing square.

Step 2 Measure the distance between these two points on the framing square. That is the set.

NOTE

In this example, the measurement is 17 inches.

Step 3 Divide the measurement by the sine for the angle of the fitting.

NOTE

Since the measurement taken in Step 2 was 17 inches and you are using 60-degree elbows, divide 17 by 0.866 to get 19.6 or 19⅝ inches. This is the length of travel from center to center.

4.5.0 Interpolation

When precise measurements involving trigonometry are made, it is sometimes necessary to find an unknown value between two given values on a trig functions table. Calculating an unknown value in this way is called interpolation. It is used when a trig functions calculator is unavailable.

When a value is interpolated, increasing and decreasing functions are calculated differently. The functions of sines and tangents are called increasing functions because they increase when the size of the angle increases. The functions of cosines are called decreasing functions because they decrease when the size of the angle decreases.

These calculations can be used for the following:

- Interpolating numerical values from angles
- Interpolating angles from numerical values

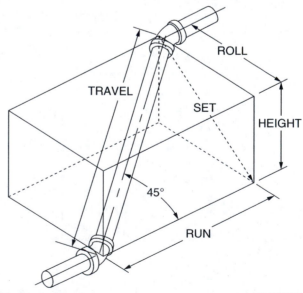

Figure 21 ◆ Rolling offset.

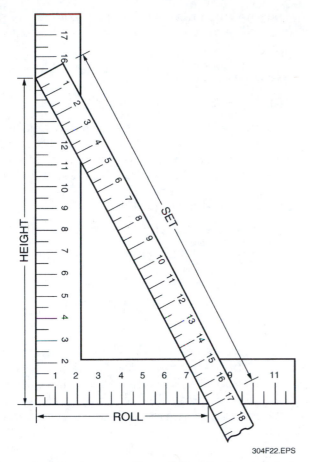

Figure 22 ◆ Laying out rolling offset, using framing square.

304F22.EPS

4.5.1 Interpolating Numerical Values from Angles

Follow these steps to calculate the numerical values of the functions of angles.

Step 1 Find the numerical value of the next larger angle in the proper column.

Example: sin 38 degrees = 0.6157

 NOTE

For the following steps, assume that the angle to be interpolated is sin 37.4 degrees. Use *Table 4* to find the numerical values for the calculations.

Step 2 Find the numerical value of the next smaller angle in the proper column.

Example: sin 37 degrees = 0.6018

Step 3 Subtract the smaller value from the larger value.

Example: 0.6157 – 0.6018 = 0.0139

Step 4 Multiply the difference by the fraction of the degree to be interpolated.

Example: 0.0139 × 0.4 = 0.00556

 NOTE

If calculating an increasing function, go to Step 5. If calculating a decreasing function, go to Step 6.

Step 5 Add the answer from step 4 to the value of the smaller angle found in Step 2.

Example: 0.00556 + 0.6018 = 0.60736 rounded to 0.6074

Step 6 Subtract the answer from Step 4 from the value of the smaller angle found in Step 2.

4.5.2 Interpolating Angles from Numerical Values

Follow these steps to calculate an angle when the numerical value is known.

 NOTE

For the following steps, assume that the numerical value of the function of the unknown angle is 0.8942. Use *Table 4* to find the numerical values for the calculations.

Step 1 Find the next larger numerical value in the proper column.

Example: cos 26 degrees = 0.8988

Step 2 Find the next smaller numerical value in the proper column.

Example: cos 27 degrees = 0.8910

Step 3 Subtract the smaller value from the larger value.

Example: 0.8988 – 0.8910 = 0.0078

Step 4 Subtract the value of the unknown angle from the value of the smaller angle.

Example: 0.8988 – 0.8942 = 0.0046

Step 5 Divide the number found in Step 4 by the number found in Step 3.

Example: 0.0046/0.0078 = 0.5897 rounded to 0.6

Step 6 Add the answer from Step 5 to the value of the smaller angle found in Step 1.

Example: cos 26 degrees + 0.6 = 26.6 degrees

5.0.0 ◆ CALCULATING TAKEOUTS USING TRIGONOMETRY

Takeout, takeoff, or takeup is the length of space that a fitting occupies in a run of pipe. Manufacturers' information and many pipefitting reference books list the dimensions of fittings, but these references are not always available. The pipefitter must be able to calculate takeouts when determining the cut lengths of pipe to be placed between fittings. This section explains calculating takeouts for other elbows.

5.1.0 Takeouts

Offsets often turn through odd-sized angles. The odd-angled elbows are usually fabricated from standard, 90-degree long radius elbows. The formula for finding the takeout of any welding elbow is the following:

tan (½ turn) × radius of elbow

This formula can be divided into three parts:

- *Tangent* – The tangent of an angle can be found in a trig functions table. If you know the dimensions of the right triangle, the tangent can be figured by dividing the side opposite the angle by the side adjacent to the angle.
- *½ turn* – To figure the takeout for any elbow, divide the number of degrees in the elbow by 2, and then find the tangent of this angle.
- *Radius of elbow* – The radius of a standard, long radius 90-degree welding elbow is the same as the takeout of the elbow. To find the radius, multiply the nominal size by 1½.

This formula can be used to figure the takeout for a 6-inch, 30-degree welding elbow. *Figure 23* shows a 30-degree welding elbow.

takeout = tan (½ turn) × radius
takeout = tan (½ × 30 degrees) × (1½ × 6 inches)
takeout = tan 15 degrees × 9 inches
takeout = 0.2679 × 9 inches
takeout = 2.41 inches or 2¹³⁄₃₂ inches

5.2.0 Odd Angles

To determine the inside and outside cutpoints on the elbow for an odd-degree miter, you can use the radius of the inside and outside of the elbow, and divide 360 degrees by the degree of the turn. Since the formula for the length of a circle is pi times the diameter, pi times twice the radius of the inside of the elbow times the part of a circle will allow you to calculate the length along the centerline of the inside arc.

L(arc) = 2πr (degree of bend/360)

The same formula used on the outside arc of the fitting will give you the cutpoint on that side of the miter. For example:

You are to cut a six-inch, schedule 40 LR elbow to make a 22½ degree miter fitting.

360/22.5 = 16

The outside diameter of a six-inch, schedule 40 fitting is 6.625 inches. So, since the centerline radius of the elbow is nine inches, the inside radius must be 3.3125 inches less and the outside radius 3.3125 inches more.

For the outside cut point:

Length of arc = 2(12.3125) × 3.1416/16

Length of arc = 4.835 inches on the outside arc

For the inside cut point:

Length of arc = 2(9-3.3125) × 3.1416/16

Length of arc = 2.2335

The lengths of the arc segments for such elbows are available as precalculated tables in several published sources. The other way to lay out such a fitting is to draw a full-size template (*Figure 24*), lay out the angles, and transfer the intersections to the elbow.

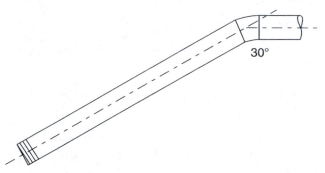

Figure 23 ◆ 30-degree welding elbow.

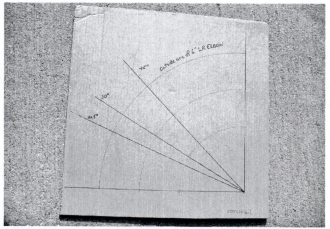

Figure 24 ◆ Full-size template.

1. The fraction equivalent of 6.125 is _____.
   a. 6⅛
   b. 6¼
   c. 6½
   d. 6⅞

2. The decimal equivalent of ⁹⁄₁₆ inch is _____.
   a. 0.375
   b. 0.4125
   c. 0.5000
   d. 0.5625

3. One atmosphere equals _____ pounds per square inch.
   a. 0.8842
   b. 14.696
   c. 29.92
   d. 34

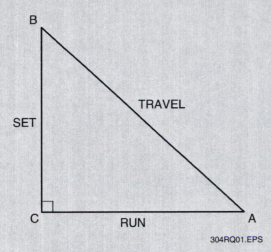

Figure 1

4. If the set in *Figure 1* is 4 feet, and the travel is 35 feet, the run is _____. (Round off to the nearest tenth.)
   a. 31 feet
   b. 34.8 feet
   c. 39 feet
   d. 51 feet

5. If the set in *Figure 1* is 6 feet and the run is 8 feet, the travel is _____. (Round off to the nearest tenth.)
   a. 4 feet
   b. 8 feet
   c. 10 feet
   d. 12 feet

6. The sine of a reference angle is the ratio of the _____.
   a. opposite side to the adjacent
   b. opposite angle to the reference angle
   c. side opposite to the hypotenuse
   d. side adjacent to the hypotenuse

7. The cosine of the reference angle is the ratio of the _____.
   a. opposite side to the adjacent
   b. opposite angle to the reference angle
   c. side opposite to the hypotenuse
   d. side adjacent to the hypotenuse

8. If the set is 4 feet and the travel is 35 feet, angle A is _____.
   a. 3.6 degrees
   b. 6.6 degrees
   c. 11.2 degrees
   d. 15 degrees

9. If side AC is 14 feet and the hypotenuse is 19 feet, the value of cos angle A to four places is _____.
   a. 0.3571
   b. 0.5000
   c. 0.7368
   d. 0.8055

10. The ratio of the two sides in a right triangle is directly related to the size of the _____ in the triangle.
    a. apex
    b. left angle
    c. reference angle
    d. right angle

11. If the set is 25 feet, and angle A is 22.5 degrees, the travel is _____ feet.
    a. 35.2
    b. 45.5
    c. 65.3
    d. 71

12. If angle B is 30 degrees and the hypotenuse is 48", the run is _____ inches.
    a. 12
    b. 16
    c. 24
    d. 36

13. If angle A is 45 degrees and side run is 48",
    side AB is _____ inches.

    a. 48
    b. 56.6
    c. 67.9
    d. 72

14. If the set is 20 inches and angle B is 75 degrees,
    the travel is _____ inches.

    a. 45.2
    b. 62.8
    c. 77.3
    d. 74.6

15. If angle A is 30 degrees, and side BC is 45
    inches, side AC is _____ inches.

    a. 77.9
    b. 84.5
    c. 90.1
    d. 48

16. If angle A is 30 degrees, angle B is _____
    degrees.

    a. 22.5
    b. 30
    c. 60
    d. 90

17. If angle A is 45 degrees and side AC is 48",
    side BC is _____ inches.

    a. 45
    b. 48
    c. 52
    d. 56.6

18. If the run is 24 inches and the travel is 48
    inches, angle A is _____ degrees.

    a. 30
    b. 45
    c. 60
    d. 90

19. If the set is 6 feet and the run is 8 feet, angle
    A is _____ degrees.

    a. 10
    b. 22.5
    c. 36.9
    d. 45

20. If angle A is 15 degrees and side BC is 20
    inches, side AC is _____ inches.

    a. 35
    b. 48.3
    c. 69.5
    d. 74.6

21. If the set is 25 feet and angle A is 22.5
    degrees, the run is _____ feet.

    a. 16
    b. 25
    c. 45.3
    d. 60.4

22. If the run is 3 and the travel is 7", angle A is
    _____ degrees.

    a. 64.6
    b. 70.1
    c. 72
    d. 84.6

23. If the set is 2 feet and the travel is 11 feet,
    angle A equals _____ degrees.

    a. 10.5
    b. 15
    c. 22.4
    d. 25.4

24. The height of a rolling offset is 6 feet, the roll
    is 8 feet, and the angle is 45 degrees, the
    travel is _____ feet.

    a. 10
    b. 11.2
    c. 14.1
    d. 17.2

25. To determine the cutpoint for the inside arc
    of an elbow, multiply the diameter of the
    inside arc by _____.

    a. the outside diameter
    b. ⅔π
    c. (360/degree of angle)
    d. πr²

Summary

Pipefitters use a lot of different tools, and not all of those are made out of metal. The mental and mathematical tools allow you to know the shape and dimensions of assemblies before you build them. The trigonometric ratios, sine, cosine, and tangents tell you the set, run, angles, and travel of an offset from incomplete information, such as the difference of elevation and the angle of the offset, or the travel and the run. Mathematics will also allow you to cut custom fitting angles.

Notes

Adjacent side: The side of a right triangle that is next to the reference angle.

Cosine: Trigonometric ratio between the adjacent side and the hypotenuse, written as adjacent divided by the hypotenuse.

Hypotenuse: The longest side of a right triangle. It is always located opposite the right angle.

Opposite side: The side of a right triangle that is located directly across from the reference angle.

Ratio: A comparison of one value to another value.

Reference angle: The angle to which the sides are related as adjacent and opposite.

Sine: Ratio between the opposite side and the hypotenuse, the opposite divided by the hypotenuse.

Tangent: Ratio between the opposite side and the adjacent side, opposite divided by the adjacent.

Resources & Acknowledgments

Additional Resources

This module is intended to be a thorough resource for task training. The following reference works are suggested for further study. These are optional materials for continued education rather than for task training.

Pipe Fitter's Math Guide, 1989. Johnny Hamilton. Clinton, NC: Construction Trade Press.

Applied Construction Math, Latest Edition. Upper Saddle River, NJ: Prentice Hall Publishing.

Figure Credits

Topaz Publications, Inc., 304F05 (photo), 304F16, 304F24

The NCCER makes every effort to keep these textbooks up-to-date and free of technical errors. We appreciate your help in this process. If you have an idea for improving this textbook, or if you find an error, a typographical mistake, or an inaccuracy in NCCER's Contren® textbooks, please write us, using this form or a photocopy. Be sure to include the exact module number, page number, a detailed description, and the correction, if applicable. Your input will be brought to the attention of the Technical Review Committee. Thank you for your assistance.

Instructors – If you found that additional materials were necessary in order to teach this module effectively, please let us know so that we may include them in the Equipment/Materials list in the Annotated Instructor's Guide.

Write: Product Development and Revision
National Center for Construction Education and Research
3600 NW 43rd St, Bldg G, Gainesville, FL 32606

Fax: 352-334-0932

E-mail: curriculum@nccer.org

Craft _____ Module Name _____

Copyright Date _____ Module Number _____ Page Number(s) _____

Description _____

_____

_____

_____

(Optional) Correction _____

_____

_____

(Optional) Your Name and Address _____

_____

_____

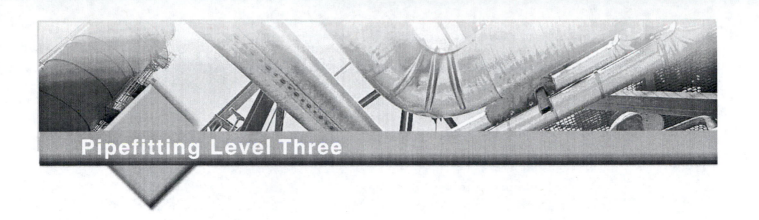

08305-07

Motorized Equipment II

08305-07
Motorized Equipment II

Topics to be presented in this module include:

Overview

Pipefitters use many different machines to move materials. It is also frequently necessary that the fitter be lifted off the ground to gain access to the spool connections. Manlifts and scaffolds are the way that pipefitters get to where the work is.

Objectives

When you have completed this module, you will be able to do the following:

1. Identify and explain types of manlifts.
2. Explain manlift safety rules and hazards.
3. Inspect scissors-type and telescoping boom manlifts.
4. Explain the use of cable lifts.
5. Identify and explain the use of drain cleaners.

Trade Terms

Cornice
Counterweight
Hydraulic
Lanyard
Load rating
Outrigger
Parapet wall
Safety factor
Wire rope

Required Trainee Materials

1. Pencil and paper
2. Appropriate personal protective equipment

Prerequisites

Before you begin this module, it is recommended that you successfully complete *Core Curriculum*; *Pipefitting Level One*; *Pipefitting Level Two*, and *Pipefitting Level Three*, Modules 08301-07 through 08304-07.

This course map shows all of the modules in the third level of the *Pipefitting* curriculum. The suggested training order begins at the bottom and proceeds up. Skill levels increase as you advance on the course map. The local Training Program Sponsor may adjust the training order.

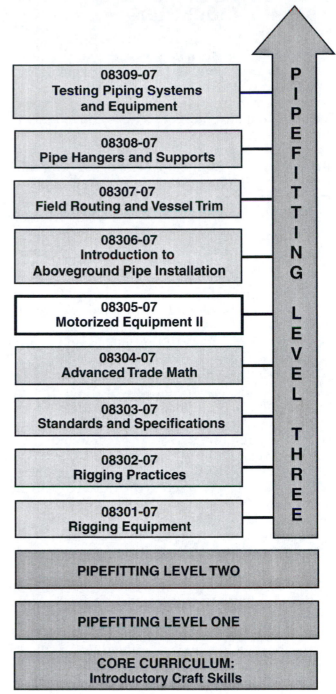

305CMAP.EPS

1.0.0 ◆ INTRODUCTION

Pipefitters use various types of motorized equipment throughout their careers. Company policies, as well as local, state, and federal regulations, dictate the type of certification and training required to operate various types of motorized equipment. This module introduces you to motorized equipment that pipefitters use and work with in the vicinity of the job site. Before operating any piece of motorized equipment, you must read and fully understand the manufacturer's operating procedures for a specific piece of equipment. You must also obtain authorization from your immediate supervisor to operate a piece of equipment.

2.0.0 ◆ MANLIFTS

Manlifts are used to lift workers and tools to overhead work areas. Manlift controls allow the operator to adjust the working platform to almost any position. Manlifts are self propelled and can be powered by batteries, gas or diesel engines, or **hydraulic** power. Pipefitters must be able to operate manlifts safely and effectively. Manlifts are available in many different models, sizes, working heights, and lifting capacities. Two common types of manlifts are the scissors-type and the telescoping boom.

Scissors-type manlifts are driven beneath the work area and then raised to provide access to the work area. They are available in rough-terrain or indoor models with two- or four-wheel drive. *Figure 1* shows a scissors-type manlift. Note that not all scissors lifts have **outriggers**.

Telescoping boom manlifts are more versatile and have more working positions than scissors manlifts. The work platform on a telescoping boom manlift is smaller than that of a scissors-type manlift and is referred to as a basket. The basket is attached to a boom mounted to a turntable that allows the work platform to rotate in a full circle. Some telescoping boom manlifts are designed for direct overhead work; others are designed to lift the basket up vertically and over horizontally to enable better access in congested work areas. *Figure 2* shows telescoping boom manlifts.

2.1.0 Manlift Safety

Before operating any manlift, you must know what the machine can and cannot do. Never operate a manlift until after you have read and fully understand the manufacturer's operation manual. Practice operating the manlift under close supervision of a qualified operator before you

operate the manlift alone. Also check your company policies regarding necessary certifications before operating a manlift.

2.1.1 General Safety Rules

Follow these general safety rules when working with any type of manlift:

- Do not stand on midrails or handrails.
- Do not place ladders, steps, or similar items on the work platform to increase reach.
- Do not attach wire, cable, or similar items to the platform.
- Maintain a clearance of at least 15 feet between any part of the machine and any electrical line or power source. Always allow room for platform movement and electrical line swaying.
- Use safety props when the platform is extended for maintenance or inspection purposes.
- Wear approved headgear at all times when working from a manlift or on the ground in areas where manlifts are being used.
- Wear a safety harness with a **lanyard** attached to the basket or the rails of the work platform.
- Read and obey all safety warnings and cautions located on the manlift.
- Observe all weight limits for the manlift you are using. Do not overload the manlift.

2.1.2 Operating Safety Rules

 WARNING!
Do not circumvent or defeat any safety devices.

Follow these basic safety rules when operating any manlift:

- Do not operate any manlift unless you have been authorized and qualified to do so.
- Do not operate the machine unless it has been serviced and maintained according to the manufacturer's specifications and schedule.
- Do not allow anyone to tamper with, service, or operate the manlift from the ground while the platform is being used.
- Do not allow anyone within 6 feet of the manlift when the platform is being raised or the machine is moving.
- Beware of overhead obstructions when raising the work platform.

- Fasten the platform chain gates before operating the manlift.
- Do not raise the platform unless the handrails are in place and secure.
- Keep both feet on the platform deck when working from the platform.
- Hold onto the platform handrail when traveling or when raising the platform.
- Allow only one person to be responsible for all machine operations when two or more people are on the platform.
- Do not move or work from the manlift on soft or uneven surfaces.

- Do not elevate the platform when wind speed is above 30 miles per hour.
- Do not drive the manlift and operate the work platform at the same time.
- If the manlift is equipped with outriggers, do not raise the platform unless the outriggers are fully extended horizontally and are lowered and in full contact with the ground or floor.
- Ensure that the machine is level before operating the work platform.
- Do not use the platform or basket to lift materials that exceed the weight limits of the manlift.

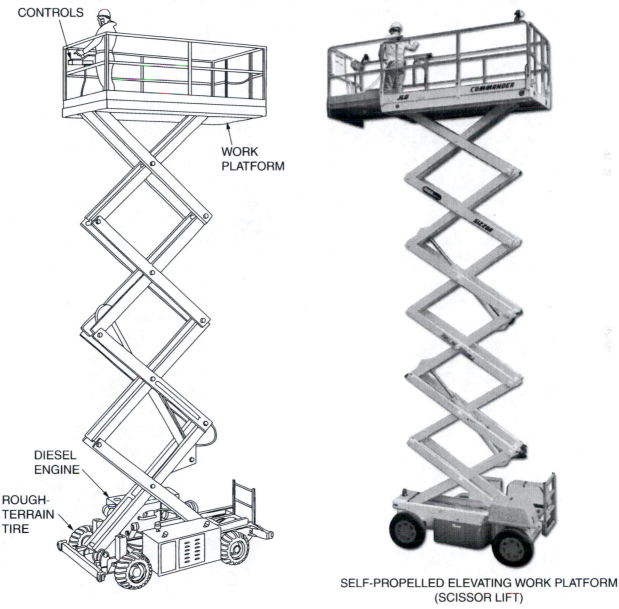

CONTROLS

WORK PLATFORM

DIESEL ENGINE

ROUGH-TERRAIN TIRE

SELF-PROPELLED ELEVATING WORK PLATFORM
(SCISSOR LIFT)

305F01.EPS

Figure 1 ◆ Scissors-type manlift.

2.1.3 Driving Safety Rules

Follow these safety guidelines when driving any manlift:

- Be alert and attentive when driving, keeping your eyes in the direction of travel.
- Do not travel until the outriggers, beams, jack, and platform are completely retracted.
- Do not travel on grades steeper than 10 percent or on side slopes greater than 5 percent.
- Use high engine speed when traveling up an incline and low engine speed when traveling down an incline.
- Do not drive over any type of temporary floor covering unless you have verified that the floor covering is of adequate strength to support the load.

- Allow at least 6 feet to stop when traveling at normal speeds and 3 feet to stop when traveling at low speeds.
- Do not shift the control lever through neutral and into the opposite direction. Always shift to neutral, stop the manlift, and then move the lever to the desired position.
- Travel in reverse for short distances only, and always give a warning signal.

2.1.4 Scissors-Type Manlift Safety Rules

Follow these safety rules when using scissors-type manlifts:

- Do not allow people, materials, or objects between the extended scissors of the manlift.

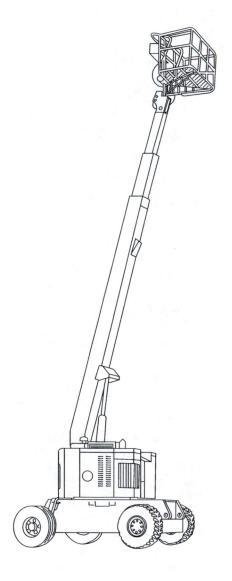

BOOM SUPPORTED WORK
PLATFORM (BOOM LIFT)

305F02.EPS

Figure 2 ◆ Telescoping boom manlifts.

- Before raising the lift, ensure that the outriggers are fully extended and lowered to within 1 inch of the floor.
- Do not move or drive the scissors manlift unless the platform is completely lowered.

2.1.5 Telescoping Boom Manlift Safety Rules

Follow these safety rules when using telescoping boom manlifts:

- Ensure that the boom is fully lowered and that the turntable is secured before moving or driving the manlift.
- Position the boom in the direction of travel.
- Do not push or pull objects by retracting or extending the boom.
- Place a physical barrier or barricade around the manlift before raising the work platform.
- Retract boom to lowest possible length and lower boom before traveling

2.2.0 Inspecting Manlifts

Before operating the manlift, a walk-around inspection must be performed. This inspection should be performed according to the manufacturer's recommendations. The following guidelines apply to all manlifts:

- Inspect all standing surfaces for cleanliness. Ensure that no oil, fuel, spills, rags, papers, or tools are on the manlift.
- Keep all information and operating placards clean and readable.
- Ensure that the machine operating log is current and that no entries have been left uncleared.
- Ensure that all items needing daily lubrication have been serviced.
- Ensure that the brakes are in good working condition and will stop the machine.

2.2.1 Inspecting Scissors-Type Manlifts

Figure 3 shows the scissors-type manlift inspection points. The lift is raised in *Figure 3* so that the parts can be seen. Do not raise the lift before the actual inspection is made.

Follow these guidelines to inspect scissors-type manlifts:

- Ensure that the chain gate is properly secured and has no visible damage.
- Ensure that the handrail assembly is properly secured, has no loose or missing parts, and has no visible damage.

- Ensure that the work platform has no loose or missing parts, visible damage, or excessive dirt and debris.
- Ensure that the scissors arms have no visible damage, abrasions, or distortions.
- Ensure that the pivot pins have no loose or missing hardware, no damage or wear to pin heads to cause the pin to rotate, and no wear to the pin or bushing.
- Ensure that the lift cylinder has no rust, nicks, scratches, or foreign material on the shaft.
- Ensure that the frame has no visible damage and no loose or missing parts.
- Ensure that the outriggers have no loose, damaged, or missing fasteners and lock pins.
- Ensure that the hydraulic cylinders have no hose damage or leakage.
- Ensure that the tires are properly inflated and have no visible damage.
- Ensure that the wheels have no loose or missing lug nuts.
- Ensure that the sliding wear pad blocks have plenty of lubrication and no excessive wear.
- Ensure that the hydraulic oil supply is full and that the filter element is in good condition.
- Ensure that the drive hubs have no visible damage or leakage.
- Ensure that the tie rods and linkages have no visible damage and no loose or missing parts.
- Ensure that the front spindle assembly has no excessive wear or damage.
- Ensure that all switches on the ground and platform control boxes are working properly and that the directions are secure and readable.
- Ensure that the cooling fans are free of obstructions.
- Ensure that the battery is properly charged, has no visible damage or corrosion, and has tight cables.
- Ensure that the muffler and exhaust systems are properly secured and have no leakage.
- Ensure that the engine oil is at the proper level and that the filler cap is secure.
- Ensure that the hydraulic pump and valve bank have no leakage.
- Ensure that the fuel tank is full and has no visible damage or leakage.
- Ensure that the electrical and hydraulic lines have no visible damage or cuts.

2.2.2 Inspecting Telescoping Boom Manlifts

Figure 4 shows the top view of the telescoping boom manlift inspection points.

Follow these guidelines to inspect telescoping boom manlifts:

- Ensure that the work platform has no loose or missing parts, visible damage, or excessive dirt and debris.

- Ensure that the frame has no visible damage and no loose or missing parts.
- Ensure that the tires are properly inflated and have no visible damage.
- Ensure that the wheels have no loose or missing lug nuts.
- Ensure that the hydraulic oil supply is full and that the filter element is in good condition.

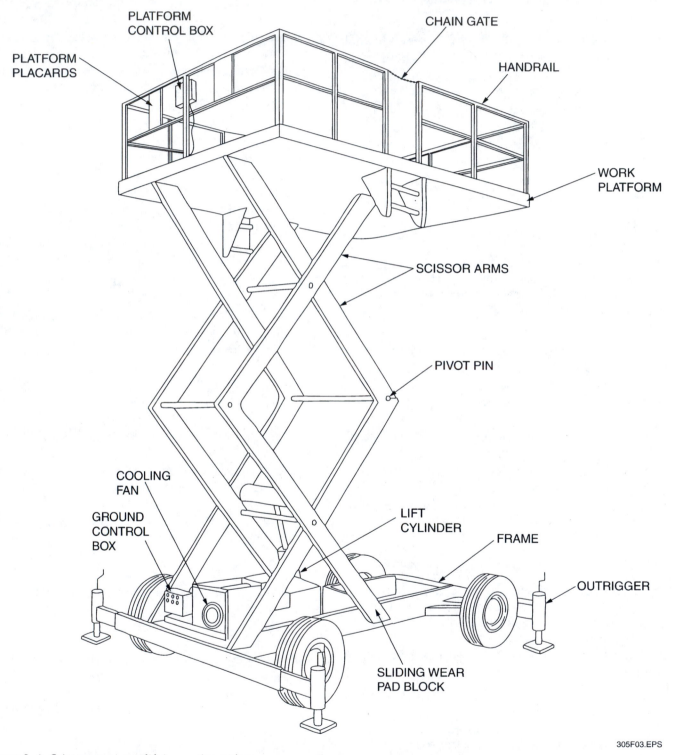

305F03.EPS

Figure 3 ◆ Scissors-type manlift inspection points.

- Ensure that the drive hubs have no visible damage or leakage.
- Ensure that the tie rods and linkages have no visible damage and no loose or missing parts.

- Ensure that the tie rod end studs are locked and that the lock pins are properly installed on extending-axle machines.

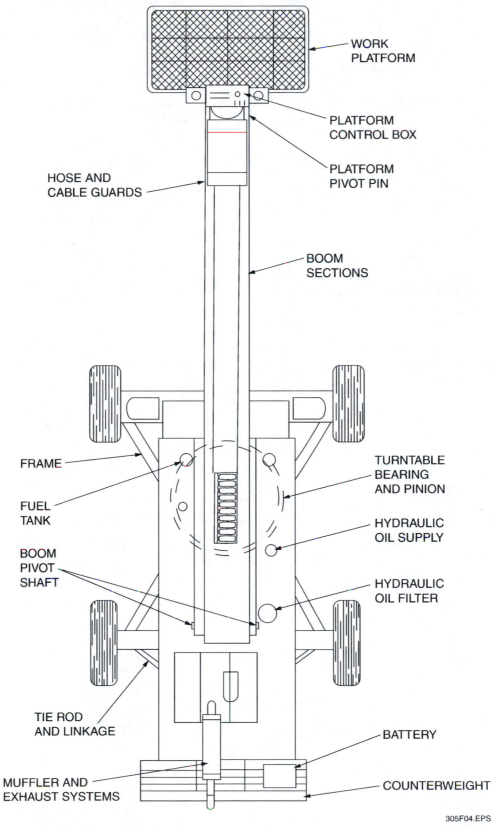

WORK PLATFORM

PLATFORM CONTROL BOX

PLATFORM PIVOT PIN

HOSE AND CABLE GUARDS

BOOM SECTIONS

FRAME

FUEL TANK

BOOM PIVOT SHAFT

TURNTABLE BEARING AND PINION

HYDRAULIC OIL SUPPLY

HYDRAULIC OIL FILTER

TIE ROD AND LINKAGE

BATTERY

MUFFLER AND EXHAUST SYSTEMS

COUNTERWEIGHT

305F04.EPS

Figure 4 ◆ Telescoping boom manlift inspection points.

- Ensure that all switches on the ground and platform control boxes are working properly and that the directions are secure and readable. *Figure 5* shows a typical platform control box.
- Ensure that the battery is properly charged, has no visible damage or corrosion, and has tight cables.
- Ensure that the muffler and exhaust systems are properly secured and have no leakage.
- Ensure that the engine oil is at the proper level and that the filler cap is secure.
- Ensure that the hydraulic pump and valve bank have no leakage.
- Ensure that the fuel tank is full and has no visible damage or leakage.
- Ensure that the electrical and hydraulic lines have no visible damage or cuts.
- Ensure that the power track has no loose, damaged, or missing parts.
- Ensure that the control valve and engine compartment have no loose or missing parts, no leakage, and no unsupported wires or hoses.

- Ensure that the **counterweight** is properly secured.
- Ensure that all cowling, doors, and latches are properly secured, are in proper working condition, and have no loose or missing parts.
- Ensure that the turntable bearing and pinion has no visible damage and has no loose or missing hardware.
- Ensure that the boom sections have no visible damage and that the wear pads are secure.
- Ensure that the platform pivot pin is properly secured.

3.0.0 ◆ CABLE LIFTS

A cable lift is powered, suspended scaffolding. The cable lift is suspended by **wire rope** and powered by an electric or pneumatic motor. Small cable lifts designed for one person are known as single-stage cable lifts. Fly decks can be added to each side of the single-stage cable lift to accommodate as many as two people working from the lift. Two cable lifts can be joined by fixed or adjustable platform sec-

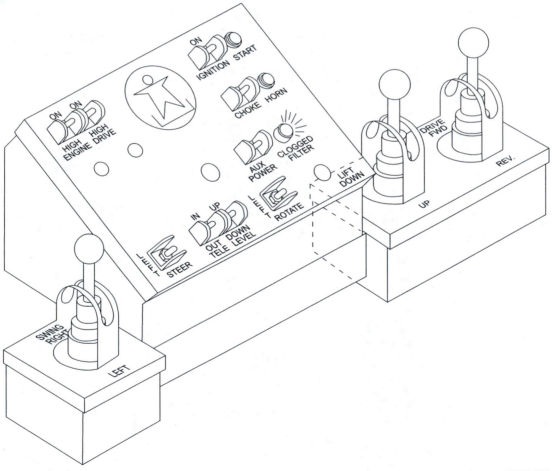

305F05.EPS

Figure 5 ◆ Platform control box.

tions to increase the working area of the cable lift up to 48 feet. When working from a suspended scaffold, wear a safety harness that is securely tied to the structure either directly or by means of a wire rope. The safety harness cannot be tied to the cable lift or to the wire rope suspending the cable lift. *Figure 6* shows cable lifts.

Persons using these types of lifts must be trained by the equipment manufacturer to properly set up and use the cable lift.

3.1.0 Cable Lift Setup

The rigging device and the adjoining structure must be capable of supporting the maximum gross load of the cable lift, which is normally 1,000

TWO-PERSON CABLE LIFT

MODULAR PLATFORM

305F06.EPS

Figure 6 ◆ Cable lifts.

pounds, with a **safety factor** of 4. The rigging device and cable lift must be set up by trained, qualified rigging personnel who have been authorized to set up the cable lift. A separate rigging method must be supplied for the safety line of the worker on the cable lift. The cable lift can be suspended from an I-beam, a **cornice** or **parapet wall**, or the roof of a building.

3.1.1 I-Beam Rigging

Manufacturers of cable lifts supply special clamps for attaching the cable lift wire rope to I-beams. These devices should not be used on a vertical beam or in a position that would place the suspension rope parallel to the beam. The cable lift beam clamp can be stationary or equipped with rollers so that it is movable along the beam. When using rolling beam clamps, ensure that the width of the I-beam flange is consistent and that the I-beam has no cutouts or notches that would allow the roller to roll off. The end of the beam must also be blocked, clamped, or constructed to prevent the roller from rolling off. The required **load rating** and the width of the I-beam flange dictate the type of I-beam clamp to use. *Figure 7* shows the types of I-beam clamps.

3.1.2 Cornice Hook Rigging

A cornice hook is used to suspend a cable lift from a cornice or parapet wall of a building. Ensuring that the cornice or parapet is able to support the maximum load rating with a safety factor of 4 is critical when using this type of rigging hardware. The cornice hook fits over the cornice or parapet wall, and a block of wood is placed between the point of the hook and the wall. A tieback is attached to the hook and must be firmly secured to a structurally sound portion of the building. The tieback must have a strength equivalent to the suspension rope and must also have a load-rating safety factor of four. The suspension wire rope must be shackled to the cornice hook and held away from the wall so that it passes through the wire rope guide of the cable lift without obstruction. It may be necessary to use a standoff attachment with the hook to hold the wire rope away from the wall. *Figure 8* shows cornice hooks.

3.1.3 Portable Roof Outrigger Rigging

A portable roof outrigger supports a vertically hung, suspended wire rope for a cable lift from a flat roof or floor of a building. An outrigger is required to be either fastened very strongly to the roof or counterweighted to a safety factor of at least four times the expected load. If it is fastened to the roof, the beam, the fastening, and the point of connection must be designed and approved by an engineer. If it is counterweighted, the beam and the counterweights must be designed for that purpose, and the weights must be securely fastened to the outrigger beam. The length of the beam over the roof itself must be at least one and one half times the length extending off the edge. Care must be taken to ensure that the roof or floor is capable of withstanding the weight of the rated load and the weight of the outrigger and counterweight. *Figure 9* shows two types of portable roof outriggers.

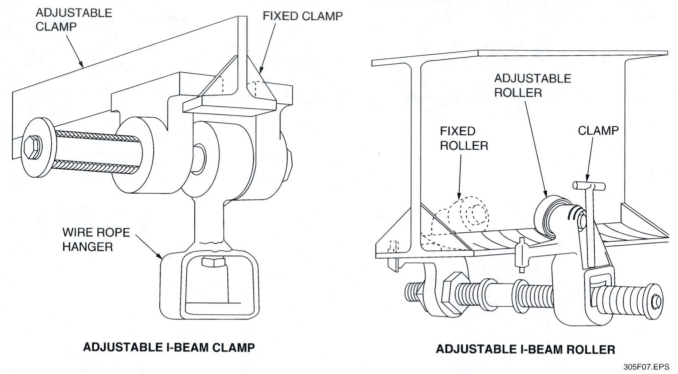

Figure 7 ◆ I-beam clamps.

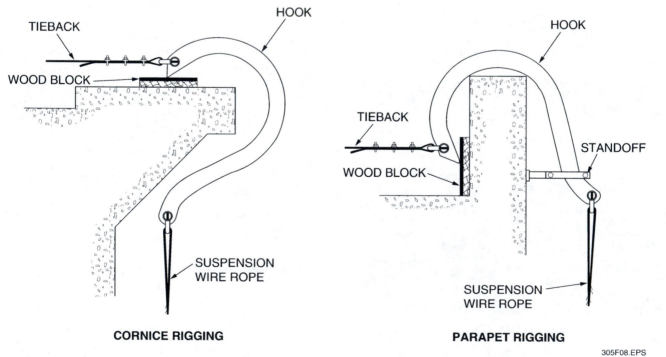

Figure 8 ◆ Cornice hooks.

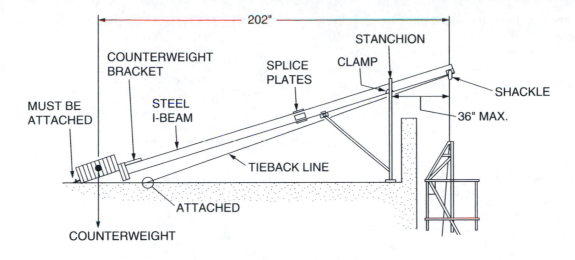

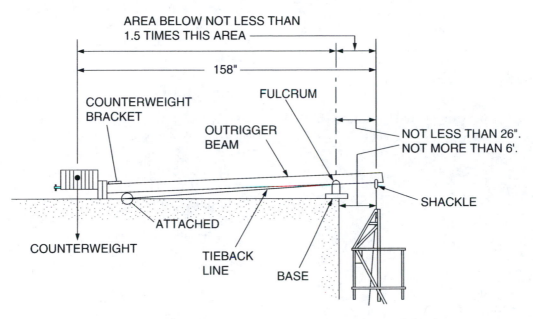

Figure 9 ◆ Portable roof outriggers.

305F09.EPS

3.2.0 Cable Lift Safety

If used improperly, cable lifts can cause accidents resulting in injuries or even death. You must be properly trained on the specific equipment that you are using and certified at your job site before operating any cable lift. The Occupational Safety and Health Administration's Standards for the Construction Industry govern the use of powered scaffolds. The Scaffold Industry Association, Inc., also provides safety guidelines for the use of powered, suspended scaffolds. All of the safety guidelines that apply to scaffolds and the provisions specific to the use of suspended, powered scaffolds apply to cable lifts. The wire rope, tiebacks, wall hooks, beam clamps, and safety harnesses must be inspected by qualified personnel before they are used. OSHA sets the minimum guidelines for the type of wire rope, hooks, and beam clamps that are used. No persons should set up or operate a cable lift unless they have been specifically trained on all local, state, and federal regulations pertaining to scaffolding, suspended scaffolding, and personal protective equipment.

4.0.0 ◆ HYDRAULIC TORQUE WRENCHES

Hydraulic torque wrenches are often used for assembling and disassembling large flanges with high bolt torque applications. These automatic wrenches can effectively decrease bolting time to reduce maintenance costs and can provide up to 20,000 foot-pounds of torque for heavy-duty bolting operations.

The hydraulic torque wrench has a reaction arm that rests against the flange to transfer the torque to the flange, making the wrench easy to handle. Always ensure that the reaction arm is resting against the flange before applying pressure to the wrench. Serious injury could occur if there is a gap between the reaction arm and the flange.

The hydraulic torque wrench is used with an electric hydraulic pump that has a calibrated pressure gauge built in to control the amount of torque that the wrench exerts on the bolt. Before using the hydraulic torque wrench, ensure that the pressure gauge has a current calibration sticker. If the gauge has not been calibrated recently, it may be inaccurate and give false readings. Dial in the correct torque setting on the hydraulic pump, and the wrench is ready to operate.

Most pumps come with 25 feet of hydraulic hose to attach to the wrench, but extra hose can be added to suit the job application. Always read and follow the manufacturer's operating instructions when using a hydraulic torque wrench. *Figure 10* shows a hydraulic torque wrench.

5.0.0 ◆ DRAIN CLEANERS

Drain cleaners clear obstructions from drain pipes by rotating a cable or rodding through the pipeline. Drain and sewer cleaning machines are available in a variety of sizes for different applications. Five types of drain cleaners include handheld drain cleaners, sectional drain cleaners, drum-type drain cleaners, rodding machines, and end tools.

5.1.0 Handheld Drain Cleaners

Handheld drain cleaners can be operated either manually or with a 7.2-volt drill motor. These drain cleaners are used in plumbing applications with small lines up to 2½ inches. They have a totally closed cable container, are lightweight, and are easy to use. The hand spinner can hold up to 25 feet of cable, and the drill model can hold up to 50 feet of cable. *Figure 11* shows handheld drain cleaners.

5.2.0 Sectional Drain Cleaners

Sectional drain cleaners clear blockages by spinning short sections of cable through the pipeline at high speeds and low torque. The cable is used with a special end tool to drill through the obstruction. The machine and cable are transported separately to the work area to make transportation of the machine easier. Cables are loaded into the machine one cable at a time during operation. *Figure 12* shows sectional drain cleaners.

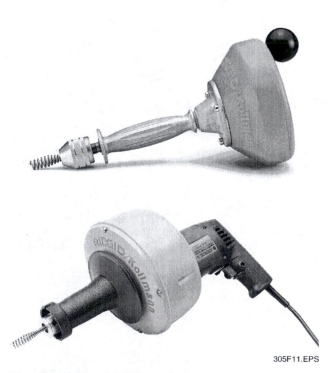

305F10.EPS

Figure 10 ◆ Hydraulic torque wrench.

305F11.EPS

Figure 11 ◆ Handheld drain cleaners.

5.3.0 Drum-Type Drain Cleaners

Drum-type drain cleaners clear obstructions in drains and sewer lines by spinning a cable at low speeds and high torque. The drum-type drain cleaner and its cable are a single unit. These units are somewhat heavier than the sectional cleaners, but setup is fast and easy because the cable is already loaded into the machine. *Figure 13* shows a drum-type drain cleaner.

5.4.0 Rodding Machines

Rodding machines clear tough obstructions in long, straight runs of large-diameter drains and sewers. Rodding machines are used with pipe that ranges in diameter from 8 to 24 inches. Rodding machines typically use a 5-horsepower, gasoline-powered engine with a 4-speed transmission to drive sectional sewer rods through the pipeline at 80 to 200 rpm. A throttle control allows the operator to adjust the speed of the rod within each gear. When the throttle is released, the engine returns to an idle and the clutch disengages to allow the operator to add more rod sections. *Figure 14* shows a rodding machine.

In addition to rodding machines and drain cleaners, video equipment such as the Rigid®

SeeSnake *(Figure 15)* can be fed through a piping system, allowing the operator to view obstructions or breaks in the line.

305F13.EPS

Figure 13 ◆ Drum-type drain cleaner.

305F14.EPS

Figure 14 ◆ Rodding machine.

ELECTRIC

PNEUMATIC

305F12.EPS

Figure 12 ◆ Sectional drain cleaners.

305F15.EPS

Figure 15 ◆ SeeSnake.

5.5.0 Drain Cleaner End Tools

All drain cleaners operate in basically the same manner. The cable is inserted into the pipe, and the drain cleaner is started in the forward direction. Once all of the cable is run through the pipe, the machine is run in reverse to remove the cable. A special end tool is attached to the cable, depending on the type of clog in the pipe. A wide variety of end tools are available for different applications. *Figure 16* shows drain cleaner end tools.

Follow these guidelines when selecting an end tool to clean a drain:

- Use the drop-head auger for cleaning back-to-back mounted fixtures, such as sinks, where the cable must be fed to enter the pipe.

- Use the straight auger for exploring and breaking up stoppages or for removing samples of the stoppage to the surface to determine what type of end tool is needed.
- Use the funnel auger for breaking up the remains left by the straight auger.
- Use the hook auger for hooking and breaking up heavy and dense root stoppages.
- Use the retrieving auger for finding a cable that has broken or is lost.
- Use the grease cutter for clearing lines that have become clogged with grease or detergents.
- Use the four-blade saw-tooth cutter for unclogging blockages caused by hardened, glazed material, such as chemical deposits.

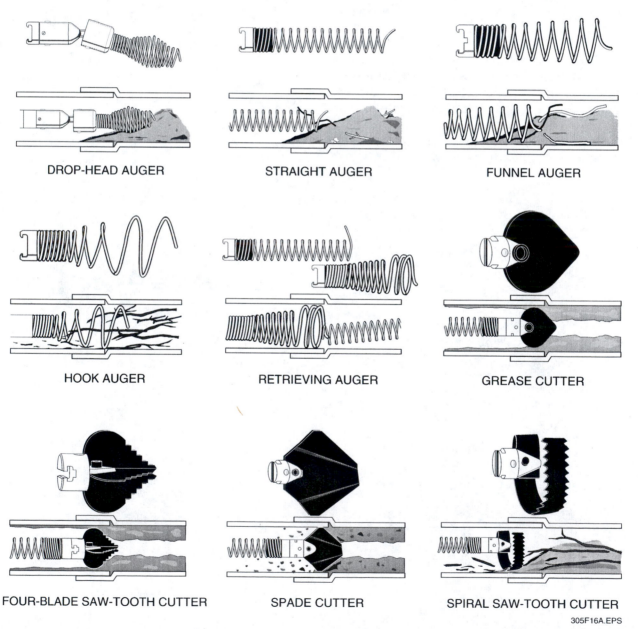

DROP-HEAD AUGER STRAIGHT AUGER FUNNEL AUGER

HOOK AUGER RETRIEVING AUGER GREASE CUTTER

FOUR-BLADE SAW-TOOTH CUTTER SPADE CUTTER SPIRAL SAW-TOOTH CUTTER

305F16A.EPS

Figure 16 ◆ Drain cleaner end tools. (1 of 2)

- Use the spade cutter for opening floor drains after an auger has been used.
- Use the spiral saw-tooth cutter for breaking up any type of stoppage caused by roots, rags, or sticks.
- Use the saw-tooth cutter for clearing lines blocked heavily by roots.
- Use the spiral-bar cutter for clearing main sewer lines blocked by roots, leaf debris, sticks, sawdust, and cloth.
- Use the shark-tooth cutter for removing material stuck to the internal pipe walls.

- Use the grease "C" cutter for clearing grease blockage in lines leading from garbage disposal units or waste pipe.
- Use the expanding finish cutters for final removal of material stuck to walls and certain fibrous roots.
- Use the chain knocker for cleaning scale from the internal pipe walls and boiler tubes.
- Use the flue brush for finish-cleaning required for boiler tubes and heat exchangers.

Much pipe cleaning, for reasons of efficiency, is done with high-pressure water jetting equipment, but such equipment is both loud and dangerous, and therefore not suitable for some applications.

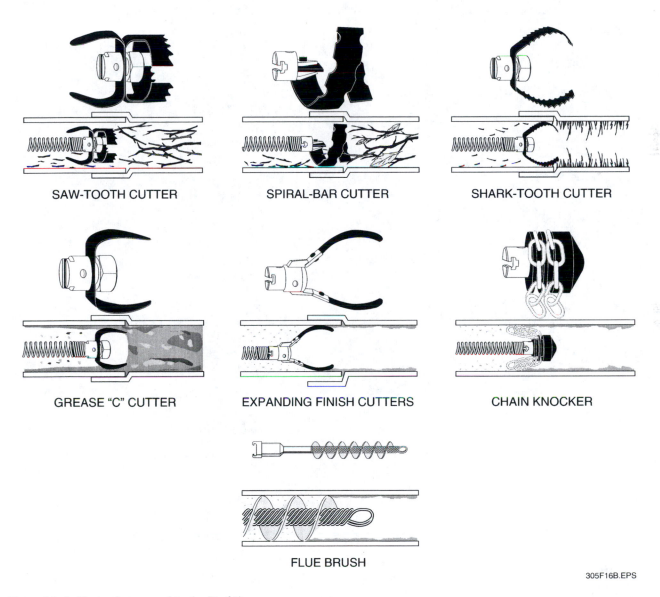

SAW-TOOTH CUTTER   SPIRAL-BAR CUTTER   SHARK-TOOTH CUTTER

GREASE "C" CUTTER   EXPANDING FINISH CUTTERS   CHAIN KNOCKER

FLUE BRUSH

305F16B.EPS

Figure 16 ◆ Drain cleaner end tools. (2 of 2)

1. The two most common types of manlift are _____.
   a. rough-country forklift and pallet lifter
   b. scissors type and telescoping boom
   c. moveable scaffold and pallet truck
   d. pallet jack and cable lift

2. It is safe to place a ladder on a scissors lift to reach higher places.
   a. True
   b. False

3. Do not raise any manlift without _____.
   a. checking for overhead obstructions
   b. turning off the motor
   c. releasing the brakes
   d. removing the handrail

4. Do not move or work from a manlift on _____ surfaces.
   a. soft or uneven
   b. hard or flat
   c. raised
   d. concrete

5. Do not drive a manlift on slopes greater than _____ percent.
   a. 1
   b. 3
   c. 5
   d. 10

6. Cable lifts are suspended on _____.
   a. hemp rope
   b. nylon rope
   c. chains
   d. wire rope

7. The length of an outrigger on a roof must be _____ off the roof.
   a. half as long as the part
   b. the same as the part
   c. one and a half times as long as the part
   d. less than the part

8. When using a hydraulic torque wrench, make sure the _____ is resting against the flange.
   a. hose
   b. pressure gauge
   c. reaction arm
   d. other hand

9. The drain cleaner that works by spinning a cable at low speeds and high torque is the _____.
   a. sectional type
   b. handheld
   c. drum type
   d. rodding machine

10. To clear a tough obstruction in a long, straight run of large-diameter sewer or drain, use a _____.
    a. plunger
    b. rodding machine
    c. closet rod
    d. handheld

Summary

Motorized equipment is a useful tool at the job site. This module has introduced you to some of the equipment that you will use and work with as a pipefitter. Always remember to read and follow the manufacturer's operating procedures and fully understand the equipment before attempting to operate it. Never push a piece of equipment beyond its capabilities. Follow all safety rules to ensure the safety of yourself and your co-workers.

Notes

Trade Terms
Introduced in This Module

Cornice: The crowning, overhanging lip at the top of a wall.

Counterweight: A device that counterbalances an opposing load.

Hydraulic: Operated by fluid pressure.

Lanyard: A rope or line used to fasten a safety harness to an existing structure.

Load rating: The usable lifting capacity of a piece of hardware, including the safety factor.

Outrigger: A steel beam or supporting leg extending from a crane or other piece of equipment to provide stability by widening the base.

Parapet wall: A low retaining wall at the edge of a roof.

Safety factor: The load or tension at which a piece of rigging equipment will fail divided by the actual load to be placed on the equipment. The safety factor for rigging equipment should never be less than four to one.

Wire rope: A rope formed of twisting strands of wire.

Resources & Acknowledgments

Additional Resources

This module is intended to be a thorough resource for task training. The following reference work is suggested for further study. This is optional material for continued education rather than for task training.

Scaffolding, Latest Edition. Upper Saddle River, NJ: Prentice Hall Publishing.

Figure Credits

JLG Industries, 305F01 (photo), 305F02 (photo)

Spider, a division of SafeWorks, LLC, 305F06–305F09

Enerpac, 305F10 (top)

HYTORC, Division of UNEX Corporation, 305F10 (bottom)

Ridge Tool Company (RIDGID®), 305F11–305F16

NCCER makes every effort to keep these textbooks up-to-date and free of technical errors. We appreciate your help in this process. If you have an idea for improving this textbook, or if you find an error, a typographical mistake, or an inaccuracy in NCCER's Contren® textbooks, please write us, using this form or a photocopy. Be sure to include the exact module number, page number, a detailed description, and the correction, if applicable. Your input will be brought to the attention of the Technical Review Committee. Thank you for your assistance.

Instructors – If you found that additional materials were necessary in order to teach this module effectively, please let us know so that we may include them in the Equipment/Materials list in the Annotated Instructor's Guide.

Write: Product Development and Revision
National Center for Construction Education and Research
3600 NW 43rd St, Bldg G, Gainesville, FL 32606

Fax: 352-334-0932

E-mail: curriculum@nccer.org

Craft _____ Module Name _____

Copyright Date _____ Module Number _____ Page Number(s) _____

Description _____

_____

_____

_____

(Optional) Correction _____

_____

_____

(Optional) Your Name and Address _____

_____

_____

08306-07

Introduction to Aboveground Pipe Installation

08306-07
Introduction to Aboveground Pipe Installation

Topics to be presented in this module include:

Overview

The skills of working on aboveground piping are quite different than those for buried pipe. The requirement still exists that the pipe be calculated, measured, cut, and welded or assembled; however, the pipe is not supported by the earth of a trench, and frequently the pipe must be raised above an obstruction. The geometry is similar, but now the runs are much more three-dimensional. The main form of boltup pipe connection is flanged, and the tolerances are closer. Here you will learn the tools and methods for aboveground work.

Objectives

When you have completed this module, you will be able to do the following:

1. Store pipe and materials.
2. Identify types of flanges.
3. Identify types of gaskets used with flanges.
4. Lay out and cut gaskets.
5. Explain the location of flange bolt holes.
6. Install pipe with flanged connections.
7. Lay out and install pipe sleeves and floor penetrations.
8. Read and interpret spool sheets.
9. Explain how to erect spools in a piping system.

Trade Terms

Cavitation
Dunnage
Gasket
Pressure differential

Pressure vessel
Severe service
Turbulence

Required Trainee Materials

1. Pencil and paper
2. Appropriate personal protective equipment

Prerequisites

Before you begin this module, it is recommended that you successfully complete *Core Curriculum; Pipefitting Level One; Pipefitting Level Two;* and *Pipefitting Level Three*, Modules 08301-07 through 08305-07.

This course map shows all of the modules in the third level of the *Pipefitting* curriculum. The suggested training order begins at the bottom and proceeds up. Skill levels increase as you advance on the course map. The local Training Program Sponsor may adjust the training order.

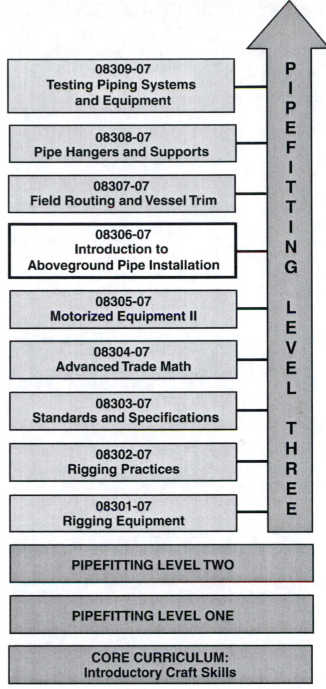

306CMAP.EPS

1.0.0 ◆ INTRODUCTION

This module introduces aboveground pipe installation. Once you are familiar with reading blueprints and fabricating piping components, you are ready to use these skills to install aboveground piping systems. This module explains how to properly store pipe and materials, identify flanges and **gaskets**, install flanged piping components, lay out and install floor penetrations, read and interpret spool sheets, and erect spools.

2.0.0 ◆ STORING PIPE AND MATERIALS

After piping materials are received from the supplier and before they are used on the job, they must be stored on the job site. Storage is a critical part of the construction process and requires careful planning. Proper storage procedures maintain the quality of the pipe and materials and save time by enabling workers to find materials easily. The following sections explain the guidelines that must be followed when storing piping materials inside and outside and explain how to protect pipe ends.

2.1.0 Inside Storage

If material is stored inside a warehouse, a floor plan of the warehouse is needed to designate where various materials are located and where they need to be placed when received. The floor space must be organized with several key factors in mind, such as safety, access to the material, distance from where the material is to be used, and any other special considerations dictated by the particular demands of the job site.

Small materials such as fittings are usually ordered in large quantities and stored on shelves or in bins. They should be carefully sorted and clearly marked so that accurate inventory records can be kept. It is extremely important that materials be checked out of the warehouse only as they are needed. Do not take more supplies or materials than the job calls for.

2.2.0 Outside Storage

Most job sites store materials outside. The storage area should be well planned and laid out in the same manner as a warehouse would be laid out, with adequate space between materials to allow access for forklift trucks and other vehicles. A consideration should be the flow of operations; if possible, the materials for some particular operation

should be stored where they are easily available to the place where that particular activity is to occur. It may be feasible to make several staging areas where the sequence of operations is mirrored in supply storage points. Follow these guidelines when storing pipe and materials outdoors:

- Place new materials in specially designated areas only.
- Label all materials clearly with identification tags or stenciled placards.
- Store all pipe and materials on appropriate hardwood **dunnage** to facilitate transportation of the materials and to keep them off the ground.
- Store all valves in an upright position so that water does not collect in them and damage the seats.
- Store all materials on the basis of their compatibility with other materials. For example, do not store stainless steel pipe with carbon steel pipe because the rust from carbon steel can cause the stainless steel to corrode.

One very important consideration that should not be overlooked is the effect that the weather will have on the materials. Materials must be properly protected from rain, snow, heat, and humidity.

2.3.0 Protecting Pipe, Fittings, and Valves

The ends of piping and piping materials must also be protected from the effects of the weather. The piping specifications often provide instructions on how the ends should be protected. If special instructions are not included in the specifications, the methods given in the following sections should be used to protect flange faces, male threaded ends, beveled ends, and socket weld and female threaded ends.

2.3.1 Flange Faces

Flange faces must always be protected when storing flanges. Cover the flange faces with a wood, plastic, plywood, or hardboard disc. The disc can be strapped, pressed, bolted, or clipped into place. The protective disc needs to cover only the gasket surface of the flange face. If the specifications require that the flange cover be sealed weather-tight, use a rubber or vinyl disc between the flange and the cover, and bolt the cover in place with no less than four bolts. *Figure 1* shows protected flange faces.

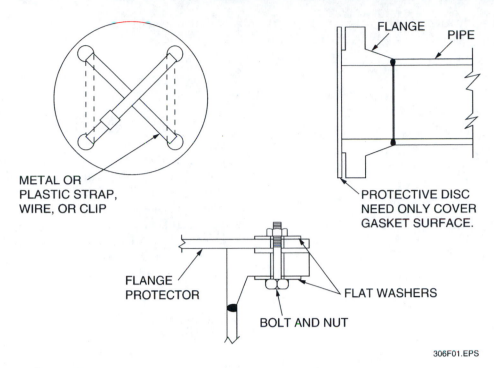

Figure 1 ◆ Protected flange faces.

2.3.2 Male Threaded Ends

Male threaded ends should be protected by press-on plastic or light metal caps. *Figure 2* shows protected male threaded ends.

2.3.3 Beveled Ends

Beveled ends on carbon steel pipe should be protected by press-on end protectors that are lined with wood or hardboard discs. Beveled ends on stainless steel pipe should be protected with plastic or metal press-on end connectors that are lined with wood or hardboard discs and polyethylene film. If the end must be sealed watertight, wrap the end protector and the pipe with at least two layers of waterproof tape. *Figure 3* shows protected beveled ends.

2.3.4 Socket Weld and Female Threaded Ends

Socket weld and female threaded ends should be protected using a press-in plastic or light metal insert. If weatherproofing is required, seal the edge of the insert with at least two layers of waterproof tape. *Figure 4* shows protected socket weld and female threaded ends.

3.0.0 ◆ FLANGED PIPING SYSTEMS

The most common method of joining pipe is with welded connections, but systems that may need to be dismantled occasionally for maintenance or other reasons are joined with flanges. A flanged joint consists of two flanges, a gasket or other device to seal the joint, and the necessary bolts or stud bolts to hold the joint together.

Like most other piping components, flanges are made from several different types of materials, but the most commonly used are forged steel and cast iron. The flanges are available in several dif-

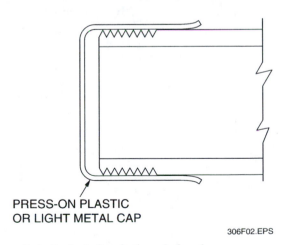

Figure 2 ◆ Protected male threaded ends.

ferent basic primary ratings. The primary rating is a pressure rating based on a pressure/temperature relationship. *Table 1* shows the basic primary ratings.

3.1.0 Types of Flanges

Several types of flanges are used in different types of piping systems and different service applications. Some of the most common flanges include the following:

• Weld neck
• Slip-on
• Slip-on reducing
• Blind
• Socket weld
• Threaded
• Lap-joint
• Cast iron

3.1.1 Weld Neck Flanges

Weld neck flanges are distinguished from other types of flanges by a long, tapered hub that is welded to the pipe end. This hub reinforces the flange and allows material to flow smoothly between the pipe and the flange. This transition between flange thickness and pipe wall thickness is very helpful under severe conditions, such as

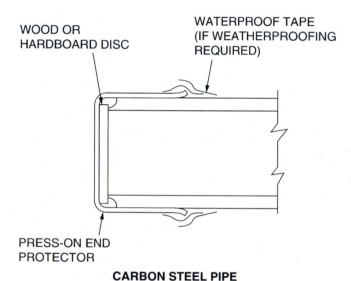

CARBON STEEL PIPE

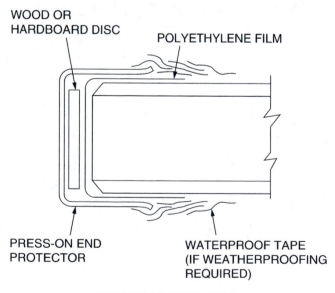

STAINLESS STEEL PIPE

306F03.EPS

Figure 3 ◆ Protected beveled ends.

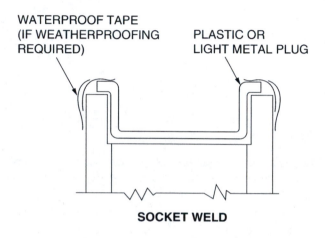

SOCKET WELD

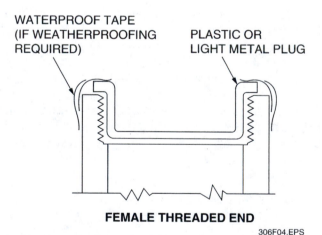

FEMALE THREADED END

306F04.EPS

Figure 4 ◆ Protected socket weld and female threaded ends.

Table 1 Basic Primary Ratings

Forged Steel		Cast Iron
Class 150	Class 900	Class 25
Class 300	Class 1,500	Class 125
Class 400	Class 2,500	Class 250
Class 600		Class 800

306T01.EPS

repeated bending caused by thermal expansion, and strengthens the endurance of the flange assemblies to equal that of a typical butt weld pipe joint. The weld neck flange is preferred over other flanges for very **severe service** conditions. *Figure 5* shows a weld neck flange. Weld neck flanges must be marked with the weight, as in standard, extra heavy, or extra-extra heavy. The markings will be stamped on the neck of the flange. The flanges of different weights have different bore sizes, that is, internal diameters, as do the pipe of the different weights, and the bores would not match if the pipe and flanges were not matched.

3.1.2 Slip-On Flanges

Slip-on flanges are preferred by many contractors because of the low initial cost, reduced accuracy required in cutting the pipe to length, and greater ease in alignment procedures. Slip-on flanges slip over the end of the pipe and are fillet-welded to the pipe both at the neck and at the face of the flange. Normal practice is to extend the flange face beyond the end of the pipe about ⅜ inch or the wall thickness of the pipe, whichever is greater.

Slip-on flanges (*Figure 6*) are used where the space is limited and a weld neck cannot be used and are also used with many low-pressure applications of Class 150 to Class 300.

3.1.3 Slip-On Reducing Flanges

The slip-on reducing flange is used for reducing the line size when a weld neck flange and reducer combination will not fit. A slip-on reducing flange should be used only when the flow is from the smaller-sized pipe to the larger-sized pipe. If the flow is in the opposite direction, the smaller opening will cause **turbulence** or an excessive **pressure differential** at that point. If the difference in sizes of the standard and reducing flanges is too great, the pressure differential can produce **cavitation**. What actually happens is that the liquid enters the smaller pipe and speeds up, and loses pressure. When the pressure drops below the vapor pressure of the liquid (the pressure at which it becomes a gas instead of a liquid) bubbles form. When the bubbles collapse, the result is a physical impact wave to the vessel or parts of the vessel. This makes the waves keep striking the inside of the pipe, or the fitting to which it is attached and damages the surface. If space limitations make a slip-on reducing flange necessary, the approval of the equipment manufacturer must be obtained. Manufacturers of equipment such as pumps, heaters, and heat exchangers must control the pressure drops in their systems, and the equipment nozzle sizes directly affect the pressure drop through the system. This is why it is not uncommon for equipment nozzle sizes to differ from pipe sizes. Slip-on reducing flanges are available in all sizes and ratings. *Figure 7* shows a slip-on reducing flange.

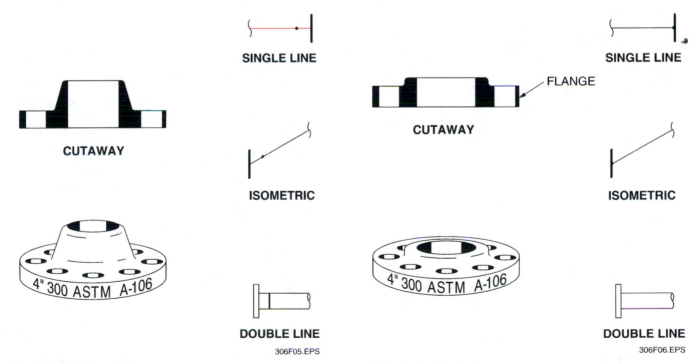

SINGLE LINE

CUTAWAY

4" 300 ASTM A-106

ISOMETRIC

DOUBLE LINE

306F05.EPS

Figure 5 ◆ Weld neck flange.

SINGLE LINE

FLANGE

CUTAWAY

4" 300 ASTM A-106

ISOMETRIC

DOUBLE LINE

306F06.EPS

Figure 6 ◆ Slip-on flange.

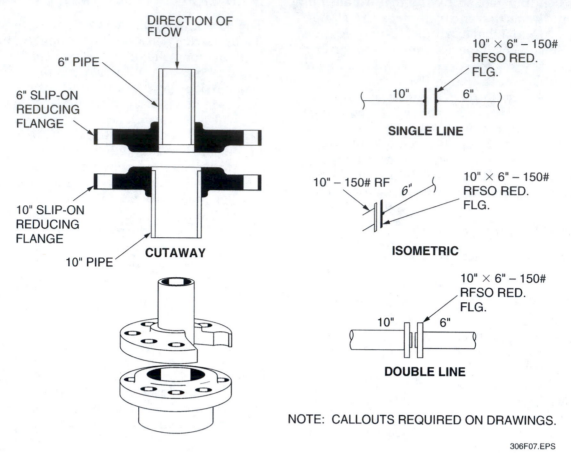

DIRECTION OF FLOW

6" PIPE

6" SLIP-ON REDUCING FLANGE

10" SLIP-ON REDUCING FLANGE

10" PIPE

CUTAWAY

10" × 6" – 150# RFSO RED. FLG.

10"   6"

SINGLE LINE

10" – 150# RF   6"   10" × 6" – 150# RFSO RED. FLG.

ISOMETRIC

10" × 6" – 150# RFSO RED. FLG.

10"   6"

DOUBLE LINE

NOTE: CALLOUTS REQUIRED ON DRAWINGS.

306F07.EPS

Figure 7 ◆ Slip-on reducing flange.

3.1.4 Blind Flanges

Blind flanges are used to close off the ends of pipes, valves, and **pressure vessel** openings. From the standpoint of internal pressure and bolt loading, blind flanges are the most highly stressed of all flange types since the maximum stresses in a blind flange are bending stresses at the center. These flanges are available in the seven primary ratings, and their dimensions can be found in manufacturer's catalogs. *Figure 8* shows a blind flange.

3.1.5 Socket Weld Flanges

Socket weld flanges are attached to pipe ends by inserting the pipe into a socket on the flange and then fillet-welding the flange to the pipe at the flange hub. These flanges are used primarily with small, high-pressure piping. Socket weld flanges are available in Class 150, Class 300, Class 600, and Class 1,500 pressure ratings. *Figure 9* shows a socket weld flange.

3.1.6 Threaded Flanges

Threaded flanges are available in all sizes and pressure ratings up to and including 12 inches and

Class 2,500. However, the use of high-pressure threaded flanges should be limited to special applications. Their major advantage is that they can be

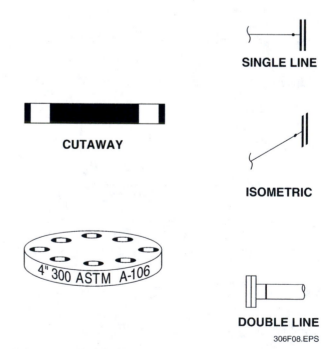

SINGLE LINE

CUTAWAY

ISOMETRIC

4" 300 ASTM A-106

DOUBLE LINE

306F08.EPS

Figure 8 ◆ Blind flange.

assembled without welding, but the threaded joint will not withstand high temperatures or bending stresses. *Figure 10* shows a threaded flange.

3.1.7 Lap-Joint Flanges

Lap-joint flanges, also known as Van Stone flanges, are used with lap-joint stub ends. Lap-joint stub ends are straight pieces of pipe with a lap on the end, which the flange rests against. These can be used at all pressures, and the flange is normally not welded to the stub end, making alignment easier. When using lap-joint flanges in vertical applications, have a welder weld steel nubs to the sides of the stub end beneath the flange. This prevents the flange from sliding down the pipe when the bolts are being removed.

The main use of lap-joint flanges in carbon steel piping systems is in services that require frequent disassembly for inspection and cleaning. When used in systems subject to high corrosion, these flanges can usually be salvaged because they do not actually come in contact with the substance flowing through the pipe.

These flanges are also used in many cases for economical reasons. When running stainless steel piping systems, you can use carbon steel lap-joint flanges with stainless steel stub ends to save money. The use of lap-joint flanges at points where severe bending stresses occur should be avoided. *Figure 11* shows a lap-joint flange and a stub end.

3.1.8 Cast Iron Flanges

The two most common ratings for cast iron flanges are Class 125 and Class 250. These flanges are generally found on equipment nozzles, such as on pumps and turbines. Mating cast iron flanges to forged steel poses some problems that are discussed in the next section on flange facings. The main point to remember is that a Class 125 cast iron flange mates with a Class 150 forged steel flange, and a Class 250 cast iron flange mates with a Class 300 forged steel flange.

3.2.0 Flange Facings

Facing refers to the sealing surface machined onto flanges. Flange facings are governed by American National Standards Institute (ANSI) standards to ensure that all flanges of each type and rating have the same facing. Typically, you should never use two different face styles at one joint, but this may be necessary when connecting pipe to manufactured equipment. The industry standard for flange facing finish is the phonographic serrated finish. The phonographic finish means that the facing of the flange has a continuous narrow, shallow spiral groove in the face. The other common facing finish is a concentric serrated finish. This is a series of shallow circular grooves sharing a common center. The normal range is between 30 and 55 grooves per lateral inch of the diameter of the face. Both finishes allow soft gasket material to fill the grooves and keep liquids from leaking. Some

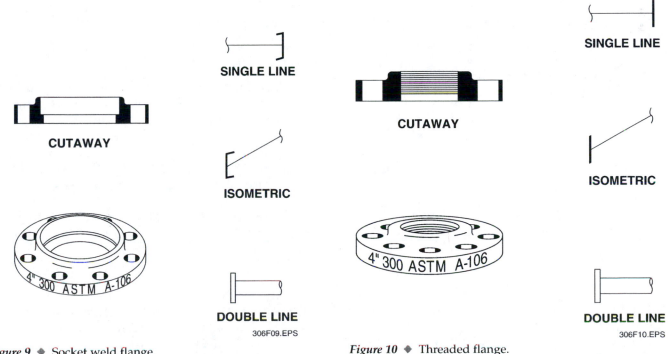

Figure 9 ◆ Socket weld flange.

306F09.EPS

Figure 10 ◆ Threaded flange.

306F10.EPS

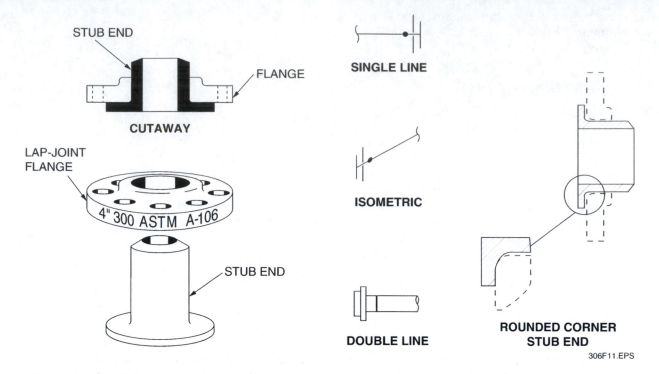

Figure 11 ◆ Lap-joint flange and stub ends.

special industrial applications may use a smooth finish, sometimes called a cold water finish, to be used without a gasket. The various types of flange facings include the following:

- Raised-face
- Flat-face
- Ring joint type
- Tongue and groove
- Male and female

Figure 12A shows flange facing finishes and *Figure 13* shows flange facings.

3.2.1 Raised-Face Flanges

The raised-face flange has a wide, raised rim around the center of the flange and is the most common facing used with steel flanges. The raised face on Class 150 and Class 300 flanges is ¹⁄₁₆ inch high, and the face on flanges of all other pressure ratings is ¼ inch high. This flange uses a wide, flat ring gasket that is squeezed tightly between the flanges. The facing is machine-finished with tiny, spiral grooves that bite into and help hold the gasket.

3.2.2 Flat-Face Flanges

The flat-face flange is flat across the entire face. This flange uses a gasket that has bolt holes cut in it and that covers the entire flange face. Flat-face flanges are mainly used to join Class 150 and Class 300 forged steel flanges with Class 125 and Class 250 cast iron flanges. The brittle nature of cast iron flanges requires that flat-faced flanges be used so that full face contact is achieved between the flanges.

If you must connect a flat-face flange to a raised-face flange, you must confirm that the flat-face flange is not cast iron because the pressure of the bolts can crack or break the cast iron flange. If the flat-face flange is cast iron, change the raised-face flange to a flat-face flange only after obtaining engineering approval.

3.2.3 Ring Joint Type Flanges

The ring joint type (RJT) flange (*Figure 13*) has a different type of facing because the contact surface of the seal is below the actual flange facing. This type of flange uses an oval, rectangular, or octagonal steel ring as a seal that fits into a ring groove that is machined into the faces of the flanges. Only one steel ring is required for each pair of mating flanges.

The RJT flange facing is the most expensive standard facing, but for use in oil and refinery services, it is usually regarded as the most efficient because internal pressure acts on the ring to increase the sealing force. One disadvantage of using this type of flange is that in order to remove a valve or a section of pipe from the system, the flanges have to be spread far enough apart to remove the ring. In tight hookups, this can cause problems.

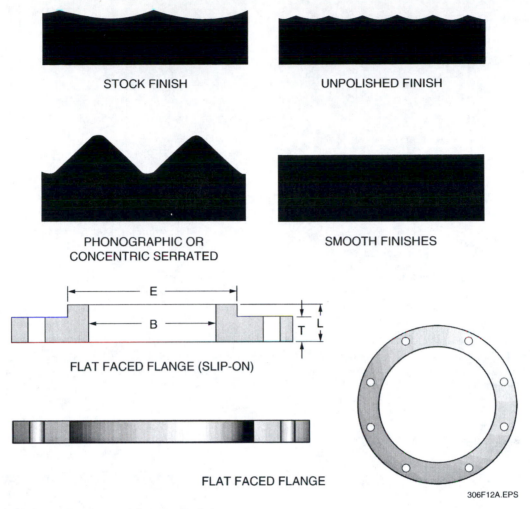

STOCK FINISH

UNPOLISHED FINISH

PHONOGRAPHIC OR
CONCENTRIC SERRATED

SMOOTH FINISHES

FLAT FACED FLANGE (SLIP-ON)

FLAT FACED FLANGE

306F12A.EPS

Figure 12 ◆ Flange facing finishes and flanges. (1 of 2)

3.2.4 Tongue and Groove Flanges

Tongue and groove flange faces are classified as either large tongue and groove or small tongue and groove. This type of flange was originally developed for hydraulic service and offers high gasket pressure in a small area between the tongue and the groove. One advantage that this type of flange offers is that it is more resistant to flange stresses and bending. Disadvantages include excessive gasket pressure with changes in temperature, mushrooming of the tongue under excessive bolt torquing, and the tight tolerances involved in machining and assembly.

3.2.5 Male and Female Flanges

Male and female flange faces fit together much like the tongue and groove flange facings. A rim on the male face fits into a machined groove in the female flange. A flat ring gasket is used between the rim and the groove in the female flange. The

same disadvantages that apply to tongue and groove flanges also apply to male and female flange facings, and male and female flanges are also classified as either large or small.

4.0.0 ◆ FLANGE GASKETS

To ensure that the composition of various types of gaskets, gasket material, and some packing meets the required specifications, the American Society for Testing and Materials International (ASTM) devised a method of identifying different types of gasket material.

Gaskets are available in a variety of types and materials (*Figure 14*). The type of gasket used must be matched to the process characteristics and operating conditions to which it will be exposed. For example, different gasket materials are capable of withstanding different temperature and pressure ranges, and certain gasket materials are compatible with different process fluids. In piping systems, gaskets are placed between two flanges

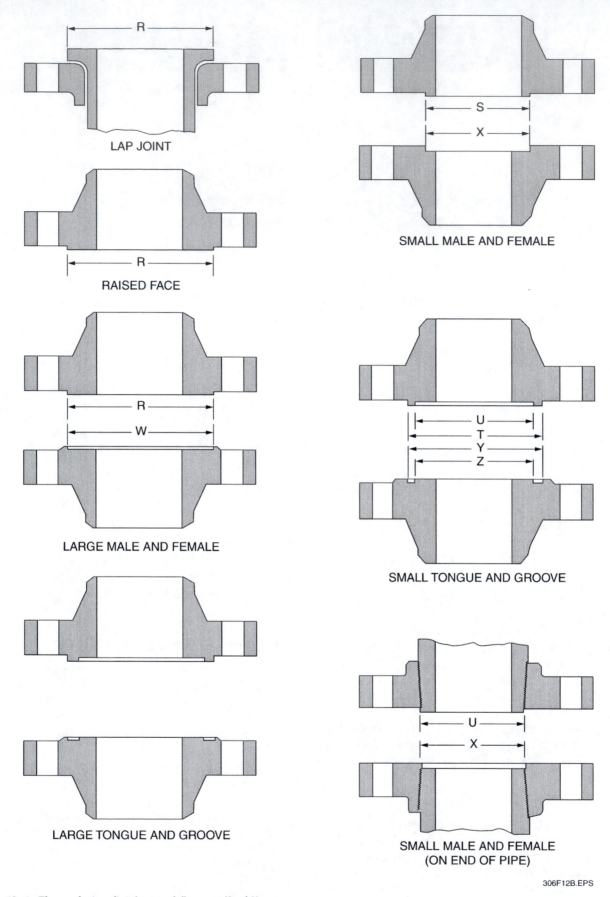

LAP JOINT

RAISED FACE

LARGE MALE AND FEMALE

LARGE TONGUE AND GROOVE

SMALL MALE AND FEMALE

SMALL TONGUE AND GROOVE

SMALL MALE AND FEMALE
(ON END OF PIPE)

306F12B.EPS

Figure 12 ◆ Flange facing finishes and flanges. (2 of 2)

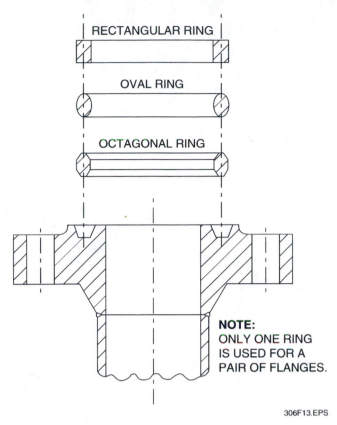

RECTANGULAR RING

OVAL RING

OCTAGONAL RING

NOTE:
ONLY ONE RING
IS USED FOR A
PAIR OF FLANGES.

306F13.EPS

Figure 13 ◆ Metal ring joint gaskets.

to make the joint leakproof. Gaskets are normally sized by the thickness of the gasket material in fractions of an inch.

Based on manufacturer's specs, adhesive materials such as Loctite® gasket replacement are used in certain applications, especially in machine assemblies, as a replacement for cut gaskets. The material is applied from a tube or pump to the clean surface of the flange, and replaces the gasket. Such applications are medium strength, and you must follow the instructions of the manufacturer absolutely. The drawbacks include having to use only recommended cleaning solutions beforehand, and that the material takes from 24 hours to a week to cure.

4.1.0 Gasket Materials

Gaskets are made of many different types of materials to meet the demands of the particular process system in which they are installed. Generally, one of the following four conditions will exist in a process system. These conditions affect the types of gaskets that can be used in the system.

- *High temperature/low pressure* – Generally includes temperatures from 500°F to 1,200°F and pressures up to 600 psi.
- *High temperature/high pressure* – Generally includes temperatures from 500°F to 1,200°F and pressures from 600 to 2,500 psi.

- *Low temperature/low pressure* – Generally includes temperatures up to 500°F and pressures up to 600 psi.
- *Low temperature/high pressure* – Generally includes temperatures up to 500°F and pressures from 600 to 2,500 psi.

Usually, the types of gaskets and gasket materials to be used on a project are listed in a special section of the piping specifications. Gasket materials are rated by ANSI specifications for applicable pressures.

Since most process systems are subject to a combination of temperature and pressure, it is extremely important to match the proper gasket material to the operating conditions of the process system. The type of gasket material used is based on several factors. Generally, the gasket is chosen to withstand the temperature, pressure, and chemical properties of the process medium. Different gasket materials have different pressure and temperature ratings. These ratings are a reflection of the ability of a gasket material to withstand forces in a process system. Various types of gaskets include the following:

- Teflon® (PTFE)
- Fiberglass
- Acrylic fiber
- Metal
- Rubber
- Cork
- Vinyl
- Ceramic
- Asbestos
- Ethylene propylene diene terpolymer (EPDM)
- Neoprene
- Nitrile
- Silicone
- Viton®
- Gylon®

4.1.1 Teflon® Gaskets

Teflon® flange gaskets, also known as PTFE, are nonporous, nongalvanic gaskets. They are usually white and are available in several pressure/temperature limits. PTFE is available in different forms and products. Standard amorphous PTFE, the original form of PTFE, is available in sheet form and can be cut into the various sizes necessary to fit the applications. Amorphous PTFE is frequently associated with a phenomenon referred to as gasket creep and cold flow which results in a loss of bolt load. When under continuous load, PTFE tends to con-

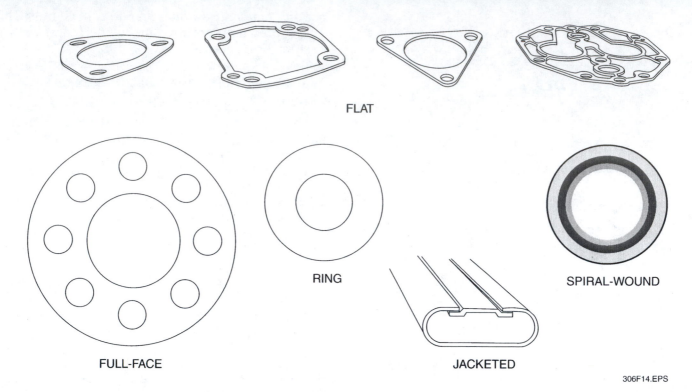

FLAT

RING

SPIRAL-WOUND

FULL-FACE

JACKETED

306F14.EPS

Figure 14 ◆ Gaskets.

tinuously flow due to a lack of strength. This often creates a necessity for plant personnel to re-torque bolts in order to compensate for this material flow.

Fillers may introduce other problems, including different characteristics from the non-stick and chemically stable PTFE. The flow was slowed somewhat, but other advantages were lost.

Finally a process was discovered to expand PTFE in a manner so as to dramatically reduce and nearly eliminate PTFE gasket creep and cold flow. Products are made from virgin PTFE resin, have no fillers, have the same chemical compatibility as the original PTFE, operate in the same temperature range (–450°F to +500°F), and have no colorizers or other additives. Expanded PTFE, due to its very high tensile strength properties, can be used in pressure environments from full vacuum to 3,000 psi. The photo in *Figure 15* represents an array of cut gaskets from one of the two expanded PTFE manufacturers.

The company that makes GORE-TEX® outerwear also makes cord gasket material (*Figure 16*), made from expanded PTFE. It is sold in rolls, to be applied to the joint by hand. The biggest advantage of PTFE is that it is chemically inert. This means that it does not react chemically with the process medium.

4.1.2 Fiberglass Gaskets

Fiberglass gaskets are treated so that they will remain soft and pliable. They sometimes include wire inserts for service for boiler handhold and manhole covers, as well as for tank heads and other high-pressure applications. Fiberglass gaskets are usually white in color. They can typically be used in applications up to 380°F and 180 psi.

306F15.EPS

Figure 15 ◆ Teflon® flange gaskets.

4.1.3 Acrylic Fiber Gaskets

Acrylic fiber flange gaskets are used with oil, steam, weak acids, and alkalines. They are usually off-white in color. Acrylic fiber gaskets can typically be used in applications up to 700°F and 1,200 psi. Many gaskets combine fibers of various materials filled with nitrile, EPDM, or neoprene (*Figure 17*).

4.1.4 Metal Gaskets

Metal gaskets are used for machine flanges and where metal-to-metal fits are required. They are good for applications in which higher temperature fluctuations and pressures are normal. When using metal gaskets, the material must be suitable for use within the process system. The following are some of the types of metal gaskets available:

- *Solid-metal gaskets* – These gaskets are made from solid metal. They are used in processes suitable for the metal or alloy used to make the gasket. Common materials are copper, Monel®, steel, and iron. Copper gaskets are not used in high-pressure high-temperature systems. Monel®, steel, and iron gaskets can be used in a variety of pressure and temperature applications.

- *Ring joint metal gaskets* – The four types of standard ring joint gaskets (*Figure 18*), defined by the ring section, are: R octagonal; R oval; RX (eccentric octagonal); and BX (rectangular). The octagonal is considered superior. The rings are made of the softest carbon steel or iron available. For lower temperatures, the rings are made of plastic to resist corrosion or to insulate the joint from electric currents. The ring joint gasket is effective at very high pressures.

- *Serrated metal gaskets* – Serrated metal gaskets are solid metal gaskets that have concentric ribs machined into their surfaces. With the contact area reduced to a few concentric lines, the required bolt load is reduced considerably. This design forms an efficient joint. Serrated gaskets are used with smooth-finished flange faces.

- *Corrugated metal gaskets* – Corrugated metal gaskets are generally used on low-pressure systems where the flanges are smooth and bolt pressure is low. The ridges of the corrugations tend to concentrate the gasket loading along the concentric ridges.

- *Laminated metal gaskets* – These gaskets are made of metal with a soft filler. The laminate can be parallel to the flange face or spiral-wound. Laminated gaskets can be used in high-pressure, high-temperature applications.

- *Corrugated metal gaskets with asbestos inserted* – These gaskets are used for a variety of pressure and temperature applications in steam, water, gas, air, oil, oil vapor, and refrigerant systems.

306F17.EPS

Figure 17 ◆ Gaskets made of a combination of materials.

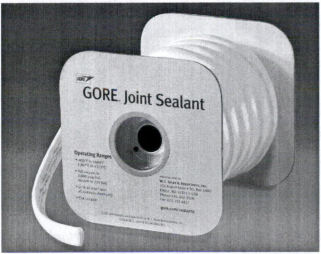

306F16.EPS

Figure 16 ◆ Cord gasket material.

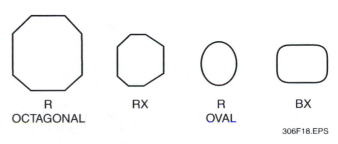

306F18.EPS

Figure 18 ◆ Standard ring joint gasket.

- *Flat metal jacket gaskets with asbestos gasket material* – These gaskets are used in the same applications as corrugated metal gaskets with asbestos inserted.
- *Corrugated metal jacket gaskets with heat-resistant, synthetic filler* – These gaskets are used for the same applications as corrugated metal with asbestos-inserted gaskets. Synthetic replacements for asbestos are often referred to as nonasbestos.
- *Spiral-wound metal gaskets with heat-resistant, synthetic filler* – Spiral-wound metal gaskets are used for a variety of pressure and temperature applications in the same type systems as corrugated metal jacket gaskets with heat-resistant, synthetic filler. All spiral-wound metal gaskets are color-coded to identify the type of metal from which they are made. Spiral-wound gaskets normally have an inner ring, an outer ring, and a section of preformed windings in the middle that serve as the sealing element. The inner and outer rings are usually spot-welded, and serve as blowout preventers for the sealing element. The preformed shapes of the sealing element are not welded, and are usually slightly thicker through the gasket.

Table 2 shows the industry standard color-code for spiral-wound gaskets.

4.1.5 Natural Rubber Gaskets

Natural rubber is seldom used in installations today. It has low solvent and oil resistance, but is excellent when used with water. Natural rubber has excellent resilience and may be used at temperatures up to 175°F. Rubber gaskets are available in a variety of pressure and temperature ratings. Rubber gaskets generally come in thicknesses from ⅟₁₆ to 1 inch. Rubber gaskets are used in low-pressure/low-temperature water, gas, air, and refrigerant systems. The following types of rubber gaskets are used:

- *Standard black or red rubber gaskets* – These gaskets are used for saturated steam up to 100 psi. They have an approximate temperature range of –20°F to 170°F.
- *Reinforced rubber gaskets* – These gaskets are strengthened by polyester fabric plies. They are commonly used for saturated steam and low-pressure steam. Reinforced rubber gaskets have an approximate temperature range of –40°F to 200°F. They are typically black.
- *High-test, reinforced rubber gaskets* – These gaskets are used for pressures up to 500 psi. They have an approximate temperature range of –40°F to 200°F, and they are black.

4.1.6 Cork Gaskets

Cork gaskets are compressible, flexible, lightweight, resilient, and nonabsorbent. They resist most oils, petroleum products, and chemicals. Cork is rarely used in industry, except in special cases, as a vibration-absorbing material because of its cellular construction. It can be used for most liquids, even when boiling, except for strong alkalis. Cork gaskets are generally available in thicknesses from ⅟₁₆ to ½ inch.

4.1.7 Vinyl Gaskets

Vinyl gaskets are specially fabricated for use as oil-resistant gaskets. They also have a good resistance to water, chemicals, oxidizing agents, ozone, and abrasion. Vinyl gaskets are black in color and have an approximate temperature range of 20°F to 160°F.

4.1.8 Ceramic Gaskets

Ceramic gaskets are used for high-temperature air applications, such as boiler systems, where other types of materials would fail due to the high temperature. Ceramic gaskets can typically be used for temperatures exceeding 1,500°F.

Table 2 Industry Standard Color-Code for Spiral-Wound Gaskets

Metal	Color
Type 304 SS	Yellow
Type 316 SS	Green
Type 347 SS	Blue
Type 321 SS	Turquoise
Monel®	Orange
Nickel	Red
Titanium	Purple
Alloy 20	Black
Carbon steel	Silver
Hastelloy B	Brown
Hastelloy C	Beige
Inconel®	Gold
Filler Material	**Color of Stripe**
Polytetrafluoroethylene (PTFE)	White
Ceramic	Light green
Flexible graphite	Gray
Mica/graphite	Salmon

306T02.EPS

4.1.9 Asbestos Gaskets

Gaskets are still sold and used that contain encapsulated asbestos. This means that inside a metal form is asbestos. These gaskets are used in high-temperature contexts. You should tell your supervisor if the metallic exterior is broken and the white asbestos fibers exposed.

In most locations, legal requirements exist for any action that may cause any possible exposure to asbestos. The area where the asbestos is to be removed will be completely isolated from the outside air. Even a single fiber of asbestos in a worker's lungs can cause a form of lung cancer. In addition to an airtight seal, asbestos workers are required to wear protective clothing and an air supply, all of which is decontaminated before leaving the sealed area. If the gasket is replaced, you will not be installing another asbestos gasket.

 WARNING!
If you do find an asbestos gasket, do not attempt to deal with it. Inform your supervisor immediately.

4.1.10 EPDM

EPDM is very good in an oxygen environment. It has good ozone resistance. EPDM has low resistance to solvent and oil. It is a good material to be used with water and chemicals and has a maximum temperature rating of 350°F. It also has excellent flame resistance.

4.1.11 Neoprene

Neoprene, which has good resilience, is only used in noncritical conditions. Neoprene has fair oil and solvent resistance, but has poor chemical resistance and is not recommended for water service. It has a maximum temperature rating of 250°F.

4.1.12 Nitrile

Nitrile is used with medium-pressure oil and solvent services. Nitrile cannot be used with acetone or methyl ethyl ketone (MEK). Nitrile has a maximum temperature rating of 250°F. Nitrile has poor ozone resistance, but is an excellent material to use with water or alcohol. Nitrile also has good chemical resistance but poor flame resistance.

4.1.13 Silicone

Silicone is a rubber-like material that is widely used. There are more grades of silicone material than any other rubber-type material. It has a maximum temperature rating of 550°F. Silicone is not recommended for use with oil or solvents. It has fair to good chemical resistance, excellent ozone resistance, and fair to good flame resistance. It has poor to excellent resilience depending on the grade used. It is not recommended for use on water systems.

4.1.14 Viton®

Viton® is a fluoroelastomer material. It has excellent chemical resistance and is the most commonly used chemical-resistant material. It has a maximum temperature rating of 400°F. Viton®, a hard material, has excellent oil and solvent resistance and also has excellent ozone resistance. Viton® is a relatively inexpensive gasket material.

4.1.15 Gylon® or Amerilon®

Under proprietary names like Gylon® and Amerilon®, there are also some new materials that combine Kevlar® and PTFE, to produce a very tough and resistant low temperature gasket. It is chemically inert, has a high resilience (45 percent recovery), and a maximum temperature rating of 500°F. Gylon® 3510 has one of the best sealability factors of any gasket material in the industry. It is one of the most commonly used gasket materials. It is used heavily in the chemical and paper industries.

4.1.16 Graphite-Impregnated Gaskets

These gaskets (*Figure 19*) are made of fiberglass and impregnated with graphite particles. They are used to provide a tight seal in some high-temperature applications. However, they are reactive to oxidation.

4.1.17 Soft Metal Gaskets

Soft metal and aluminum gaskets are designed to compress into the concentric rings machined into the surface of carbon steel and stainless steel flanges. Metal gaskets are used for oil and gas pipelines, where their non-reactive materials and high wear resistance keep them working over long periods without replacement.

4.1.18 Gasket Color Codes

Manufacturers use color codes to help identify gaskets made of different materials. These color codes will vary from one manufacturer to another. Some manufacturers use an additional color-coding scheme on the outside diameter of some gaskets so the material can be identified without breaking the seal. In such cases, the gasket body may be the same color as other types of gaskets, but the OD color will be different. The important thing is to check the color-coding scheme used by the given manufacturer before replacing a gasket.

NOTE

The use of a commercial liquid gasket remover can make removing old gasket material easier. It reduces the amount of scraping and sanding needed, thus preventing possible damage to flange surfaces.

4.2.0 Types of Flange Gaskets

There are different types of gaskets that must be matched to different types of flanges. For example, flat ring gaskets are used with raised-face flanges; full-face gaskets are used with flat-face flanges; and metal ring joint gaskets are used with RJT flanges.

Figure 19 ◆ Graphite-impregnated gasket.

4.2.1 Flat Ring Gaskets

Flat ring gaskets are used on flanges with raised faces. They have an outside diameter slightly larger than the outside diameter of the raised face. The material used may be metallic, nonmetallic, or a combination of both. The raised-face flange dimensions and class rating of the flange determine the size of flat ring gasket needed. *Figure 20* shows a flat ring gasket.

4.2.2 Full-Face Gaskets

Full-face gaskets are used with flat-face flanges and extend all the way to the outer edge of the flange. These gaskets have bolt holes that match the holes of the flanges they are used with. *Figure 21* shows a full-face gasket.

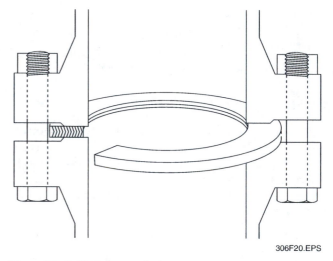

306F20.EPS

Figure 20 ◆ Flat ring gasket.

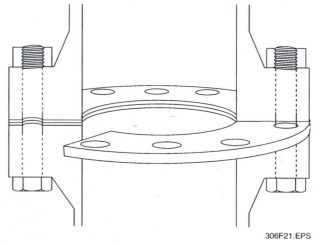

306F21.EPS

Figure 21 ◆ Full-face gasket.

5.0.0 ◆ FABRICATING GASKETS

The easiest way to make a gasket is to use the existing flange or the old gasket as a template to cut out the new gasket. If this is not possible, the following section describes the procedures for laying out and cutting a new gasket.

5.1.0 Laying Out a New Gasket

The following procedure describes how to lay out and cut a new gasket for a pipe flange.

WARNING!

Approved eye protection should be worn to protect the eyes from airborne gasket fibers or metal shards when fabricating gaskets. Hand protection should also be worn to protect the hands from injury when sharp gasket cutter blades are being used and/or sharp metal gasket edges are present.

Step 1 Select the proper gasket material for the conditions and the process.

Step 2 Take the following three measurements and draw them as concentric circles:

- Diameter of pipe opening
- Outside diameter of flange
- Diameter of the bolt hole circle

The diameter of the bolt circle is found by measuring from the edges of opposite holes, as shown in *Figure 22*. To get a center measurement, measure from the inside edge of one hole to the outside edge of the opposite hole.

Step 3 Find the radius of the bolt circle as shown in *Figure 23*. The radius of the bolt circle is equal to half the diameter.

Step 4 Draw a line through the circle's center for opposite holes.

Step 5 Lay out holes a little larger than their actual size, as shown in *Figure 24*. Gaskets with an even number of holes (4, 8, 16) can also be laid out using the swing arc method. This method uses the divider to bisect distances. Flanges with an odd number of holes can be laid out using a protractor by simply dividing 360 degrees by the number of holes.

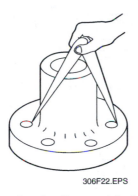

306F22.EPS

Figure 22 ◆ Measuring the diameter of a bolt circle.

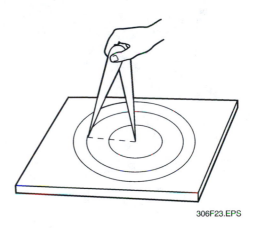

306F23.EPS

Figure 23 ◆ Radius of bolt circle.

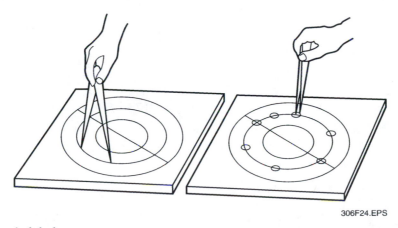

306F24.EPS

Figure 24 ◆ Laying out bolt holes.

Step 6 To check the distance, walk the dividers around the circle.

Step 7 Cut out the inside circle and outside circle of the gasket using a gasket cutter, as shown in *Figure 25*. The gasket cutter blade should not protrude more than $\frac{1}{32}$ inch more than the thickness of the gasket material. Never hammer the gasket, as hammering may cause lumps in the gasket.

Step 8 Place the gasket material on hard wood to protect the punch edge. Then, punch out the holes using the proper size hole punch as shown in *Figure 26*.

5.1.1 Bisecting Angles Using Dividers

Follow these steps to bisect an angle into equal parts using dividers and a straightedge.

Step 1 Scribe two intersecting lines, using a protractor, to construct a 60-degree angle. The point of intersection is point A.

Step 2 Place the point of one divider leg where the two lines intersect at point A.

Step 3 Scribe an arc that intersects both sides of the angle (*Figure 27*).

Step 4 Scribe two intersecting arcs from the intersection points on each side of the angle to establish point B (*Figure 28*).

Step 5 Scribe a line from point A through point B using a straightedge. This line forms two equal angles.

Step 6 Check the accuracy of each angle using a protractor.

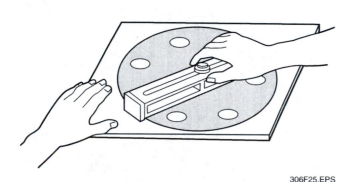

306F25.EPS

Figure 25 ◆ Using a gasket cutter.

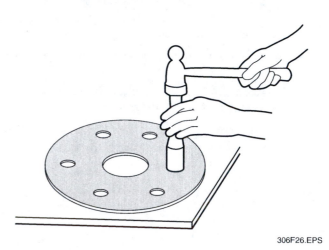

306F26.EPS

Figure 26 ◆ Using a hole punch.

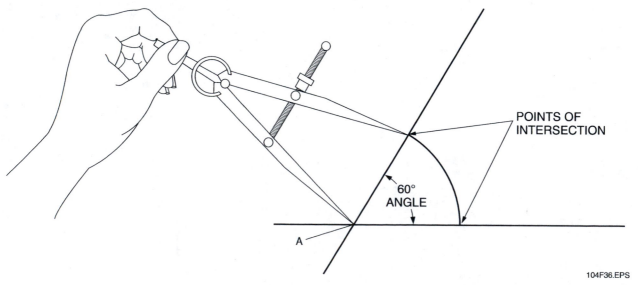

POINTS OF INTERSECTION

60° ANGLE

A

104F36.EPS

Figure 27 ◆ Scribed arc.

5.2.0 Tracing a New Gasket

The following procedure describes how to trace and cut a new metal gasket for a pipe flange.

Step 1   Spread bluing ink on the pipe flange face, as shown in *Figure 29*.

Step 2   Place the gasket material on the flange face and make an impression, as shown in *Figure 30*.

Step 3   Lift the gasket material off the flange face. You should have an impression of the gasket, as shown in *Figure 31*.

Step 4   Cut out the gasket using tin snips, as shown in *Figure 32*.

Step 5   Place the gasket material on hard wood to protect the punch edge, and then punch out the holes with the proper size hole punch.

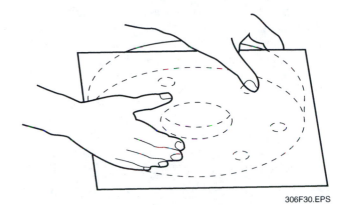

306F30.EPS

Figure 30 ◆ Making a bluing ink impression.

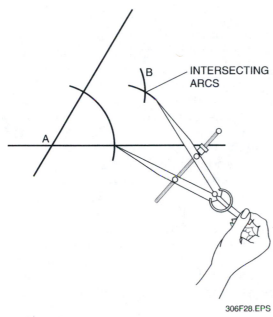

306F28.EPS

Figure 28 ◆ Intersecting arcs.

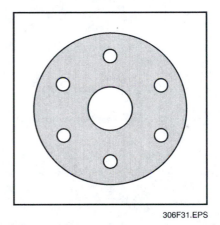

306F31.EPS

Figure 31 ◆ Bluing ink impression.

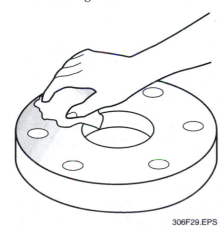

306F29.EPS

Figure 29 ◆ Bluing ink tracing.

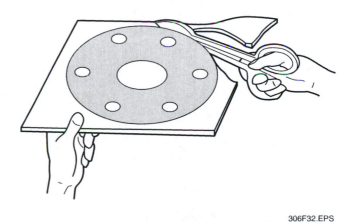

306F32.EPS

Figure 32 ◆ Cutting out the gasket with tin snips.

Step 6 Clean any loose particles or wet ink off the gasket using the appropriate cleaner. It is not necessary to clean the dried bluing ink from the surfaces.

Step 7 Check the gasket against the flange for proper fit. If it does not fit correctly, discard the gasket and start over.

5.3.0 Machine Gaskets

Machine gaskets come in many configurations. It is not practical to try to lay out a machine gasket in the same manner as a flange gasket. You must transfer a pattern from the machine to the gasket. This is done using bluing ink. Follow these steps to lay out and cut a machine gasket:

Step 1 Thoroughly clean the machine surface where the gasket fits (known as the flat) to remove the old gasket and any foreign matter. Use solvent if needed and prescribed.

Step 2 Ensure that the machine surface is completely dry.

Step 3 Cut a piece of gasket material a little larger than the area to be covered by the gasket.

Step 4 Apply an even coat of bluing ink to the machine flat.

Step 5 Lay the gasket on the machine flat.

CAUTION

Use extra care when placing the gasket on the machine flat to prevent smearing the gasket with ink. The ink should make a distinct impression on the gasket.

Step 6 Rub the gasket with your finger along the machine flat to ensure that the ink contacts the gasket at all points.

Step 7 Carefully remove the gasket from the flat.

Step 8 Allow the ink to dry.

Step 9 Cut out the gasket using scissors or a utility knife.

Step 10 Punch the bolt holes using the proper size hole punch.

Step 11 Place the gasket on the machine flat and check the fit.

6.0.0 ◆ INSTALLING PIPE FLANGES

ANSI has developed a standard that all manufacturers use as a guide for bolt hole locations on all flanges up to 24 inches. Flanges larger than 24 inches can have different diameter bolt holes and a different number of bolt holes, depending on the type of flange. When working with these larger flanges, you must determine whether the flange is an MSS SP 44, API 105, or AWWA C207-55 flange. You must also check the pressure class rating of the flange and refer to flange dimensioning charts to determine the number and size of the bolt holes. The number of bolt holes required on flanges up to 24 inches also varies with the size and rating of the flange, although there is always an even number of holes in graduations of four. The bolt pattern for each size and rating is consistent in that each type, size, and rating has the same bolt circle.

ANSI also has a standard to use as a guide for fabrication. The standard requires that flange bolts always straddle the horizontal and vertical center lines unless otherwise indicated. This is known as the two-hole method. It is possible for a job specification to require that the bolt holes line up with the vertical and horizontal lines. This method is known as the one-hole method. The general rule is to reference one center line only, depending on the type of drawing used. In elevation drawings, the usual reference is the vertical center line, and in plan drawings, the usual reference is the north-south center line. This rule ensures that all fabricators refer to the same center line. To determine the angular location of the bolt holes, divide 360 degrees by the number of bolt holes. *Figure 33* shows bolt hole locations.

The four steps in pipe flange installation are cleaning the parts, aligning the parts, installing the gaskets, and tightening the flange bolts.

6.1.0 Cleaning Parts

As with all pipe joints, flanged joints must be cleaned to ensure the best possible fit. Flanges from the factory and on other equipment often have a gum or other residue that must be scraped off the face. Flanges with grooved faces must be clean and free of dirt so that the gasket will fit properly. It is sometimes necessary to use solvents to remove the protective coating put on flanges at the factory. Always check the engineering specifications before applying any type of solvent to the flanges. Gaskets should also be cleaned before installation. Flanged joints are designed so that the face of each flange matches up squarely with a gasket to form a tight seal. Any type of obstruction or debris that could cause the flanges to match improperly or break the seal must be removed.

6.2.0 Aligning Parts

After it is cleaned, the pipe must be put in place and properly supported. The jacks and stands must then be adjusted until the flanges are properly aligned. A drift pin can be used to align the bolt holes. Never try to align the bolt holes with your fingers. *Figure 34* shows good and bad alignment. Specifications and standards will give tolerances for misalignment.

6.3.0 Installing Gaskets

After the parts are aligned, the gaskets should be selected and put in place. As shown earlier, each flange has its own gasket type. Follow these steps to install a full-face gasket between flanges:

Step 1   Check the specification to determine the type of gasket needed.

Step 2   Find the proper gasket, based on the specification. Lubricate the gasket if necessary.

Step 3   Place the gasket between the flange faces.

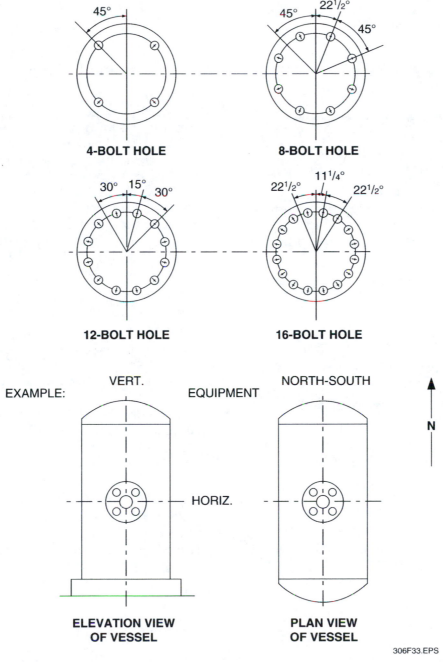

Figure 33 ◆ Bolt hole locations.

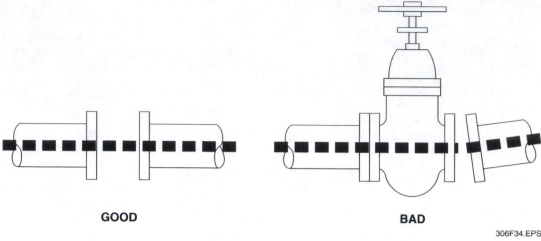

GOOD　　　　　　　　　　　　**BAD**

306F34.EPS

Figure 34 ◆ Good and bad alignment.

Step 4　Hold the gasket in place, and insert the bottom bolts into the flanges.

 WARNING!

Never use your fingers to hold the gasket in place. Use a drift pin to avoid pinching your fingers.

Step 5　Insert all the bolts through the flange bolt holes.

Step 6　Place the nuts on the bolts, and tighten the nuts finger-tight.

Follow these steps to install a gasket between two raised-face flanges.

Step 1　Check the specification to determine the type of gasket needed.

Step 2　Find the proper gasket, based on the specification.

Step 3　Insert the bottom bolts into the flanges. Lubricate the gasket if necessary.

Step 4　Place the gasket between the flange faces.

Step 5　Insert all the bolts through the flange bolt holes.

Step 6　Place the nuts on the bolts, and tighten them finger tight.

6.4.0 Tightening Flange Bolts

Flange bolts must be tightened carefully to avoid warping the flange. If the bolts are not tightened correctly, the joint may leak or crack. Pressure must be applied evenly, in the correct amount and in the correct tightening sequence, to make a good seal. *Figure 35* shows the proper tightening sequence for flange bolts.

A torque wrench is used to apply equal pressure at all points around the flange. With the aid of a torque wrench, a pipefitter can properly adjust and tighten a flanged joint. Follow these guidelines when tightening flange bolts:

- Make sure that the threads of the flange, nut, and bolt are clean and lightly oiled before torquing.
- If the bolts of a flange have already been tightened, loosen and retighten them. Loosen each bolt one full turn, using an open-end or socket wrench.
- Make sure that the torque wrench is correctly adjusted. Most companies calibrate precision tools at regular intervals.
- Always place the torque wrench on the nut, not on the bolt. Use an open-end or socket wrench to hold the bolt while torquing.
- Always use a smooth, even motion when using the torque wrench. A hurried, uneven motion does not give a correct measurement.

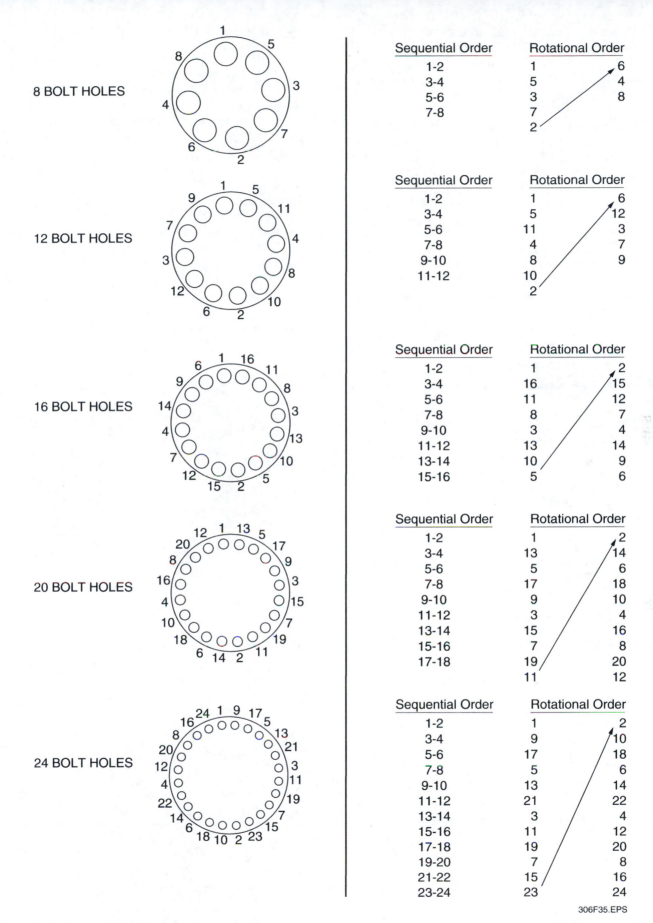

8 BOLT HOLES

Sequential Order	Rotational Order	
1-2	1	6
3-4	5	4
5-6	3	8
7-8	7	
	2	

12 BOLT HOLES

Sequential Order	Rotational Order	
1-2	1	6
3-4	5	12
5-6	11	3
7-8	4	7
9-10	8	9
11-12	10	
	2	

16 BOLT HOLES

Sequential Order	Rotational Order	
1-2	1	2
3-4	16	15
5-6	11	12
7-8	8	7
9-10	3	4
11-12	13	14
13-14	10	9
15-16	5	6

20 BOLT HOLES

Sequential Order	Rotational Order	
1-2	1	2
3-4	13	14
5-6	5	6
7-8	17	18
9-10	9	10
11-12	3	4
13-14	15	16
15-16	7	8
17-18	19	20
	11	12

24 BOLT HOLES

Sequential Order	Rotational Order	
1-2	1	2
3-4	9	10
5-6	17	18
7-8	5	6
9-10	13	14
11-12	21	22
13-14	3	4
15-16	11	12
17-18	19	20
19-20	7	8
21-22	15	16
23-24	23	24

306F35.EPS

Figure 35 ◆ Proper tightening sequence.

Each piping system may use different torque settings, depending on the sizes of the flanges and bolts, the pressure or temperature of the system, and the type of gasket used. Consult your supervisor or the piping drawings for the proper torque setting.

The bolt holes of all flanges can be identified by a simple numbering system. There are no numbers on the flanges, but each hole, starting at the top and moving clockwise, is referred to by a number. To properly torque a flange joint, you must use a crossover method of tightening the bolts. Follow these steps to tighten flange bolts:

Step 1 Set the torque wrench at 25 percent of the required torque.

> **CAUTION**
>
> Do not torque bolts to the full pressure the first time around the system because this causes unequal pressure on the flanges.

Step 2 Tighten the top, center bolt to this torque.

Step 3 Tighten the bolt directly opposite this bolt, the bottom center bolt, to this torque.

Step 4 Tighten the side bolt to this torque.

Step 5 Tighten the bolt directly opposite of the bolt tightened in Step 4 to the set torque.

Step 6 Continue this process until all the bolts are tightened to this torque.

Step 7 Reset the torque wrench to 50 percent of the required torque.

Step 8 Tighten all bolts to this torque, following the crossover method.

Step 9 Reset the torque wrench to 75 percent of the required torque.

Step 10 Tighten all bolts to this torque, following the crossover method.

Step 11 Reset the torque wrench to 100 percent of the required torque.

Step 12 Tighten all bolts to this torque, repeating the crossover, or starting at number 1 and moving around the flange clockwise.

Step 13 Repeat Step 12 for a final tightening until there is no movement.

6.5.0 Grooved Pipe and Fittings

Roll grooved and cut grooved systems are another common way to join pipe above ground. The joints consist of a camming coupling (*Figure 36*) and grooved pipe that is used in pipe or tubing applications where the joint may need to be taken apart. The groove is cut or rolled three-quarters of an inch from the end of the pipe or fitting (*Figure 37*). The grooved pipe is pressed firmly into the coupling, and the two clamps are pulled into place against the sides of the coupling.

Long used in fire protection applications, these are now used in petroleum industry applications also. In addition to the cut and rolled grooves, several companies have also developed an equivalent of the Ridgid ProPress® joining system, used with Schedule 5 and Schedule 10 pipe. The pipe is pushed together, one end inside the other, and formed into a groove with a special tool (*Figure 38*) in the field.

306F36.EPS

Figure 36 ◆ Camming coupling.

306F37.EPS

Figure 37 ◆ Roll-grooved tubing.

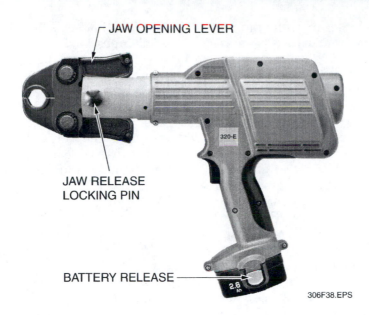

JAW OPENING LEVER

JAW RELEASE
LOCKING PIN

BATTERY RELEASE

306F38.EPS

Figure 38 ◆ ProPress tool.

The ProPress tool is a hydraulic clamp that forms the pipe to the fittings. The fittings have a raised bead around the leg of the fitting. The pipe slips over the bead, and the tool compresses the pipe over the bead, to form a tight fit.

To form a ProPress joint, perform the following steps:

Step 1 Choose the correct size jaw for the tool.

Step 2 Be sure the tool is unplugged or the battery removed and the locking pin on the tool is fully open; that is, the pin is pulled out until it stops. Hold the tool over the lever end of the jaws and slide the tool into position.

Step 3 Push the locking pin in to the locked position.

Step 4 Grasp the tool by the handgrip and the levers of the jaws (see *Figure 38*), opening the jaws. Slide the jaws over the pipe at the bead on the fitting.

Step 5 When the jaws are in place, move your hand off the levers. Press the trigger. The jaws will close fully (*Figure 39*) and retract.

NOTE

The tool will close fully if the trigger is pulled, whether the trigger is released or not. Be very careful that no fingers are between the jaws when you pull the trigger.

The tool jaws should be cleaned at regular intervals as instructed in the documentation supplied by the manufacturer. Read and obey all instructions in the manual.

7.0.0 ◆ LAYING OUT AND INSTALLING PIPE SLEEVES AND FLOOR PENETRATIONS

Pipe sleeves are often installed through floors during construction when the floor is poured so that the concrete sets around the pipe sleeve. As a pipefitter, you will be expected to lay out and install pipe sleeves and floor penetrations through existing floors. Pipe sleeves are generally fabricated out

306F39.EPS

Figure 39 ◆ Holding the ProPress tool in use.

of carbon steel pipe that is larger than the pipe it protects. The pipe can be supported in the sleeve by a single bolt riser clamp or by lugs welded to the pipe that rest against the top of the pipe sleeve. Your job specifications dictate the size of the pipe sleeve, depending on the size of pipe being run and the service of the pipe. Never run an unprotected pipe through a hole in a concrete floor. *Figure 40* shows a pipe sleeve and a pipe sleeve installation.

Floor penetrations and wall penetrations are specified and detailed in construction drawings. However, some understanding of the standard approach to penetrations is necessary. First, the pipe is protected as it passes through the wall. This function is achieved by the wall sleeve first. In most localities and applications, the wall sleeve is a piece of pipe, larger than the pipe that is to pass through it. In addition, the pipe normally has a flat ring around the middle of the outside of it inside the wall. This ring, called a waterstop, is designed both to hold the sleeve in place, and to keep water from running through the wall between the sleeve and the wall material. If the wall or floor has been drilled to insert the wall sleeve, the space around the outside of the sleeve will be grouted.

In some locations, especially in desert contexts, the inside of the sleeve is filled with mineral wool or certain proprietary fireproof packings and simply grouted around the pipe *(Figure 41)*. The pipe is supported in place mechanically, with braces or supports, and no further provision is made. In most cases, however, either flexible link seals are used, or a spool is made that passes through the wall and flexible couplings are used to provide for differential movement.

Flexible link seals are made under a number of brand names, such as Link-Seal® or PipeSeal®, but the systems work the same. These are available either as waterseals or as firestops. The seals are composed of thick rubber pieces, slightly curved, with holes down the length of half of each end, rather like a hinge *(Figure 42)*. The pieces fit together so the holes align, and a bolt is pushed through to make a link with a metal pressure plate on each end, under the bolt head or the nut. The

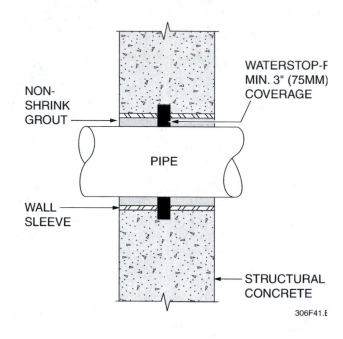

Figure 41 ◆ Sleeved penetration detail.

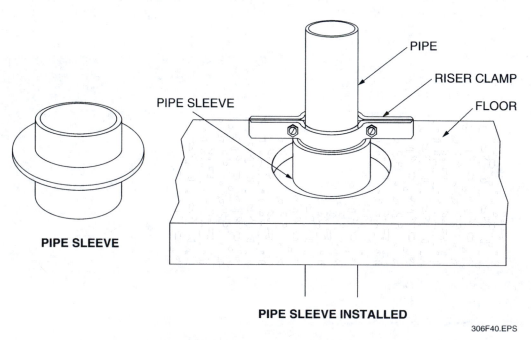

Figure 40 ◆ Pipe sleeve and pipe sleeve installation.

links are assembled until there are enough to go all the way around the pipe. The last bolt completes the ring, and the ring of links is slid inside the wall sleeve (*Figure 43*). Tightening the bolts causes the links to expand and fill the space around the pipe. The bolts are tightened sequentially and gradually several times around the pipe. Flexible link systems are usually supplemented by grout or firestopping foams.

The floor penetration can be laid out using the column line control points (*Figure 44*).

306F42.EPS

Figure 42 ◆ Flexible link seals.

As you can see in Section C-C, the 4-inch pipeline 245 and the 6-inch pipeline 250 penetrate the concrete floor to the lower level. To find the exact locations of these floor penetrations, refer to the drawing. *Figure 45* shows a part of this drawing.

Using the drawing, you can lay out the floor penetrations. The following procedure explains how to lay out the floor penetration for line 250. Follow these steps to lay out and install floor penetrations and pipe sleeves.

Step 1 Look at the north arrow on the drawing and locate north in the building in which you are installing the floor penetration.

Step 2 Locate the column nearest the floor penetration. In this case, the column is 20.

Step 3 Mark a chalkline on the floor between the center of column 20 and the center of column 19.

Step 4 Identify the distance west of column 20 at which the floor penetration will be located. Note that the center of the pipe going through the floor is located 8 feet, 6 inches from the center of column 20.

Step 5 Measure from the center of column 20 8 feet, 6 inches down the chalkline, and make a mark across the line.

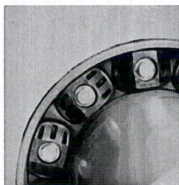

306F43.EPS

Figure 43 ◆ Putting flexible link seals together.

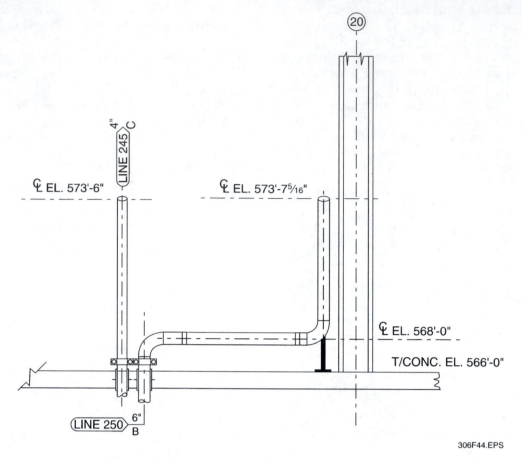

306F44.EPS

Figure 44 ◆ Laying out the floor penetration location.

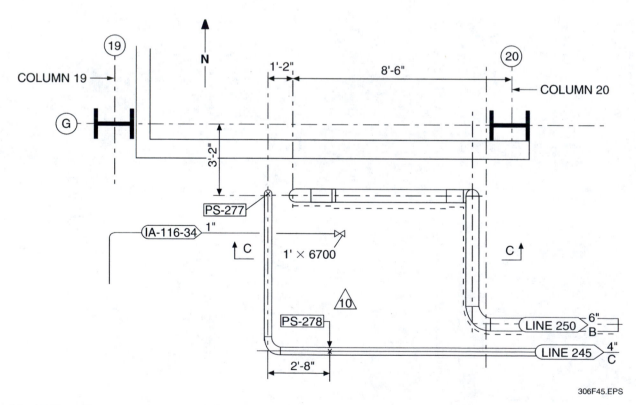

306F45.EPS

Figure 45 ◆ Floor penetration on pipeline.

Step 6 Identify the distance south of column 20 at which the floor penetration will be located. Note that the center of the pipe going through the floor is located 3 feet, 2 inches from the chalkline between the center of column 20 and the center of column 19.

Step 7 Draw a line perpendicular to the chalkline at the mark made in Step 5, using a framing square.

Step 8 Measure down the perpendicular line, and make a mark 3 feet, 2 inches south of the chalkline.

Step 9 Mark a chalkline between the mark made in Step 5 and at least 1 foot past the mark made in Step 8. *Figure 46* shows the chalkline layout.

Step 10 Use the 3–4–5 method to ensure that the chalklines intersect at a true 90-degree angle.

Step 11 Draw an 18-inch line perpendicular to the north-south chalkline at the center point of the floor penetration, using a framing square (see *Figure 46*).

Step 12 Place a catch pan and a barricade around the area on the floor directly underneath the spot where the floor penetration will be made. This is to protect the workers in the area below when you drill the hole in the concrete. Check your job specifications for any other required precautions, such as special warning signs or a lookout person stationed below.

Step 13 Check your job specifications to determine what size hole needs to be drilled for a 6-inch pipe.

Step 14 Set up a core drill for drilling the hole in the concrete.

Step 15 Identify the size pipe sleeve needed.

Step 16 Select a drill bit to match the pipe sleeve chosen. The drill bit selected for a pipe sleeve must be slightly larger in diameter than the pipe sleeve. The pipe sleeve must be able to fit inside the hole and must be long enough to extend all the way through the concrete.

WARNING!
Inspect the core drill and bit carefully to make sure they are in good operating condition. If they are not, you could be injured by parts that chip and fly off during operation.

Step 17 Put on safety glasses, ear protectors, dust mask, and gloves.

WARNING!
Drilling in concrete generates noise, dust, and possible flying objects. Make sure others in the area also wear protective equipment.

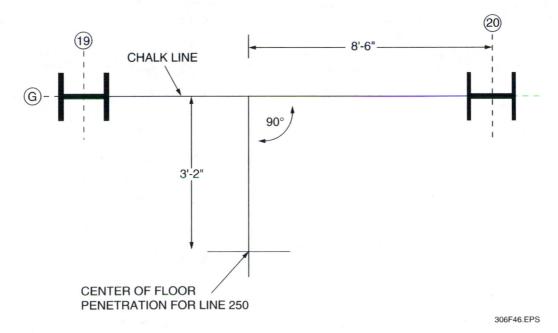

Figure 46 ◆ Chalkline layout.

306F46.EPS

Step 18 Assemble the drill and drill bit.

Step 19 Drill the hole for the pipe sleeve. The drill bit may be lubricated with water during the drilling process.

Step 20 Insert the pipe sleeve into the floor penetration according to job specifications.

8.0.0 ◆ READING AND INTERPRETING SPOOL SHEETS

Isometric drawings (ISOs) are detailed line drawings that refer to a pipeline on a plan drawing. A spool is a prefabricated section of piping. Spool sheets are detailed drawings of spools that make up a pipeline and refer to the ISO of that line. The spools shown on an ISO are separated by field welds or flange connections. Spool sheets can be drawn as isometric drawings or as orthographic drawings showing the plan and elevation views of the spool. Information that can be found on a spool sheet includes the following:

- Instructions the welder needs to fabricate the spool
- Cut lengths of pipe
- Any fittings and flanges needed to fabricate the spool
- Materials required
- Specifications of the spool
- Isometric drawing reference number
- Number of spools required

Figure 47 shows an isometric drawing.

Three spools make up the pipeline shown in *Figure 47*. They are separated by field welds (FW) and are numbered with the line number and then 1, 2, and 3 following the line number. This is how most companies number the individual spools in a system, but you may find some variations at different job sites. If spools are fabricated by an outside contractor at your job site, they will send the assembled spools to you with the spool numbers either painted on the actual spool or tagged to the spool. Spools are normally numbered in order according to the direction in which the material flows in the system.

Figure 48 shows the spool sheet for the spool numbered 395-S35-411-002-1 from the ISO shown in *Figure 47*.

Figure 48 shows a spool that consists of a length of pipe with a 90-degree ell welded to one end and a 45-degree ell welded to the end of the 90. A Class 300, raised-face, weld neck flange is welded to the other end of the 45. The spool drawing gives you the center-to-center dimensions of the fittings and

the dimensions from the end of the pipe to the center of the 90. You can see from the bill of materials that 16-inch A106 Grade B seamless carbon steel pipe and fittings are used and the 90-degree ell is a long-radius 90. The bill of material also lists the total length of pipe used in the spool. The 2 BE on item 3 means that the pipe must be beveled on each end. The title block gives information regarding the pipeline number, spool number, number of spools required, job specification reference, weight, and also who drew and checked the spool sheet.

Figure 49 shows the spool sheet for the spool numbered 395-S35-411-002-2 from the ISO shown in *Figure 47*.

Spool sheet 395-S35-411-002-2 contains much of the same information as the first spool sheet in the title block. This spool shows a 90-degree, long-radius ell welded to a straight length of pipe. A Class 300, raised-face, weld neck flange is welded to the other end of the straight pipe. The end-to-center dimension of 10 feet, 5½ inches is given on the drawing, but the bill of material gives the actual length of the pipe to be 7 feet, 10½ inches. Subtracting 7 inches for the length through the flange gives you 9 feet, 10½ inches. Dimensions for flanges can be obtained from *The Pipe Fitter's Blue Book* or the flange manufacturer. Remembering the formula for figuring takeouts of 90-degree, long-radius ells will allow you to check the cut length of the pipe. The takeout of a long-radius ell is 1½ times the nominal pipe size. Multiply 16 inches times 1½ to get 24 inches, which is the takeout of the long radius 90-degree elbow. Subtract 24 inches from the end-to-center dimension of 9 feet, 10½ inches shown on the spool sheet, and you will see that the actual length of the pipe is 7 feet, 10½ inches. Note that the straight length of pipe must be beveled on each end.

Figure 50 shows the spool sheet for the spool numbered 395-S35-411-002-3 from the ISO shown in Figure 47.

The title block of spool sheet 395-S35-411-002-3 contains the same information as in the first spool sheet. This spool shows a 45-degree ell welded to a straight length of pipe. The end-to-center dimension of 2 feet, 4⁵⁄₁₆ inches is given on the drawing, but the bill of material gives the actual cut length of the pipe to be 1 foot, 6⁵⁄₁₆ inches, so no calculations are necessary. Remembering the formula for figuring takeouts of 45-degree ells will allow you to check the cut length. The takeout of a long-radius, 45-degree ell is ⅝ times the nominal pipe size. Multiply 16 inches times ⅝ to get 10 inches, which is the takeout of the 45-degree elbow shown in *Figure 50*. Note that the straight length of pipe must be beveled on each end.

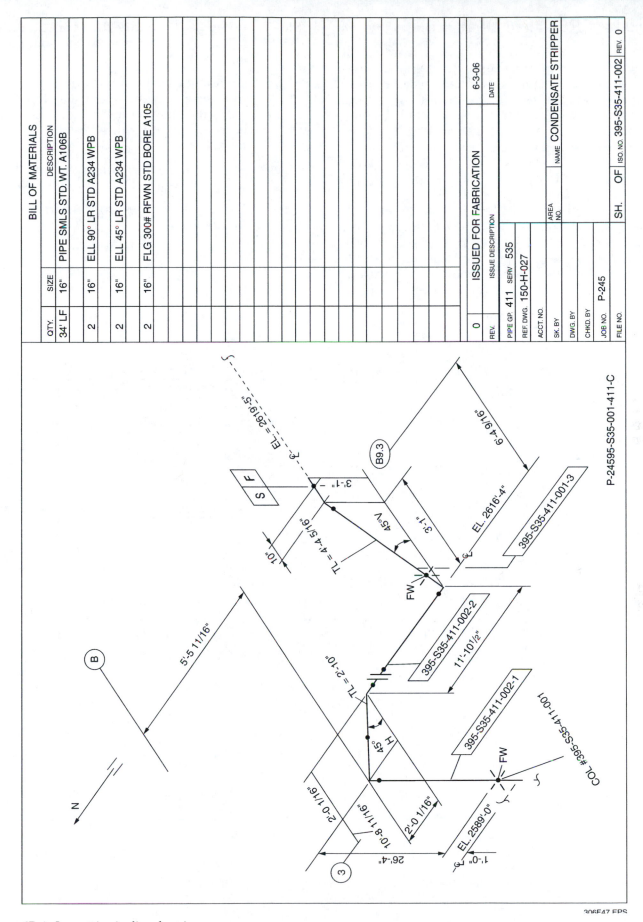

BILL OF MATERIALS

QTY.	SIZE	DESCRIPTION
34' LF	16"	PIPE SMLS STD. WT. A106B
2	16"	ELL 90° LR STD A234 WPB
2	16"	ELL 45° LR STD A234 WPB
2	16"	FLG 300# RFWN STD BORE A105

ISSUED FOR FABRICATION

REV.	ISSUE DESCRIPTION	DATE
0		6-3-06

PIPE GP. 411 SERV. 535

REF. DWG. 150-H-027

ACCT. NO.

SK. BY

DWG. BY

CHKD. BY   P-245

JOB NO.

FILE NO.

AREA NO.

NAME CONDENSATE STRIPPER

SH.   OF   ISO. NO. 395-S35-411-002 REV. 0

P-24595-S35-001-411-C

Figure 47 ◆ Isometric pipeline drawing.

306F47.EPS

9.0.0 ◆ ERECTING SPOOLS

Many companies fabricate spools from spool sheets on site in a pipe shop, while other companies have all their spools fabricated by an outside contractor and shipped to the site. Either way, the pipefitters must identify the spools and use the ISOs and plan drawings to fabricate the specific piping system that the spool was fabricated for.

When erecting spools, you must communicate with riggers and any other craftspersons who will be working with you to erect the spool, protect all other lines and equipment in the installation area, and treat all other installations in the area as live installations. The following section is a general procedure for erecting a spool. You must erect spools based on the specific guidelines and policies of your job site. Follow these steps to erect a spool.

Step 1 Locate and identify the spool according to the spool number on the spool sheet and the ISO.

Step 2 Compare the spool with the spool sheet to ensure that the spool was fabricated correctly. Ensure that the spool piece has been primed if required by the job specifications.

Step 3 Determine the location of the installation on the plan drawing and in the field.

Step 4 Determine the testing requirements of the spool.

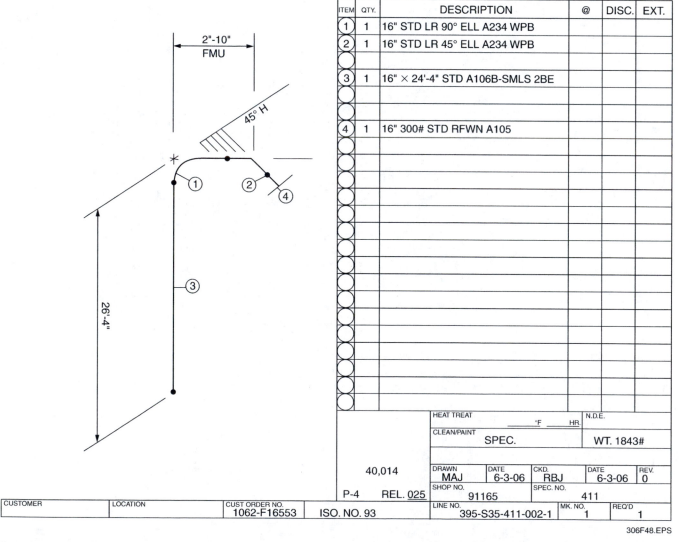

Figure 48 ◆ Spool sheet.

Step 5 Identify the method by which to connect the spool into the system.

Step 6 Determine if more than one spool can be connected on the ground and erected together.

Step 7 Determine the weight of the spool. Sometimes the weight of the spool is included on the spool sheet. If it is not, you must calculate the weight, using a weight chart for the piping being used.

Step 8 Obtain all erection materials and tools.

Step 9 Determine the access and egress to the installation point.

Step 10 Determine and obtain the supports and hangers needed for the installation. Hanger locations are shown on the plan drawing or ISOs.

Step 11 Install the hangers and supports that will permanently hold the spool in place.

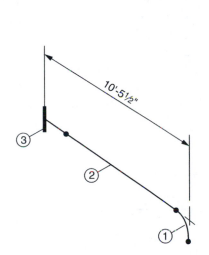

BILL OF MATERIALS

ITEM	QTY.	DESCRIPTION	@	DISC.	EXT.
1	1	16" STD LR 90° ELL A234 WPB			
2	1	16" × 7'-10 1/2" STD A106B-SMLS 2BE			
3	1	16" 300# STD RFWN A105			

HEAT TREAT				°F	HR.	N.D.E.	

CLEAN/PAINT			
	SPEC.		WT. 704#

	DRAWN	DATE	CKD.	DATE	REV.
40,014	MAJ	6-3-06	RBJ	6-3-06	0
	SHOP NO.		SPEC. NO.		
P-4 REL. 025	91165		411		

CUSTOMER	LOCATION	CUST ORDER NO. 1062-F16553	ISO. NO. 93	LINE NO. 395-S35-411-002-2	MK. NO. 2	REQ'D 1

306F49.EPS

Figure 49 ◆ Spool sheet.

Step 12 Determine if any temporary supports need to be erected to support the pipe while it is being joined.

Step 13 Prep the ends of the spool for connection. This includes removing protective caps and plugs, cleaning beveled ends with a grinder, or buffing ends to be welded.

Step 14 Inspect the exterior and interior of the spool for cleanliness, and clean the spool if necessary.

Step 15 Determine the method of transporting the spool to its installation point. Check your company policies regarding who is responsible for rigging operations and erecting scaffolding.

Step 16 Determine the tools required to lift the spool to the installation point based on the weight, the configuration of the

spool, and the location of the installation. These tools include chokers, slings, and appropriate rigging equipment. If the spool is stainless steel, remember that you must use nylon slings.

Step 17 Obtain the required rigging equipment.

Step 18 Determine the sequence for erecting the spool. If you are unsure about the sequence of spool erection, check with your supervisor or foreman.

Step 19 Transport the spool to its installation point.

Step 20 Plumb, level, and square the spool in position to be connected.

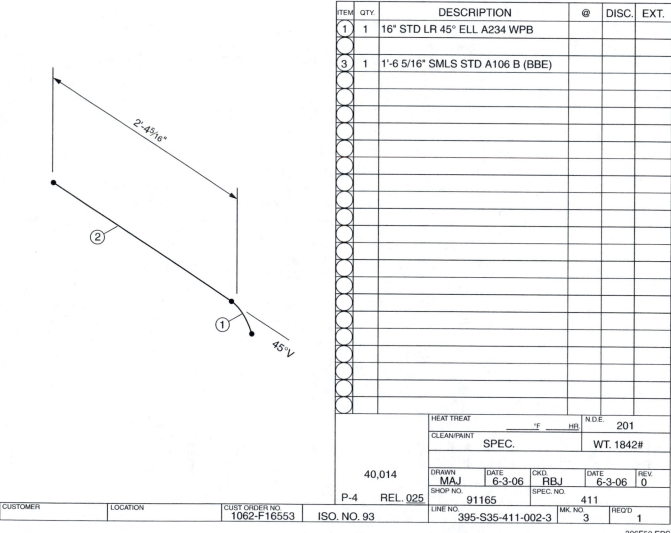

BILL OF MATERIALS

ITEM	QTY.	DESCRIPTION	@	DISC.	EXT.
1	1	16" STD LR 45° ELL A234 WPB			
3	1	1'-6 5/16" SMLS STD A106 B (BBE)			

HEAT TREAT		°F	HR.	N.D.E. 201	
CLEAN/PAINT	SPEC.			WT. 1842#	

40,014	DRAWN MAJ	DATE 6-3-06	CKD. RBJ	DATE 6-3-06	REV. 0
P-4 REL. 025	SHOP NO. 91165		SPEC. NO. 411		

CUSTOMER	LOCATION	CUST ORDER NO. 1062-F16553	ISO. NO. 93	LINE NO. 395-S35-411-002-3	MK. NO. 3	REQ'D 1

306F50.EPS

Figure 50 ◆ Spool sheet.

1. Proper storage procedures maintain the quality of the pipe and materials and _____.
   a. keep the workers busy
   b. make more work
   c. save time
   d. are difficult

2. Small materials such as fittings are usually ordered in _____.
   a. metal crates
   b. large quantities
   c. small quantities
   d. hardware stores as needed

3. Store all valves in a(n) _____.
   a. open basket
   b. shipping crate
   c. horizontal position with the handwheel to the side
   d. vertical position with the handwheel up

4. Flange protective covers must cover the _____.
   a. entire flange
   b. boltholes
   c. gasket surface
   d. bore

5. Male threaded ends should be protected by _____.
   a. press-on caps
   b. press-in plugs
   c. wooden plugs
   d. duct tape

6. Pipe systems that may need to be dismantled occasionally are frequently joined by _____.
   a. welding
   b. threaded joints
   c. brazing
   d. flanges

7. Weld neck flanges must be the same weight as the pipe, so the _____ will match.
   a. bolt holes
   b. gasket
   c. bores
   d. centers

8. Slip-on flanges are used where the space is limited, and in many _____ applications
   a. high-pressure
   b. extremely high flowrate
   c. severe service
   d. low-pressure

9. A slip-on reducing flange should only be used when the flow is from the _____ pipe.
   a. lower
   b. larger
   c. smaller
   d. higher

10. Socket weld flanges are mostly used with _____ piping.
    a. small low-pressure
    b. small high-pressure
    c. large high-pressure
    d. large low-pressure

11. When using lap-joint flanges in vertical applications, _____.
    a. weld the pipe first, then slide the flange on
    b. after the flange is on the stub end, weld lugs on the stub end
    c. bolt the flange tighter
    d. don't use gaskets

12. The most common ratings for cast iron flanges are _____.
    a. Class 125 and Class 250
    b. Class 150 and Class 300
    c. Class 400 and Class 500
    d. Class 2,500 and Class 3,000

13. You should *not* connect a cast iron flange to a _____.
    a. flat-faced flange
    b. raised-face flange
    c. slip-on flange
    d. weld neck flange

14. The most common flange facing finish is the _____.
    a. phonographic serrated
    b. concentric serrated
    c. coldwater finish
    d. rough finish

15. An RJT flange has the disadvantage that in order to remove a fitting, you must _____.
    a. push the flanges together
    b. cut the weldneck
    c. spread the flanges enough to remove the ring
    d. saw the ring in two

16. Temperatures below 500°F and pressures below 600 psi are considered _____.
    a. high temperature/low pressure
    b. high temperature/high pressure
    c. low temperature/low pressure
    d. low temperature/high pressure

17. Gaskets are chosen to suit the temperature, pressure, and chemical characteristics of the process medium.
    a. True
    b. False

18. Teflon® gaskets are chemically _____.
    a. sensitive
    b. inert
    c. explosive
    d. reactive

19. Acrylic fiber gaskets filled with nitrile could be used with _____.
    a. acetone
    b. MEK
    c. severe service applications
    d. oil

20. Corrugated metal gaskets are generally used on _____.
    a. high pressure applications
    b. concentric serrated flanges
    c. smooth flanges and low-pressure applications
    d. phonographic serrated flanges

21. Cork gaskets can be used for most liquids *except* for _____.
    a. strong alkalis
    b. acids
    c. boiling water
    d. oil

22. Graphite–impregnated gaskets are used for _____.
    a. pure oxygen atmospheres
    b. high temperatures
    c. extremely high pressures
    d. fire protection systems

23. The three concentric circles drawn for a flange layout are the outside diameter (OD) of the flange, the opening diameter of the pipe, and the _____.
    a. outside diameter of the pipe
    b. hole diameter
    c. diameter of the bolt hole circle
    d. size of the stream of fluid

24. On a six-hole flange, the radius is equal to the distance between the bolt holes.
    a. True
    b. False

25. The torque wrench must be set at _____ percent on the second pass on tightening the bolts of a flange.
    a. 20
    b. 40
    c. 50
    d. 75

Summary

As you start installing aboveground pipe, you must call on all of your previously learned skills. The installation of pipe requires you to be proficient in your measurement, calculation, and fabrication techniques. Remember to always put safety first and be aware of your working surroundings when installing pipe in the field. Your safety and the safety of your co-workers depend on the way you perform your job.

Notes

Trade Terms Introduced in This Module

Cavitation: The result of pressure loss in liquid, producing bubbles (cavities) of vapor in liquid.

Dunnage: Hardwood blocking and pallets placed underneath materials to keep them off the ground and to allow access for forklifts or for placement of chokers and slings.

Gasket: A device that is used to make a pressure-tight connection and that is usually in the form of a sheet or a ring.

Pressure differential: The difference in pressure between two points in a flow system. It is usually caused by frictional resistance to flow in the system.

Pressure vessel: A metal container that can withstand high pressures.

Severe service: A high-pressure, high-temperature piping system.

Turbulence: The motion of fluids or gases in which velocities and pressures change irregularly.

Additional Resources

This module is intended to be a thorough resource for task training. The following reference works are suggested for further study. These are optional materials for continued education rather than for task training.

Pipeline Mechanical. National Center for Construction Education and Research. Upper Saddle River, NJ: Prentice Hall.

The most famous aboveground pipeline has a website at http://www.alyeska-pipe.com/default.asp.

Figure Credits

Inertech, Inc., 306F15

W.L. Gore & Associates, 306F16

Photo courtesy of **Garlock Sealing Technologies,** 306F17, 306F19

Victaulic Company, 306F36, 306F37

Ridge Tool Company (RIDGID®), 306F38, 306F39

Cetro Building Materials Group, 306F41

Link-Seal® Modular Seals, manufactured by **Pipeline Seal and Insulator, Inc.,** 306F42, 306F43

CONTREN® LEARNING SERIES — USER UPDATE

NCCER makes every effort to keep these textbooks up-to-date and free of technical errors. We appreciate your help in this process. If you have an idea for improving this textbook, or if you find an error, a typographical mistake, or an inaccuracy in NCCER's Contren® textbooks, please write us, using this form or a photocopy. Be sure to include the exact module number, page number, a detailed description, and the correction, if applicable. Your input will be brought to the attention of the Technical Review Committee. Thank you for your assistance.

Instructors – If you found that additional materials were necessary in order to teach this module effectively, please let us know so that we may include them in the Equipment/Materials list in the Annotated Instructor's Guide.

Write: Product Development and Revision
National Center for Construction Education and Research
3600 NW 43rd St, Bldg G, Gainesville, FL 32606

Fax: 352-334-0932

E-mail: curriculum@nccer.org

Craft

Module Name

Copyright Date

Module Number

Page Number(s)

Description

(Optional) Correction

(Optional) Your Name and Address

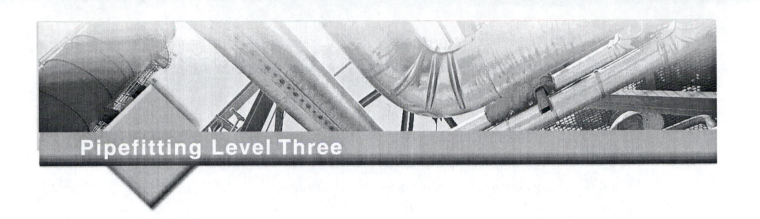

08307-07

Field Routing
and Vessel Trim

08307-07
Field Routing and Vessel Trim

Topics to be presented in this module include:

Overview

Field routing and vessel trim require pipefitters to decide how the pipe has to be assembled. Here is the practice of all the skills, putting together pipe, assembling the various components that make a vessel or piping run work. Here you learn about the different kinds of sensors and valves that are required for reactor vessels and storage tanks.

Objectives

When you have completed this module, you will be able to do the following:

1. Secure the work area.
2. Determine field run specifications.
3. Determine the required rigging equipment based on weight, location, and configuration.
4. Determine the load weight for erection equipment.
5. Determine the support needs.
6. Select and install erection materials.
7. Fabricate the field run of piping.
8. Erect vessel trim.

Trade Terms

Bridle   Grasshopper
Field routing   Vessel trim

Required Trainee Materials

1. Pencil and paper
2. Appropriate personal protective equipment

Prerequisites

Before you begin this module, it is recommended that you successfully complete *Core Curriculum*; *Pipefitting Level One*; *Pipefitting Level Two*; and *Pipefitting Level Three*, Modules 08301-07 through 08306-07.

   This course map shows all of the modules in the third level of the *Pipefitting* curriculum. The suggested training order begins at the bottom and proceeds up. Skill levels increase as you advance on the course map. The local Training Program Sponsor may adjust the training order.

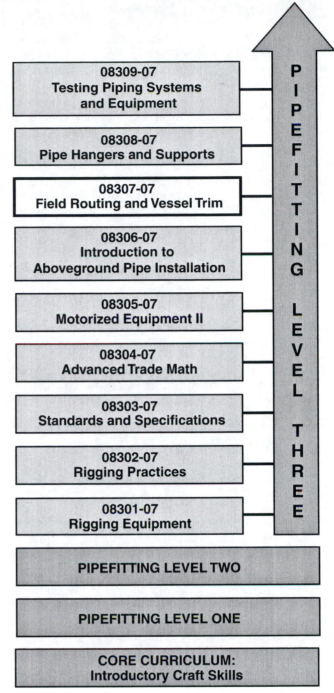

307CMAP.EPS

1.0.0 ◆ INTRODUCTION

This module introduces **field routing** and **vessel trim**. To perform field routing and vessel trim activities, you must be familiar with reading blueprints, fabricating piping components, and installing aboveground piping. This module explains how to secure the work area; determine spool specifications; determine the load weight for erection equipment; determine the method for transporting and erecting a spool; determine and obtain the required rigging for a lift based on weight, location, and configuration; determine support needs; obtain erection materials; plumb, level, and square components; and erect vessel trim.

2.0.0 ◆ SECURING THE WORK AREA

When preparing to field route pipe, it is important to secure the work area to provide a safe area in which to store and field-fabricate spool pieces. Follow these steps to secure the work area.

Step 1 Clean a space large enough to store the pipe components and to field-fabricate the spool.

Step 2 Barricade the work area to secure it from through traffic.

Step 3 Check with the supervisor and field engineer to determine the nominal pipe sizes and in-line equipment required.

Step 4 Transport the pipe to the work area, using a **grasshopper** (*Figure 1*) or suitable transportation.

Step 5 Place the pipe on dunnage.

Step 6 Inspect the inside and the outside of the pipe for cleanliness. If the pipe has been stored in an outside storage area, the inside of the pipe may contain rust and other debris that should be removed at this time.

307F01.EPS

Figure 1 ◆ Grasshopper pipe dolly.

Step 7 Transport the in-line equipment, such as valves, flanges, and elbows, to the work area using a fork truck or other suitable transportation.

Step 8 Store the in-line equipment on a wooden pallet.

Step 9 Route the necessary air hoses, electrical cords, and welding leads to the work area. Be sure to route air hoses, electrical cords, and welding leads overhead.

> **WARNING!**
> Good housekeeping is very important when setting up the work area. If it is not maintained, personnel in the work area, as well as those in the immediate area, could be in danger of flying debris, welding flash, or tripping hazards.

Step 10 Check out the necessary grinders, wire brushes, jack stands, and tools needed for field fabrication.

Step 11 Place blinds in strategic locations to protect employees from grinding debris or welding flash.

Step 12 Request that the safety engineer inspect the work area.

Step 13 Obtain from the safety engineer any grinding, welding, or burning permits necessary to perform field-fabrication duties.

3.0.0 ◆ DETERMINING FIELD RUN SPECIFICATIONS

When performing field-routing activities, you must determine the field run specifications, or dimensions of the system components, since the system is not already assembled. Follow these steps to determine the field run specifications.

Step 1 Locate point A (the starting point).

Step 2 Locate point B (the end point).

Step 3 Locate the north/south east/west columns and determine the best route for the piping system.

Step 4 Check the proposed route carefully to determine if there are any existing pipe racks or hangers that can be used.

Step 5 Look for equipment or other obstructions in the proposed route. If the equipment or obstructions cannot be removed, choose an alternate route.

Step 6 Look for the best location for getting into and out of the area of the field-routing pipe connections.

Step 7 Draw a draft of the proposed route and confer with the supervisor and field engineer to determine if there are any conflicting systems or components in the route. *Figure 2* shows a field-routing draft.

Step 8 Locate the face of point A and point B, using a plumb bob or level.

Step 9 Measure from point A to the first turn.

NOTE

When you are figuring the field route for piping, keep in mind that you should minimize the number of fittings and the amount of offset. The fewer the fittings and the smaller the offset, the less work and money will be needed.

Step 10 Write this measurement on your draft sketch.

Step 11 Continue measuring and writing the measurements on your draft sketch until point B is reached.

Step 12 Measure from the center line of point A and point B to the center of the end hanger to obtain the elevation, and write these measurements on your draft sketch.

Step 13 Recheck the measurements and discuss them with the supervisor and field engineer.

Step 14 Check the equipment at point A and point B to verify the type of connections required.

Step 15 Check the pipe and in-line equipment to ensure that they are in compliance with specifications.

NOTE

The draft sketch with the field run measurements can be divided to enable a crew to work from each end.

3.1.0 Determining the Load Weight for Erection Equipment

To prevent overloading the erection equipment, the pipefitter must know the weight of the pipe or spool piece before selecting the erection equipment. Pipe weighs a given amount per foot, depending on the wall thickness and nominal size of the pipe. Therefore, if the weight per foot and the total number of feet are known, the total weight of the load can be determined.

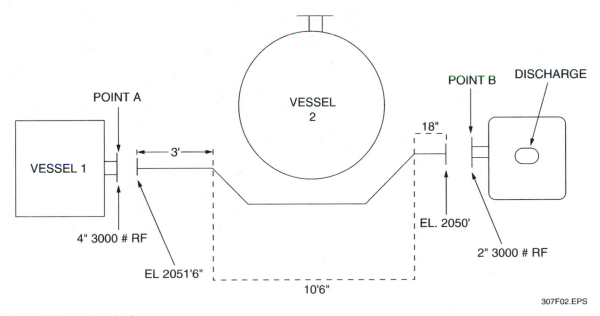

Figure 2 ◆ Field-routing draft.

307F02.EPS

Some spool pieces are field-fabricated with valves and flanges installed. To determine the weight of these spool pieces, the entire assembly must be measured and the proper weight added for each valve, depending on the valve size and material. Additional weight should also then be added for each flange. *Table 1* lists the weights of carbon steel pipe.

The numbers in the far left column represent nominal pipe size in inches. The designations and numbers across the top represent wall thickness, weight, or schedule. To use the table, find the point at which the nominal size and wall thickness of the pipe being used meet. The number in this block is the weight, in pounds, of 1 foot of pipe. To find the total weight, multiply the total number of feet by the weight per foot. Follow these steps to determine the load weight ratings for erection equipment.

Step 1 Measure the length of pipe to be lifted.

Step 2 Check the nominal pipe size.

Step 3 Check the wall thickness.

Table 1 Carbon Steel Pipe Weights

Nominal Pipe Size (inches)	Wall Thickness								
	STD	XS	XXS	10	40	60	80	120	160
	Weight Per Foot in Pounds								
2	3.65	5.02	9.03	-	3.65	-	5.02	-	7.06
2.5	5.79	7.66	13.7	-	5.79	-	7.66	-	10.01
3	7.58	10.25	18.58	-	7.58	-	10.25	-	14.31
3.5	9.11	12.51	22.85	-	9.11	-	12.51	-	-
4	10.79	14.98	27.54	-	10.79	-	14.98	18.98	22.52
6	18.97	28.57	53.16	-	18.97	-	28.57	36.42	45.34
8	28.55	43.39	72.42	-	28.55	35.66	43.39	60.69	74.71
10	40.48	54.74	104.1	-	40.48	54.74	64.40	89.27	115.7
12	49.56	65.42	125.5	-	53.56	73.22	88.57	125.5	160.3
14	54.57	72.09	-	36.71	63.37	85.01	106.1	150.8	189.2
16	62.58	82.77	-	42.05	82.77	107.5	136.6	192.4	245.2
18	70.59	93.45	-	47.39	104.8	138.2	170.8	244.1	308.6
20	78.60	104.1	-	52.73	123.1	166.5	208.9	296.4	379.1
22	86.61	114.8	-	58.07	-	197.4	250.8	353.6	451.1
24	94.62	125.5	-	63.41	171.2	238.3	296.5	429.5	542.1
26	102.6	136.2	-	85.73	-	-	-	-	-
28	110.6	146.9	-	92.41	-	-	-	-	-
30	118.7	157.5	-	99.08	-	-	-	-	-
32	126.7	168.2	-	105.8	229.9	-	-	-	-
34	134.7	178.9	-	112.4	244.6	-	-	-	-
36	142.7	189.6	-	119.1	282.4	-	-	-	-
42	166.7	221.6	-	-	330.4	-	-	-	-

307T01.EPS

NOTE

For example, to find the weight of a 10-inch schedule 80 carbon steel pipe that is 20 feet long, read across the table from the 10 in the nominal pipe size column until you reach the schedule 80 column. The number you find there is 64.40. This is the weight, in pounds, of 1 foot of pipe. Multiply this number by 20 to find the total weight of the pipe. The pipe weight is 1,288 pounds.

Step 4   Add additional weight for accessories such as valves and flanges. The weight of valves and flanges can be obtained from the shipping bill of lading.

Step 5   Add the total weight of the assembly, and select the appropriate size erection equipment.

3.2.0 Determining Support Needs

When determining the support needs for a field-routed pipe run, it is important for the pipefitter to determine the placement of the pipe hangers based on the size, weight, and type of pipe being run. When selecting pipe hangers and supports, you must also take into consideration the insulation requirements for the pipe. In order for a hanger system to do its job, it must support the pipe at regular intervals. Evenly spacing the hangers ensures that no individual hanger is overloaded. *Table 2* lists the recommended maximum hanger spacing intervals for carbon steel pipe.

In some places, because of local conditions, these intervals may not be correct. For example, in California, because of earthquake risks, hangers may need to be spaced more closely. Be sure to check with your engineer or supervisor about specific local requirements.

Table 2 serves only as a guide. Always check the field-routed pipe run for correct pipe hanger spacing. If the pipe sags, add more hangers. The pipe must run straight without sagging.

4.0.0 ◆ SELECTING AND INSTALLING ERECTION MATERIALS

The requirements of a particular job will probably determine the assembly options used. While you may or may not be in on the decision-making, you should understand the options and the reasons for the choices. The run may be built as a single spool from point A to point B, and simply hung and bolted in place, or it may be in two or more pieces.

The reason for such a decision might include the capacity and availability of rigging equipment, such as cranes, or the size, weight, and configuration of the run. An offset may compel the engineer to make a separate piece, because the angled load might make it very difficult to rig the main spool vertically. It is more complicated and difficult to do welds ten feet in the air over a vessel, for example, so it might be better to put the welded joints together on the ground and raise the higher-elevation sections as a unit.

5.0.0 ◆ FABRICATING THE FIELD RUN

After you have secured the work area, determined the field run specifications, and obtained all necessary rigging equipment and erection materials, you are ready to fabricate the proposed field run of piping. To fabricate the field run of piping, you must transport and erect the piping, install test blinds, and install temporary hydrotest spools.

5.1.0 Transporting and Erecting Piping

When transporting and erecting piping, pipefitters must communicate with riggers and any other craftpersons who will be working with them to transport or erect the piping, protect all other lines and equipment in the installation area, and treat all other installations in the area as live installations. Piping must be transported or erected according to the specific guidelines and policies of the job site. Follow these steps to determine the method of transporting and erecting piping.

Table 2 Recommended Maximum Hanger Spacing Intervals for Carbon Steel Pipe

Pipe Size	Rod Diameter	Maximum Spacing
Up to 1¼"	⅜"	8'
1½"and 2"	⅜"	10'
2½" to 3½"	½"	12'
4" and 5"	⅝"	15'
6"	¾"	17'
8" to 12"	⅞"	22'

307T02.EPS

Step 1 Go over the proposed route carefully and check for any tripping hazards or obstacles that could hinder transporting the spool.

Step 2 Obtain all rigging materials and tools.

Step 3 Rig the piping so that it can be transported as level as possible.

Step 4 Alert personnel working in the area that you are transporting piping through the area.

Step 5 Determine the method of transporting the piping to its installation point. Piping may be hand-carried by two or more pipefitters or transported with heavy equipment. The method used depends upon the weight and length of the piping.

Step 6 Determine the location of the installation.

Step 7 Determine the testing requirements of the piping.

Step 8 Identify the method for connecting the piping into the system.

Step 9 Determine if more than one piece of piping can be connected on the ground and erected at the same time.

Step 10 Determine the weight of the piping.

Step 11 Examine how to get into and out of the installation point, including scaffolds, manlifts, or ladders. These items should be put into place at this time.

Step 12 Determine what temporary supports need to be erected to support the pipe while it is being joined.

Step 13 Prepare the ends of the piping for connection. This includes removing protective caps and plugs, cleaning beveled ends with a grinder, or threading piping.

Step 14 Inspect the exterior and interior of the piping for cleanliness. Clean the piping if necessary.

Step 15 Determine the tools required to lift the piping to the installation point based on the weight and length of the piping and the location of the installation. Tools include chokers, slings, and appropriate rigging equipment. If the piping is stainless steel, you must use nylon slings.

Step 16 Obtain the required rigging equipment.

Step 17 Determine the sequence for erecting the piping. You will have to get the run into position, and keep it there until it is supported fully. If you are unsure about the sequence of piping erection, check with your supervisor.

Step 18 Transport the piping to its installation point.

Step 19 Lift the piping in place.

Step 20 Plumb, level, and square the piping in position.

Step 21 Place the temporary supports along the length of the piping to support the piping while it is being joined.

Step 22 Connect the piping.

Step 23 Determine and obtain the supports and hangers needed for the installation.

Step 24 Install the hangers and supports that will permanently hold the piping in place.

Step 25 It may be possible to install the permanent hangers in advance, allowing you to attach the hangers to the pipe run as you hang it. That would be the most efficient procedure.

5.2.0 Installing Test Blinds

Test blinds are round, flat metal components that stop the process flow in a piping system. The pipefitter must place test blinds between the flanges of a system to isolate the system while performing a hydrotest. The type of material and thickness of the test blind determines the pressure of the system being tested. *Figure 3* shows a typical test blind.

Follow these steps to install test blinds.

Step 1 Determine the type of material of the system being tested. The test blind material should be of the same kind as that of the piping system being tested—either stainless steel or carbon steel. If the system is stainless steel to carbon steel, use a stainless steel test blind.

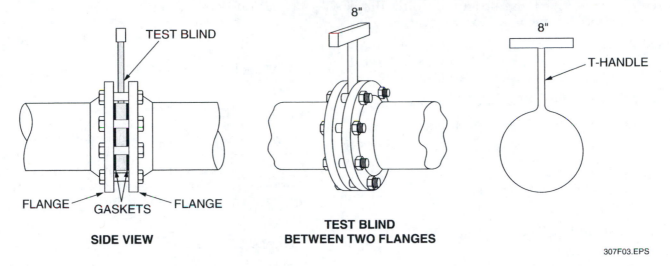

TEST BLIND

FLANGE — GASKETS — FLANGE

SIDE VIEW

8"

**TEST BLIND
BETWEEN TWO FLANGES**

8"

T-HANDLE

307F03.EPS

Figure 3 ◆ Test blind.

Step 2 Check the rating of the flanges. The flange rating is stamped on the outside edge of the flange. The flange rating ranges from 150 pounds to 2,500 pounds.

Step 3 Obtain the test blind from storage.

Step 4 Obtain two gaskets with the correct pressure rating. Refer to the flange rating to obtain gaskets with the correct pressure rating.

Step 5 Obtain extra-long flange bolts. You must use longer flange bolts when installing a test blind to allow enough space for the test blind and additional gaskets. Normally, bolts that are 1-inch longer are acceptable.

Step 6 Obtain two flange spreaders. *Figure 4* shows a typical flange spreader.

Step 7 Verify that the system is depressurized, drained, and vented. Remember: some materials are dangerous in very small quantities.

Step 8 Remove the flange bolts at the location where the test blind is to be installed.

Step 9 Place the two flange spreaders at the 3-o'clock and 9-o'clock positions if possible. This enables you to install the test blind with the handle at the 12-o'clock position, which makes the test blind and gasket installation much easier.

Step 10 Install the extra-long flange bolts in the flanges from the 3-o'clock and 9-o'clock positions down to the 6-o'clock position.

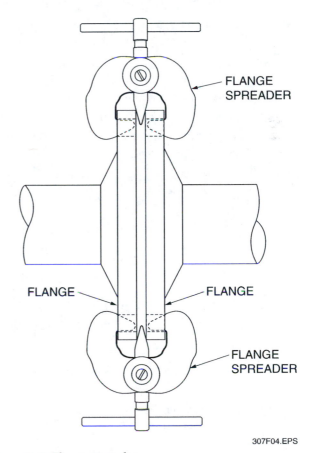

FLANGE SPREADER

FLANGE — FLANGE

FLANGE SPREADER

307F04.EPS

Figure 4 ◆ Flange spreader.

Leave the flange bolts loose to allow enough room to spread the flanges.

Step 11 Tighten the flange spreaders evenly until there is enough space to install the test blind with a gasket on each side.

Step 12 Install the test blind and gaskets.

Step 13 Remove the flange spreaders.

Step 14 Install the remaining flange bolts.

Step 15 Tighten the flange bolts using the correct tightening sequence.

5.3.0 Installing Temporary Hydrotest Spools

The pipefitter must place temporary hydrotest spools in the place of in-line equipment to isolate them while performing a hydrotest. The spool must be able to withstand the same temperatures and pressures as the rest of the system. Follow these steps to install temporary hydrotest spools.

Step 1 Determine the type of material of the system being tested.

Step 2 Check the rating of the flanges.

Step 3 Obtain the temporary spool from storage.

Step 4 Transport the temporary spool to the location where it will be installed.

Step 5 Obtain the required erection materials. The required erection materials include appropriate rigging and bolts.

Step 6 Obtain two gaskets with the correct pressure rating. Remove the in-line equipment if applicable.

Step 7 Rig the temporary spool with the appropriate rigging.

Step 8 Lift the temporary spool in place. *Figure 5* shows a typical temporary spool installation process.

Step 9 Install the lower half of the bolts in each end of the flanges of the temporary spool and the existing flanges.

Step 10 Install a new gasket at each end of the connection.

Step 11 Install the remaining bolts in each end of the connection.

Step 12 Tighten the bolts using the correct tightening sequence.

Step 13 Loosen the rigging on the temporary spool. The rigging may be left in place if the temporary spool will be removed soon. Check with your supervisor or field engineer if you are not sure.

6.0.0 ◆ VESSEL TRIM

Vessel trim is piping components attached to the exterior of a vessel that support instrumentation for monitoring the temperature, pressure, and level of fluid or gas inside the vessel. The components of vessel trim normally include the following:

- Vents
- Drains
- Valves to relieve pressure
- Instruments

6.1.0 Vents

Vents are located at the highest point of a vessel to relieve the air buildup during the filling process and release suction during the draining process. Nonpressurized vessels may have an open vent that enables the liquid level to go up and down without causing any pressure or suction problems. Vents are usually small-bore pipe and can be threaded, socket-welded, or butt-welded, depending on the application.

6.2.0 Drains

Drains are normally located at the lowest point of a vessel. Drains are usually medium-bore pipe and can be threaded, socket-welded, or butt-welded, depending on the application. Where smaller pipes are designed to move fluid out of a vessel for any reason it is common to ensure flow in one direction only with a vacuum breaker. The vacuum breaker works like a check valve. If pressure is lost upstream of the breaker, it closes.

6.3.0 Valves to Relieve Pressure

Valves used to relieve pressure in a pipeline, tank, or vessel are known as pressure safety valves (PSVs) and pressure-relief valves (PRVs). These valves are installed in pipelines to prevent excess pressure from rupturing the line and causing an accident. Both types are adjustable and operate automatically. They operate in the closed position until the pressure in the line rises above the preset pressure limit of the valve. At this point, the valve opens fully to relieve the pressure and remains open until the pressure drops, at which point it snaps shut. Relief valves, which are normally attached to the highest point of the vessel, are designed to relieve pressure in vessels containing liquid. Safety valves are normally used in steam, air, or other gas services. These valves are normally set at 1.5 times the system operating pressure and are connected to the vessel by either threaded connections or flanges, depending on the application. *Figure 6* shows a typical relief valve.

6.4.0 Instruments

Instruments are installed on vessels to provide a means for monitoring level, temperature, and pressure readings. Some instruments are provided for local readings; others are provided for remote readings.

As a pipefitter, you may or may not be assigned to install sensors or other forms of instrumentation. However, it is useful to you to understand the basic types and their accompanying equipment.

Many new varieties of sensors have appeared on the market, allowing level, pressure, and temperature to be measured much more efficiently. Some of these are mentioned in the following sections.

Level glasses, also sometimes called sight glasses, are used less frequently since the introduction of various electronic level sensors. Level glasses are still useful, in that they do provide a visual and continuous level measurement. They are vulnerable to breakage, especially in high-pressure environments, and the requirement of installing the glass in the pipe run makes them expensive, in terms of labor time. The glasses are

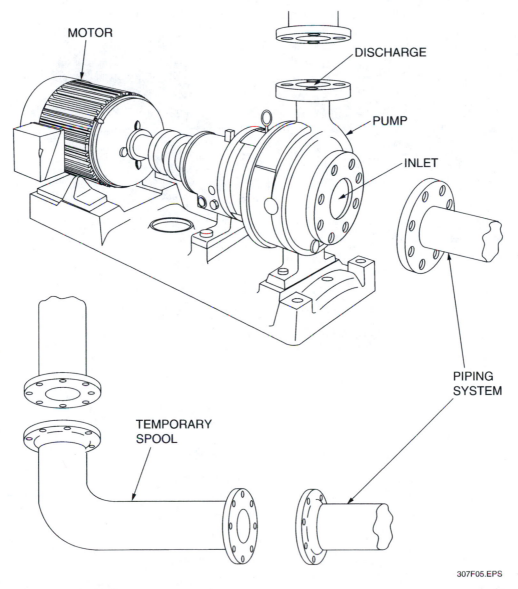

Figure 5 ◆ Temporary spool installation process.

307F05.EPS

available in flanged models; the metallic ends are fabricated, the flanges are welded on, the ends are turned true on a lathe, and the glass or clear plastic is installed and glued in place. However, the development of PLC process controllers makes the level glass less desirable and the electronic level measurement systems more useful. The electronic systems available include the following:

- Level sensors
- Pressure sensors
- Temperature sensors

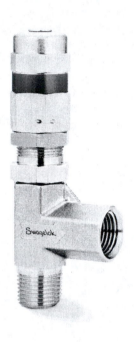
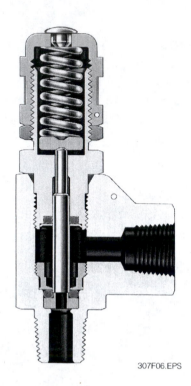

307F06.EPS

Figure 6 ◆ Typical pressure-relief valve.

6.4.1 Level Controllers

A level sensor produces a response when a specified point level is reached. This type is called a level switch. The other type continuously reads and reports the level in the vessel, either to a controller or to a data collector of some sort.

The technologies stem from various physical characteristics of processes in the vessels. The oldest technology is the use of a float to either trigger a response to a high or low point. A variant on this is the use of a magnetic float (Figure 7) that surrounds a stem with a switch inside it. When the float reaches a certain point, it causes the switch to close, and the completion of the circuit triggers either a pump or a signal to a controller or panel.

Electrical capacitance is used to measure liquid level in an application that is mounted inside the vessel. The sensor (Figure 8) reads either to a programmable linear controller or a level gauge display outside the tank. This particular technology is also capable of reporting the separation boundary of two liquids in a tank.

There are two types of acoustic sensors. Some are point sensors, using a change in the speed of sound in different materials to trigger a response. Non-contact acoustic level sensors rely on producing a directed sound at the liquid, and the time between the broadcast and the return is measured to determine the distance. This is the principle employed in sonar, the system used for detecting submarines. The unit that produces sound is usually an EMAT, an electromagnetic acoustic transducer, which turns electric current into sound, and which is mounted somewhere in the top of the tank. Some acoustic sensors are capable of reading level from outside the tank.

Another device (Figure 9) uses lasers to measure the level, again measuring the return time. The lasers are relatively expensive and can be disrupted by particulate matter in the vapors above the liquid, so their use is limited to cases where the liquid would be visible. A variation of this device uses radar to detect liquid level.

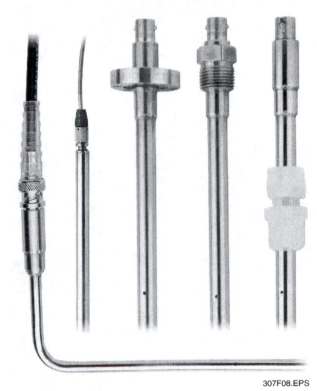

307F08.EPS

Figure 8 ◆ Liquid level sensor.

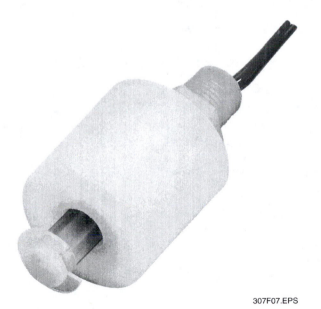

307F07.EPS

Figure 7 ◆ Magnetic float.

307F09.EPS

Figure 9 ◆ Level switch.

Level controllers are mounted on the platform side of the vessel or where they can be accessed from a ladder. At predetermined upper and lower limits, the float inside the vessel actuates controls within the level controller that send an impulse to open or close a control valve which adjusts the flow at the inlet or outlet of the vessel. Level controllers are usually attached to a pipe connected to the vessel. The controlling factor is setting the level control **bridle** at the actual elevation of the normal liquid level for the vessel. The bridle connections are located in the surge section of the vessel behind the last baffle.

Figure 10 shows a pneumatically operated system. The air supply is connected to the level controller. The level controller measures the liquid level and transmits a pneumatic signal to a remote level indicator and to the control valve. This signal controls the opening and closing of the diaphragm control valve to maintain the desired level in the vessel. *Figure 10* shows a typical level controller arrangement.

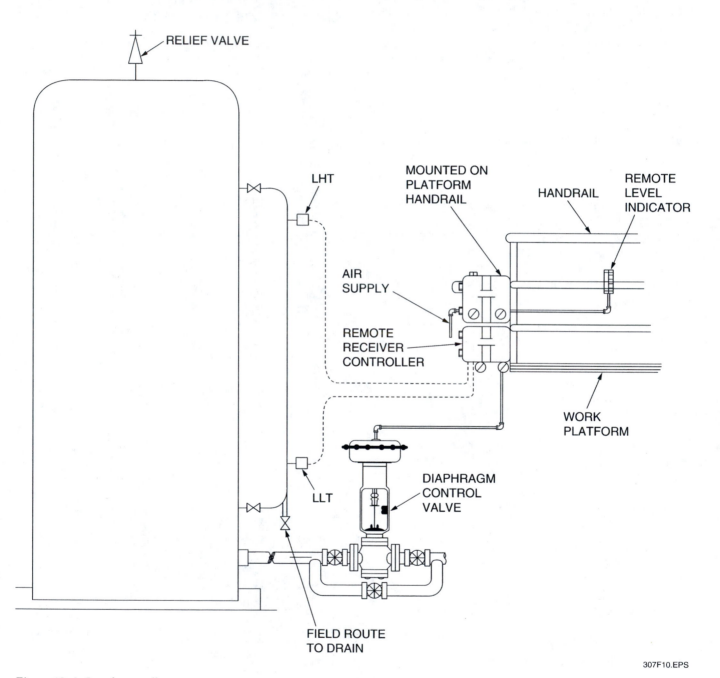

307F10.EPS

Figure 10 ◆ Level controller arrangement.

6.4.2 Level Glasses

Level glasses give a visual indication of the liquid level for both routine observations and level controller adjustments. Level glasses are usually attached to a pipe connected to the vessel. Level glasses come in a variety of lengths and can be installed in staggered positions as shown in *Figure 11*. The glasses are staggered so that when the liquid level exceeds the range of the glass, it is immediately indicated in the next glass. The combination of glasses must cover the complete float range required by the process.

6.4.3 Pressure Sensors

Some recent fiber-optic applications measure the pressure from outside the vessel by sensing a strain gauge, called a Bragg grating, applied to the surface of the vessel. The older technologies, such as the Bourdon tube pressure gauge *(Figure 12)* or the transducer *(Figure 13)* may be directly in contact with the contents of the vessel, or they may be isolated by a membrane system. Snubbers or pulsation dampers are frequently used to protect the gauge from extremely quick changes of pressure. Finally, pressure gauges are usually provided with one or more shutoff valves upstream of the gauge to allow the gauge to be repaired or recalibrated.

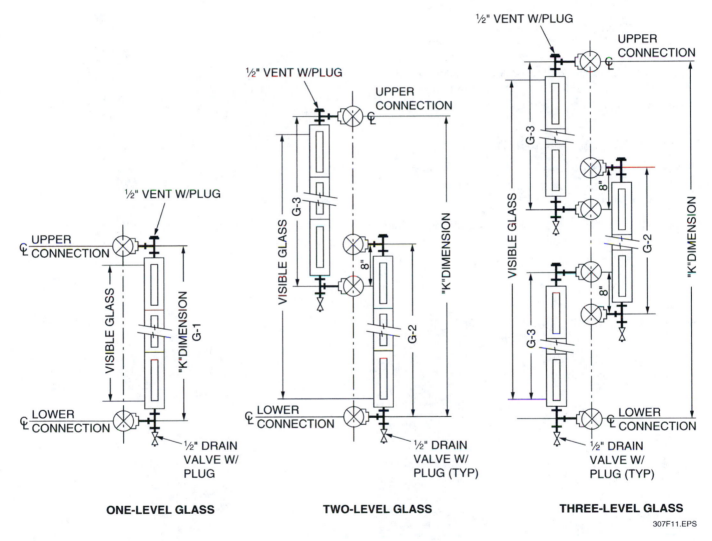

ONE-LEVEL GLASS   TWO-LEVEL GLASS   THREE-LEVEL GLASS

307F11.EPS

Figure 11 ◆ Level glass arrangement.

6.4.4 Temperature Elements

Temperature elements monitor the temperature of a vessel. The physical location of the element sometimes determines whether the reading is for liquid or vapor.

Many processes require temperature control for efficiency or for safety. Since the chemicals inside the vessel may degrade the effective operation of a temperature sensor, a thermowell is frequently used to house the temperature sensor. This is essentially a tube mounted through the vessel wall, made of a conductive material, with the temperature sensor directed inside to read. The results may either be directly read off a panel meter, or used to initiate a response through a PLC. The thermowell allows the sensor to be removed or replaced without interrupting the process.

Another way of measuring temperature in a vessel is the MIMS (mineral-insulated-metal-sheathed) thermocouple, which essentially incorporates the sensor and thermowell in a single unit. Some of these are very robust units, but if the unit fails the entire thermocouple is removed and replaced.

6.5.0 Erecting Vessel Trim

Special precautions must be taken when erecting vessel trim. Follow these guidelines when erecting vessel trim:

- Handle instrumentation and equipment with care because these items are fragile.
- Do not alter vessel flanges or other parts of the vessel.

- Ensure that all gasket materials are selected according to specifications.
- Preassemble all components before erection.
- Label all flanges upon arrival of the vessel.
- Verify the orientation and configuration of all valve handles and components to ensure accessibility.
- Hydrotest bridles before installation where applicable.
- Where pressure surges are likely, use a snubber (*Figure 14*) to suppress the surge.

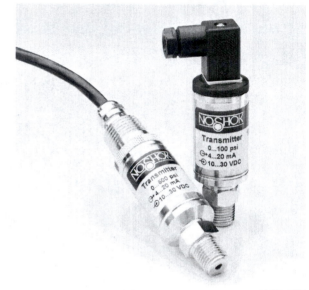

307F13.EPS

Figure 13 ◆ Transducer.

307F12.EPS

Figure 12 ◆ Bourdon tube pressure gauge.

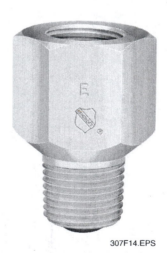

307F14.EPS

Figure 14 ◆ Pressure snubber.

1. The first step in field routing is to _____ the work area.

   a. leave
   b. secure
   c. move the pipe to
   d. measure

2. If the pipe has been stored outside, it must be checked for _____.

   a. rust inside
   b. vermin inside
   c. coatings outside
   d. end orientation

3. The first step of determining field run specifications is to _____.

   a. draw the field-routing draft
   b. figure the takeout for the elbows
   c. locate the beginning and end points
   d. hang the run

4. When you have chosen a route to run the pipe, you should check for _____.

   a. pipe stretchers
   b. existing supports or hangers
   c. trenches
   d. valves to attach to

5. To prevent overloading the erection equipment, you must know the _____ of the spool.

   a. color
   b. shape
   c. weight
   d. purpose

6. If the spool is field-fabricated with valves and fittings, there is no need to figure the weight of the flanges.

   a. True
   b. False

7. To find the total weight of a run of pipe, multiply the length (in feet) by the _____.

   a. thickness of the wall
   b. number of pieces
   c. weight of one foot of pipe
   d. diameter

8. The placement and number of hangers is based on _____.

   a. the contents of the pipe
   b. the size, weight, and type of pipe
   c. the diameter of the pipe
   d. one hanger every 20 feet

9. A hanger system must support the pipe at _____.

   a. every 10 feet
   b. regular intervals
   c. anywhere possible
   d. the ends

10. When selecting hangers and supports, you must also consider the _____.

    a. insulation requirements
    b. foundation of the building
    c. color of the hangers
    d. weight of the supports

11. When transporting and erecting piping, it is important to protect _____ in the area.

    a. the columns
    b. the concrete piers
    c. the soil
    d. all other lines and equipment

12. Rig piping so that it can be carried _____.

    a. vertically
    b. sideways
    c. as level as possible
    d. as low as possible

13. Pipefitters use test blinds between the flanges of a system to _____ the system.

    a. weld
    b. hydrotest
    c. bolt up
    d. support

14. If the system is carbon steel, the test blind should be _____.

    a. plastic
    b. stainless steel
    c. carbon steel
    d. titanium steel

15. Refer to the flange rating to find the pressure rating for the _____.
    a. piping
    b. gaskets
    c. flange spreaders
    d. fittings

16. Vents are located at the _____ of the vessel.
    a. lowest point
    b. middle
    c. bottom third
    d. highest point

17. Safety valves are normally used in _____ services.
    a. water
    b. oil
    c. steam or gas
    d. food

18. One level sensor that responds when a point level is reached is the _____.
    a. transducer
    b. magnetic float
    c. sight glass
    d. Bragg grating

19. Acoustic sensors operate by measuring the _____.
    a. time between a sound going out and returning
    b. distance between the top and bottom of the vessel
    c. number of sounds sent out
    d. pitch of the sound

20. The temperature sensor may be isolated from the contents of a vessel by a _____.
    a. bolt head
    b. insulation batt
    c. thermowell
    d. thick wall

Summary

As you start field routing and vessel trim activities, you must call on all of your previously learned skills. These activities require you to be proficient in your measurements, calculations, and fabrication techniques. Remember to always put safety first and be aware of your working surroundings when performing field routing and vessel trim activities. Your safety and the safety of your co-workers depend on the way you perform your job.

Notes

Trade Terms Introduced in This Module

Bridle: A common line used for manifolding two or more instruments to one pair of connections on a vessel.

Field routing: Describes a method for installing piping systems without having drawings to give coordinates and specifications. Field routing is normally performed with small-bore pipe.

Grasshopper: A two-wheel device that is used to transport pipe and piping components.

Vessel trim: Piping, instruments, and valves connected to vessels.

Resources & Acknowledgments

Additional Resources

This module is intended to be a thorough resource for task training. The following reference works are suggested for further study. These are optional materials for continued education rather than for task training.

Parker Instrumentation has literature on pipe and vessel instrumentation at http://www.parker.com/ead/cm2.asp?cmid=177.

Instrumentation. National Center for Construction Education and Research. Upper Saddle River, NJ: Prentice Hall.

Figure Credits

Sumner Manufacturing Company, Inc., 307F01

© 2002 Swagelok Company, 307F06

Gems Sensors & Controls, 307F07, 307F09

American Magnetics, Inc., 307F08

Topaz Publications, Inc., 307F12

NOSHOK, Inc., 307F13

Ashcroft Inc., 307F14

NCCER makes every effort to keep these textbooks up-to-date and free of technical errors. We appreciate your help in this process. If you have an idea for improving this textbook, or if you find an error, a typographical mistake, or an inaccuracy in NCCER's Contren® textbooks, please write us, using this form or a photocopy. Be sure to include the exact module number, page number, a detailed description, and the correction, if applicable. Your input will be brought to the attention of the Technical Review Committee. Thank you for your assistance.

Instructors – If you found that additional materials were necessary in order to teach this module effectively, please let us know so that we may include them in the Equipment/Materials list in the Annotated Instructor's Guide.

Write: Product Development and Revision
National Center for Construction Education and Research
3600 NW 43rd St, Bldg G, Gainesville, FL 32606

Fax: 352-334-0932

E-mail: curriculum@nccer.org

Craft _____ Module Name _____

Copyright Date _____ Module Number _____ Page Number(s) _____

Description _____

_____

_____

_____

(Optional) Correction _____

_____

_____

(Optional) Your Name and Address _____

_____

_____

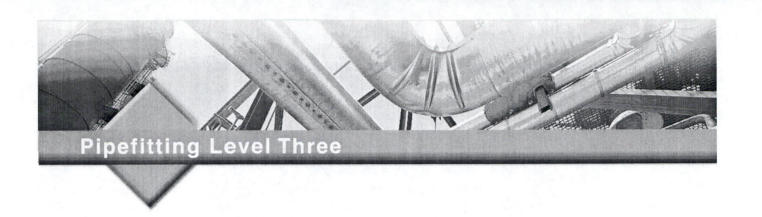

08308-07

Pipe Hangers
and Supports

08308-07
Pipe Hangers and Supports

Topics to be presented in this module include:

Overview

Pipe has to be supported. Here are the different kinds of supports and hangers that keep the pipes from breaking or bending. There are hangers for large and small pipe and for pipe that moves in different ways. The ways to adjust and install the different kinds of hangers and supports are described.

Objectives

When you have completed this module, you will be able to do the following:

1. Identify types of pipe hangers and supports.
2. Identify and interpret pipe support drawings and symbols.
3. Determine field placement of hangers.
4. Identify and install concrete fasteners.
5. Fabricate angle iron brackets to support pipe.
6. Identify and explain the types of spring can supports.
7. Identify and explain the types of variable spring can supports.
8. Identify and explain the types of constant spring can supports.
9. Explain the storing and handling procedures for spring can supports.
10. Explain how to install spring can supports.
11. Maintain spring can supports.

Trade Terms

Cold load	Load rating
Component	Pipe riser
Corrosion	Precompress
Fabricate	Run
Head room	Snubber
Hot load	Structural member
Load column	Sway brace

Required Trainee Materials

1. Pencil and paper
2. Appropriate personal protective equipment

Prerequisites

Before you begin this module, it is recommended that you successfully complete *Core Curriculum*; *Pipefitting Level One*; *Pipefitting Level Two*; and *Pipefitting Level Three*, Modules 08301-07 through 08307-07.

This course map shows all of the modules in the third level of the *Pipefitting* curriculum. The suggested training order begins at the bottom and proceeds up. Skill levels increase as you advance on the course map. The local Training Program Sponsor may adjust the training order.

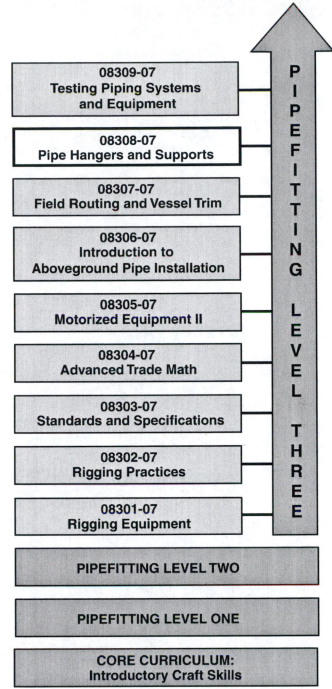

308CMAP.EPS

1.0.0 ◆ INTRODUCTION

A pipeline is not static. It can move vertically, horizontally, or in both directions. Movement within the piping system is caused mainly by thermal expansion, which is the natural tendency of metal to expand when hot, and by the weight of the substance within the pipeline. Thermal expansion depends on the temperature of the system and the material of the pipe and has nothing to do with pipe size or schedule. To control this movement, pipe hangers and supports are installed at established intervals along a pipe **run**. Pipe hangers and supports are chosen by design engineers and installed by pipefitters according to the piping drawings. Correctly installed hangers allow pipes to drain, expand, contract, and vibrate without damaging the pipeline.

2.0.0 ◆ TYPES OF PIPE HANGERS AND SUPPORTS

There are many types of pipe hangers and supports. The particular hanger or support selected for a given system depends on many factors, including the weight to be supported, the temperature of the system, the intended movement of the pipe within the hanger, and the surrounding structure. Every pipe hanger and support has a maximum **load rating**. This load rating takes into account a safety load factor of 5 to 1. This means that a hanger rated at 500 pounds must be able to support at least 2,500 pounds before failing. The load rating must never be exceeded. Manufacturers supply information about the maximum load rating of their hangers and accessories.

Some pipe hangers also have a maximum temperature rating. This is important if the hanger is to be used in pipelines with high operating temperatures. For any application, there is a wide variety of pipe hangers and supports available in many sizes and materials. The following sections explain some of the more common types of pipe hangers and supports.

2.1.0 Pipe Hangers

Pipe hangers are usually attached to **structural members**, such as steel beams or plates, or by hanger accessories embedded in concrete. Most hangers are adjustable for the correct height or elevation. Widely used pipe hangers include the following:

- Adjustable rings
- Adjustable clevis
- Double-bolt pipe clamps
- Trapeze hangers
- Job **fabricated**

2.1.1 Adjustable Rings

Adjustable rings are available in sizes to fit ½- to 8-inch pipe. Adjustable rings are recommended for supporting noninsulated, stationary pipelines with a maximum temperature rating of 650°F. They can be adjusted vertically 1 to 2 inches. Adjustable ring hangers should not be used outdoors or in highly corrosive areas. *Figure 1* shows an adjustable ring hanger.

2.1.2 Adjustable Clevis

The adjustable clevis (*Figure 2*) is available in a lightweight model for pipe that is ⅜ to 4 inches in diameter and in heavy-duty models for pipe that is ½ to 30 inches in diameter. It is recommended for suspending light loads with a maximum temperature of 650°F. The clevis can be adjusted vertically, depending on its physical size. The nut above the clevis is tightened to maintain the vertical adjustment. The nut below the clevis adjusts the height. When supporting insulated pipelines, use a shield with the clevis to prevent the insulation from being damaged by the hanger. The shield is normally either tack-welded into the clevis or held in place by two ribs in the center of the shield.

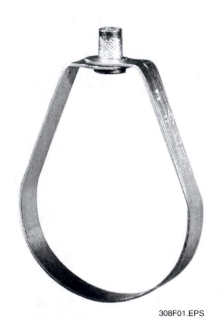

308F01.EPS

Figure 1 ◆ Adjustable ring hanger.

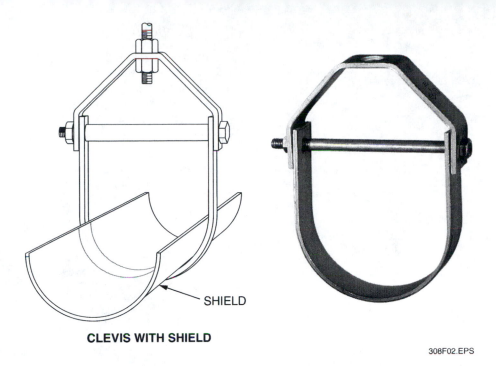

SHIELD

CLEVIS WITH SHIELD

308F02.EPS

Figure 2 ◆ Adjustable clevis and shield.

2.1.3 Double-Bolt Pipe Clamps

Double-bolt pipe clamps are available in carbon steel for use in both light- and heavy-duty applications for pipe sizes of ¾ to 36 inches. The maximum temperature rating is 750°F. They are also available in chrome-molybdenum (chrome-moly) steel that has a maximum temperature rating of 1,050°F for pipe sizes 1½ to 10 inches. *Figure 3* shows a double-bolt pipe clamp.

2.1.4 Trapeze Hangers

Trapeze hangers are available commercially or they can be field-fabricated. Field-fabricated trapeze hangers must be designed by or approved by an engineer or supervisor. They are suspended by two rods and designed for top-loading of pipe only. Vertical adjustment can be made by tightening or loosening the bolts at the bottom of the hanger. *Figure 4* shows a trapeze hanger.

2.1.5 Job-Fabricated Pipe Hangers

Job-fabricated pipe hangers are used with manufactured hangers in many applications. They are normally constructed by welding angle iron together to form braces and supports. These angle iron brackets can be welded directly to the beams in the facility or can be bolted to the beams in areas that do not permit welding. *Figure 5* shows

examples of job-fabricated pipe hangers. All such hangers must be approved by an engineer. Commercial brackets (*Figure 6*) are also used.

2.2.0 Hanger Connecting Units and Attachments

Pipe hanger connecting units and attachments attach the pipe hanger to the supporting structure.

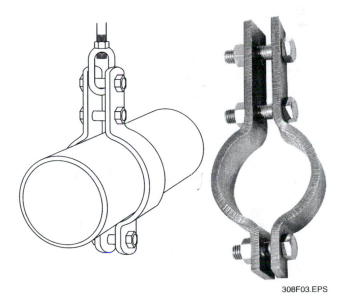

308F03.EPS

Figure 3 ◆ Double-bolt pipe clamp.

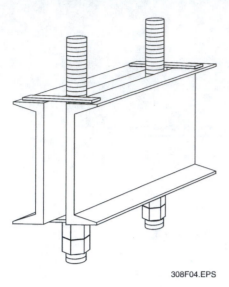

Figure 4 ◆ Trapeze hanger.

308F04.EPS

They must be able to withstand the same load that is exerted on the hanger. Some common connecting units are the following:

- Eyebolts
- Turnbuckles
- Threaded rods
- Rod attachments
- Beam clamps
- C-clamps
- Welded beam attachments

2.2.1 Eyebolts

The most basic type of connecting unit is the eye rod, or eyebolt. Eyebolts support the pipe hanger but cannot be adjusted for height unless they are threaded through a nut that is free to turn on the other side of a bearing. Eyebolts are made with an eye, or loop, at one end and threads on the other end. There are many types of eyebolts. Some have a formed and welded eye; others have a forged eye. *Figure 7* shows several types of eyebolts.

308F06.EPS

Figure 6 ◆ Commercial brackets.

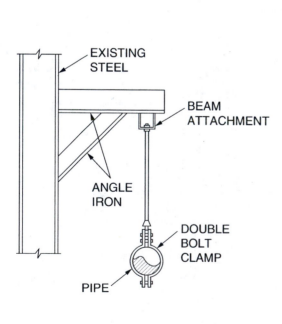

Figure 5 ◆ Job-fabricated pipe hangers.

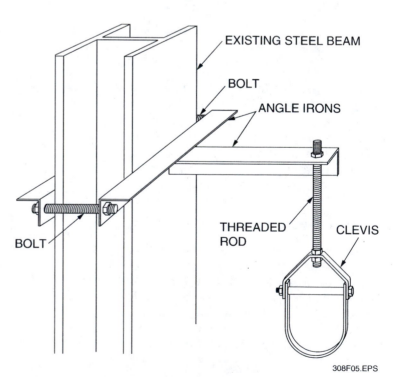

308F05.EPS

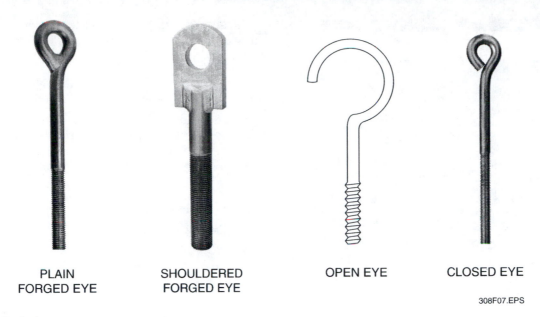

PLAIN
FORGED EYE

SHOULDERED
FORGED EYE

OPEN EYE

CLOSED EYE

308F07.EPS

Figure 7 ◆ Eyebolts.

2.2.2 Turnbuckles

Turnbuckles allow hangers to be adjusted to different heights and are used with several types of hangers. A turnbuckle consists of two threaded rods that screw into one body. One rod has right-hand threads, and the other has left-hand threads. When the body is turned, the distance between the ends of the rods is shortened or lengthened. *Figure 8* shows a turnbuckle.

2.2.3 Threaded Rods

Threaded rods are used to connect pipe hangers to various types of anchors or rod attachments. The rods can be cut to the desired length and are available in several diameters to match the hanger and rod attachment. *Figure 9* shows a threaded rod and load ratings.

2.2.4 Rod Attachments

Rod attachments are used to link pipe supports to clamps or other hangers. A wide variety of rod attachments are available. They are easy to install and allow for some vertical adjustment. *Figure 10* shows rod attachments.

2.2.5 Beam Clamps

Beam clamps come in a variety of configurations (*Figure 11*) and most commonly attach to the lower flange of standard structural steel I-beams. The maximum recommended load of beam clamps ranges between 500 and 5,000 pounds.

2.2.6 C-Clamps

Another method of fastening hangers to structural steel is with a C-clamp (*Figure 12*) used in conjunction with a threaded rod that is screwed into the bottom of the C-clamp. C-clamps are available for rod sizes of ⅜ to ¾ inch. The clamp is placed over the flange of an I-beam and tightened. The rod is then threaded into the hole on the bottom of the clamp and tightened. The lock nut of the C-clamp must not be overtightened because this could damage the clamp or strip the threads.

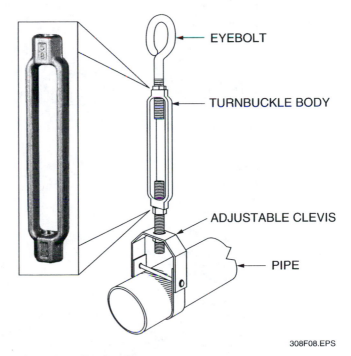

EYEBOLT

TURNBUCKLE BODY

ADJUSTABLE CLEVIS

PIPE

308F08.EPS

Figure 8 ◆ Turnbuckle.

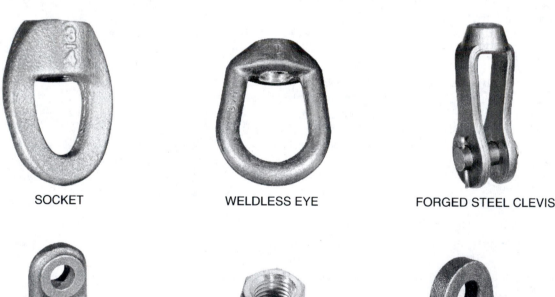

Rod Size (inches)	Load (pounds)
3/8	610
1/2	1,130
5/8	1,810
3/4	2,710
7/8	3,770
1	4,980
1/8	6,230
1¼	8,000
1½	11,630

308F09.EPS

Figure 9 ◆ Threaded rod.

2.2.7 Welded Beam Attachments

Welded beam attachments are made of carbon steel and are welded to the bottoms of beams to support the pipe hangers. They can be used with eyebolts or weldless eyes and threaded rods. If vertical adjustment is desired, use a threaded rod and nut with the weld attachment, which should be welded to the beam in an inverted position.

The size of welded beam attachment depends on the size of the threaded rod. *Figure 13* shows a welded beam attachment.

2.3.0 Pipe Supports

A pipe support is a device that is normally attached under piping to hold or support the weight of the pipe and the fluid conveyed inside. Common pipe supports include the following:

- U-bolts
- Pipe roll supports
- Pipe saddles
- Extension riser clamps
- Wall support clamps
- Job-fabricated supports

2.3.1 U-Bolts

U-bolts (*Figure 14*) are available commercially, or they can be field-fabricated. They are available in sizes that will support pipe from ½ to 36 inches and have a maximum temperature rating of 750°F. U-bolts can be used for supporting, guiding, or anchoring heavy loads, but they are mainly used for guiding pipelines. When used to guide pipelines, U-bolts do not exert any pressure on the

SOCKET   WELDLESS EYE   FORGED STEEL CLEVIS

EXTENSION PIECE   COUPLING   SIDE BEAM BRACKET

308F10.EPS

Figure 10 ◆ Rod attachments.

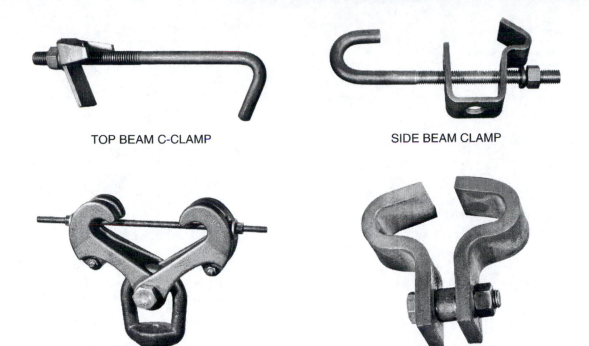

TOP BEAM C-CLAMP

SIDE BEAM CLAMP

BEAM CLAMP WITH LINKS

IRON BEAM CLAMP

308F11.EPS

Figure 11 ◆ Beam clamps.

pipe. To accomplish this, use four nuts to secure the U-bolt to the brace.

2.3.2 Pipe Roll Supports

Pipe rolls are used when end-to-end movement of a pipeline is required. They are available in a variety of styles to accommodate pipe from 1 to 30 inches and can be adjusted vertically to provide the proper pipe elevation. *Figure 15* shows a pipe roll support.

2.3.3 Pipe Saddles

Pipe saddles support pipe from the bottom. They are available to fit a wide range of pipe sizes and can be adjusted vertically to provide the proper elevation. *Figure 16* shows a pipe saddle.

2.3.4 Extension Riser Clamps

Extension riser clamps (*Figure 17*) are used to support vertical **pipe risers**. When riser clamps are used to support insulated pipes, the insulation is wrapped around the riser clamp as well as the pipe. Riser clamps have a maximum temperature rating of 650°F and come in sizes to support pipe that is ¾ to 20 inches nominal size.

2.3.5 Wall Support Clamps

If the piping system runs along a wall, simple and cheaper clamps can be used. These clamps fasten directly to the wall to hold the pipe firmly and closely against the wall. *Figure 18* shows wall support clamps.

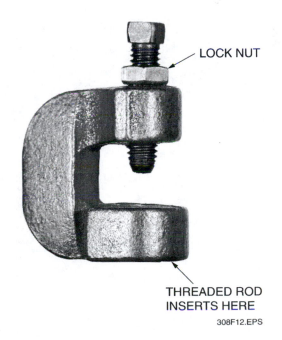

LOCK NUT

THREADED ROD INSERTS HERE

308F12.EPS

Figure 12 ◆ C-clamp.

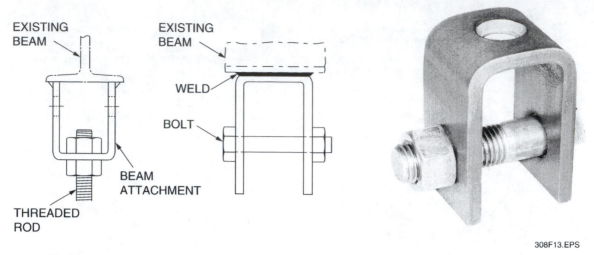

308F13.EPS

Figure 13 ◆ Welded beam attachment.

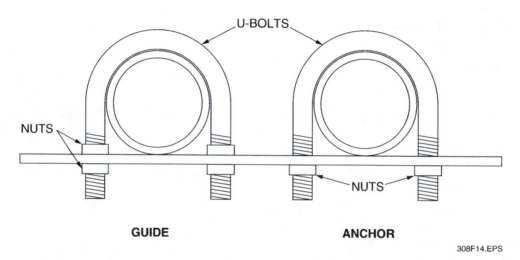

GUIDE

ANCHOR

308F14.EPS

Figure 14 ◆ U-bolts.

308F15.EPS

Figure 15 ◆ Pipe roll support.

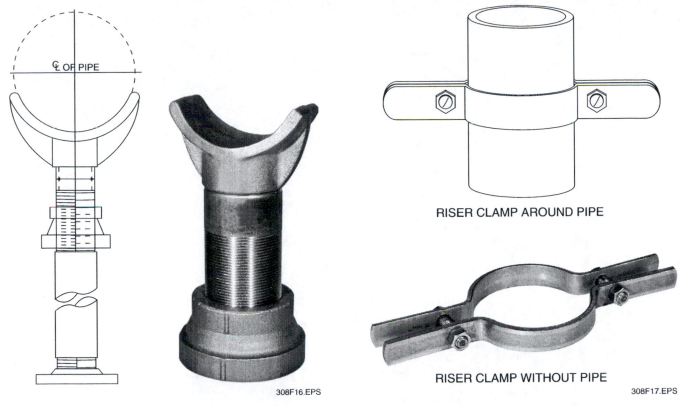

Figure 16 ◆ Pipe saddle.

RISER CLAMP AROUND PIPE

RISER CLAMP WITHOUT PIPE

308F17.EPS

Figure 17 ◆ Extension riser clamp.

308F16.EPS

HEAVY STRAP   SINGLE HOOK   LIGHT STRAP

308F18.EPS

Figure 18 ◆ Wall support clamps.

2.3.6 Job-Fabricated Supports

Pipe supports can be fabricated in the field as long as they meet design specifications. One example of a job-fabricated support is a concrete block and mortar placed under an elbow at a floor drain. Job-fabricated wall supports can also be fabricated using angle iron and a piece of steel pipe cut in half. Another job-fabricated support is a dummy leg, which is a line support welded to an elbow in the pipe run. The dummy leg can be a piece of pipe or structural beam.

Another type of job-fabricated support is a T-iron anchor that is welded to a beam and then to the pipe. If job specifications do not allow the T-iron that is welded to the pipe to be also welded to the beam, stops can be fabricated from T-iron or angle iron and welded to the T-iron anchor on each side of the beam. *Figure 19* shows job-fabricated supports. All such supports must be approved by a supervisor or an engineer.

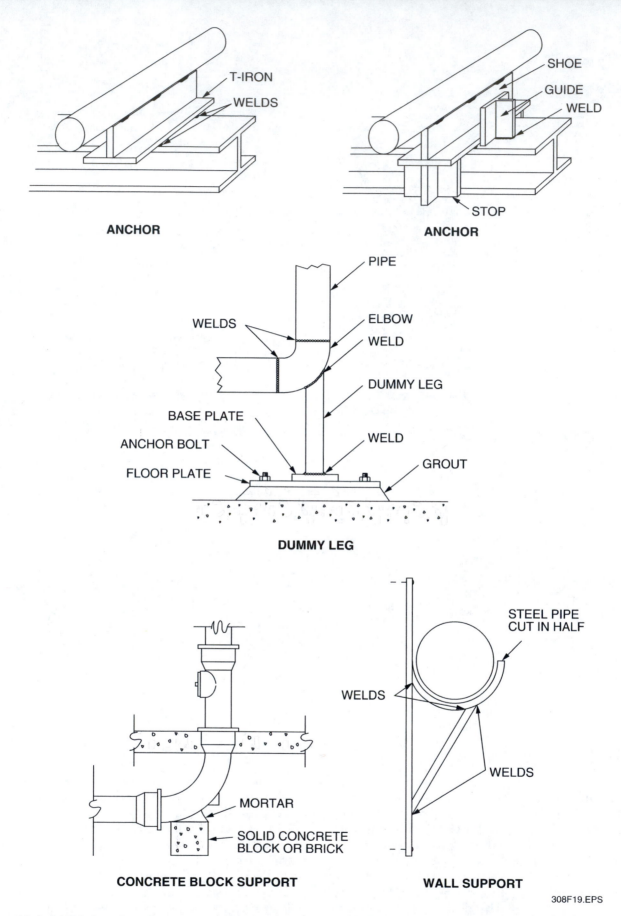

ANCHOR

ANCHOR

DUMMY LEG

CONCRETE BLOCK SUPPORT

WALL SUPPORT

308F19.EPS

Figure 19 ◆ Job-fabricated supports.

3.0.0 ◆ PIPE SUPPORT DRAWINGS AND SYMBOLS

Before a piping system is installed, engineers analyze the system and determine the type, size, and placement of all hangers and supports in the system. Piping drawings are then made, showing the placement of each type of hanger and support. The pipefitter must be able to read and interpret these drawings to install the hangers and supports in the proper places.

Symbols for pipe supports vary from job to job. On piping drawings, the hanger or support has a prefix that refers to the specific detail sheets used on that job. Check the detail sheets to determine the exact type hanger to use. *Table 1* lists examples of pipe hanger reference prefixes taken from one engineering company. Each company also uses a numerical code with the reference prefix to designate size. Check your company's standards and specifications to determine the reference prefixes used on your job site.

Each engineering company also uses pipe hanger and support symbols that are specific to a job site. These symbols appear on plan and isometric drawings and are called out by the reference prefix as explained above. *Figure 20* shows an example of pipe support drawings and symbols.

You must refer to the detail sheets of a pipe hanger symbol for all of the information needed to fabricate and install a hanger. Notice the symbol in the upper right-hand corner of *Figure 20* that designates a clevis hanger with reference number 5HR. Before installing this hanger, refer to the detail sheet for 5HR. *Figure 21* shows the detail sheet for 5HR.

Table 1 Pipe Support Prefixes

GENERAL PIPE SUPPORT LIST	
Designation	**Description**
A	Anchor
ABS	Angle base support
ASW	Adjustable spring wedge
BA	Base anchor
BS	Base support
BSA	Base support – adjustable
BSS	Base spring support
BSL	Base support – lined pipe
C	Cradle
CI	Isolation cradle
CP	Concrete pad
DA	Directional anchor
DS	Dummy support
FS	Field support
G	Guide
GSB	Gussett support – bolted
GSW	Gussett support – welded
HD	Hold down
HDS	Hold down – slotted
HR	Hanger rod
LP	Load plate
PU	Pickup
RC	Riser clamp
S	Shoe
SL	Riser clamp shear lug
SP	Slide plate
SPA	Stiffener plate
ST	Support trunnion
UG	U-bolt – guide
US	U-bolt – support

LOW-TEMPERATURE PIPE SUPPORT LIST	
Designation	**Description**
CAS	Cold anchor shoe
CBS	Cold base support
CHD	Cold hold down

HYGIENIC PIPE SUPPORT LIST	
Designation	**Description**
BSH	Base support
CLH	Clamp
FSH	Field support
HRH	Hanger rod
TSH	Toggle support

308T01.EPS

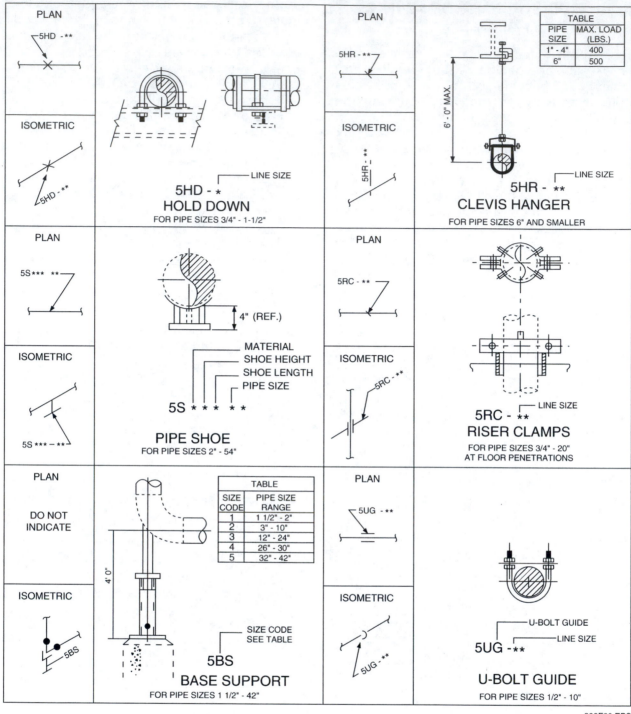

Figure 20 ◆ Pipe support drawings and symbols.

308F20.EPS

4.0.0 ◆ FIELD PLACEMENT OF HANGERS

Although piping drawings should be available for every piping system that is installed, this is not always the case. Therefore, the pipefitter often has to select the best route and install the piping. This is known as field-fabricating or field-routing a pipe run. In this case, the pipefitter has to determine the placement of the pipe hangers, based on the size, weight, and type of pipe being run. In order for a hanger system to do its job, it must support the pipe at regular intervals. Evenly spacing the hangers ensures that no individual hanger is overloaded. *Table 2* lists the recommended maximum hanger spacing intervals for carbon steel pipe.

Table 2 serves only as a guide. Always check the job-site specifications when determining pipe hanger spacing, and if the pipe sags, add more hangers. The most important thing to remember is that the pipe must run straight without sagging.

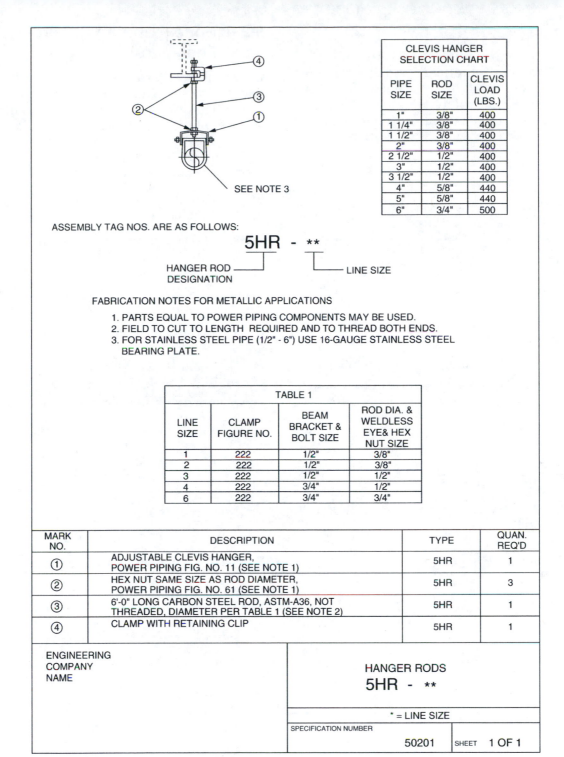

CLEVIS HANGER SELECTION CHART

PIPE SIZE	ROD SIZE	CLEVIS LOAD (LBS.)
1"	3/8"	400
1 1/4"	3/8"	400
1 1/2"	3/8"	400
2"	3/8"	400
2 1/2"	1/2"	400
3"	1/2"	400
3 1/2"	1/2"	400
4"	5/8"	440
5"	5/8"	440
6"	3/4"	500

SEE NOTE 3

ASSEMBLY TAG NOS. ARE AS FOLLOWS:

5HR - **

HANGER ROD DESIGNATION

LINE SIZE

FABRICATION NOTES FOR METALLIC APPLICATIONS

1. PARTS EQUAL TO POWER PIPING COMPONENTS MAY BE USED.
2. FIELD TO CUT TO LENGTH REQUIRED AND TO THREAD BOTH ENDS.
3. FOR STAINLESS STEEL PIPE (1/2" - 6") USE 16-GAUGE STAINLESS STEEL BEARING PLATE.

TABLE 1

LINE SIZE	CLAMP FIGURE NO.	BEAM BRACKET & BOLT SIZE	ROD DIA. & WELDLESS EYE& HEX NUT SIZE
1	222	1/2"	3/8"
2	222	1/2"	3/8"
3	222	1/2"	1/2"
4	222	3/4"	1/2"
6	222	3/4"	3/4"

MARK NO.	DESCRIPTION	TYPE	QUAN. REQ'D
①	ADJUSTABLE CLEVIS HANGER, POWER PIPING FIG. NO. 11 (SEE NOTE 1)	5HR	1
②	HEX NUT SAME SIZE AS ROD DIAMETER, POWER PIPING FIG. NO. 61 (SEE NOTE 1)	5HR	3
③	6'-0" LONG CARBON STEEL ROD, ASTM-A36, NOT THREADED, DIAMETER PER TABLE 1 (SEE NOTE 2)	5HR	1
④	CLAMP WITH RETAINING CLIP	5HR	1

ENGINEERING COMPANY NAME

HANGER RODS

5HR - **

* = LINE SIZE

SPECIFICATION NUMBER

50201   SHEET 1 OF 1

308F21.EPS

Figure 21 ◆ Detail sheet for 5HR.

5.0.0 ◆ CONCRETE FASTENERS

In a typical piping system, a clamp is placed around a pipe, and connecting units are attached to the clamp. These connecting units are then attached to a fastener that connects the hanger to the existing structure. There are many types of concrete fasteners. The following sections explain how to install these types of fasteners:

- Concrete inserts
- Nonexpanding concrete fasteners
- Adhesive anchors
- Expanding concrete fasteners
- Redheads
- Toggle bolts

5.1.0 Concrete Inserts

Concrete inserts are embedded into concrete when the concrete is poured, and the hanger extension rod is screwed into the anchor after the concrete is set. Other attachments can then be added, depending on the requirements of the system. *Figure 22* shows several types of concrete inserts.

Table 2 Recommended Maximum Hanger Spacing Intervals for Carbon Steel Pipe

Pipe Size	Rod Diameter	Maximum Spacing
Up to 1¼"	⅜"	8'
1½" and 2"	⅜"	10'
2" to 3½"	½"	12'
4" and 5"	⅝"	15'
6"	¾"	17'
8" to 12"	⅞"	22'

308T02.EPS

5.2.0 Nonexpanding Concrete Fasteners

Other concrete fasteners are made to be installed after the concrete has set. Fasteners come in either expanding or nonexpanding form. The following steps explain how to install the nonexpanding concrete fasteners shown in *Figure 23*.

Follow these steps to install a nonexpanding concrete fastener in hardened concrete:

Step 1   Identify the type of hanger or support to be used.

Step 2   Locate the area where the support is to be fastened or attached.

CAUTION

The area where the support is to be fastened should be smooth so that the item will have solid footing. Uneven footing might cause the support to twist, warp, or not tighten properly.

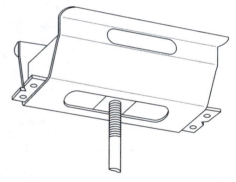

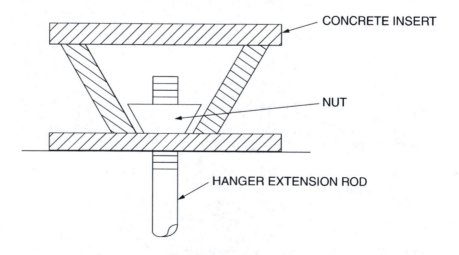

CONCRETE INSERT

NUT

HANGER EXTENSION ROD

308F22.EPS

Figure 22 ◆ Concrete inserts.

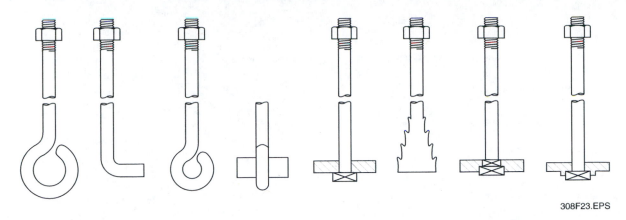

Figure 23 ◆ Nonexpanding concrete fasteners.

Step 3 Make a layout template of the holes to be drilled.

Step 4 Use the template to lay out and mark where each hole will be located.

Step 5 Set up a core drill and bit for drilling the hole in the concrete.

WARNING!
Carefully inspect the drill and bit to ensure that they are in good operating condition. If they are not, you could be injured by parts that chip and fly off during operation.

Step 6 Identify the type and size fastener needed.

Step 7 Select a drill bit to match the fastener chosen. The drill bit selected for a nonexpanding fastener must be slightly larger in diameter than the head of the fastener and the flat washer used with the fastener. The washer must be able to fit down inside the hole, just below surface level.

WARNING!
Be sure to wear ear protectors, safety glasses, and gloves because drilling in concrete generates noise, dust, and possible flying objects. Make sure others in the area also wear protective equipment.

Step 8 Drill the holes for the nonexpanding fasteners. The drill bit can be lubricated with water during the drilling process. There is always a possibility of hitting rebar when drilling in concrete. If you hit rebar, use a drill bit that is specially made to drill through rebar.

WARNING!
Drilling in concrete produces a great amount of torque. Maintain a firm grip on the drill when drilling the holes.

Step 9 Clean out the holes when finished drilling.

Step 10 Assemble and lay out the fastener and the parts for each hole.

Step 11 Position a nonexpanding fastener into the drilled hole. The fastener should extend out of the hole far enough for the threads and a little of the unthreaded bolt to be above the surface.

Step 12 Prepare a portion of filler material, either epoxy or grout.

Step 13 Fill the hole around the fastener with grout or epoxy until the hole is about three-fourths filled. The grout or epoxy will later harden and secure the fastener in the hole.

Step 14 Work the fastener up and down in the hole to settle the grout or epoxy and to eliminate air pockets.

Step 15 Finish filling the hole with grout or epoxy and work it in well, but leave enough room at the top for the washer to be installed.

Step 16 Slip the washer down over the fastener, and settle it down into the hole on top of the grout. *Figure 24* shows a properly installed nonexpanding fastener.

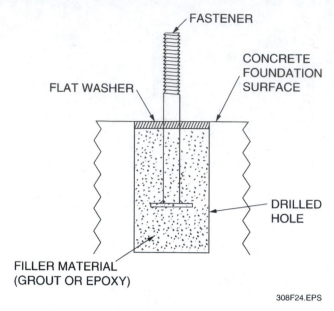

Figure 24 ◆ Properly installed nonexpanding fastener.

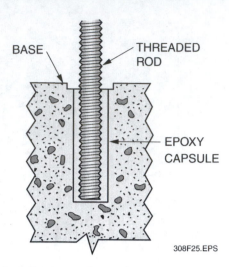

Figure 25 ◆ Adhesive anchor.

CAUTION

Make sure the fastener is straight after working it around in the hole and installing the washer. The washer centers the fastener and holds it in place until the grout or epoxy hardens. If the grout or epoxy sets and the fastener is not straight, the fastener will be unusable and will have to be removed and the installation repeated.

Step 17 Clean any excess grout or epoxy from around the hole and off the fastener threads and shaft.

Step 18 Run a matching nut down the shaft a few turns to make sure the threads are good.

Step 19 Remove the nut.

Step 20 Allow the grout or epoxy to dry fully before mounting the support to the fastener.

5.3.0 Adhesive Anchors

Another technique used for individual holes employs prepackaged capsules of epoxy and hardener. The two forms are packaged in either glass capsules or plastic. The hardener is in a separate chamber inside the resin chamber. When the hole has been drilled, the capsule is inserted, then the bolt is inserted to break the chambers. The bolt is turned to mix the resin and hardener thoroughly, and the epoxy is allowed to set (*Figure 25*). Manufacturers supply instructions as to how long the

bolt must be turned to completely mix the adhesive. With the glass capsule, the glass fragments will be brought to the surface, so the worker must take care not to wipe the top off with bare hands.

In this application, the washer is not necessary, and the hole is much smaller without sacrificing strength. In either case, follow manufacturer's directions so that the anchor will set completely. Adhesive anchors should not be used in concrete that has not set for at least seven days, nor should they be used below 40°F. Adhesive anchors must be allowed to set completely before any stress or movement is applied to the bolt.

5.4.0 Expanding Concrete Fasteners

Expanding fasteners are also installed after the concrete has been poured and has hardened. These fasteners are made in different sizes with different heads. The size of an expanding fastener is based on the size of the bolt that it will be used with. Two types of expanding concrete fasteners are wedge-type fasteners and expansion cases.

5.4.1 Wedge-Type Fasteners

Wedge-type fasteners are available in sizes ranging from ¼ to 1¼ inches. They are capable of holding materials that range in size from ¼ to 6⅞ inches thick and can be used in any position. *Figure 26* shows a wedge-type fastener.

Follow these steps to install a wedge-type fastener:

Step 1 Identify the type of hanger or support to be used.

Step 2 Locate the area where the support is to be fastened or attached.

Step 3   Make a layout template of the fastener holes to be drilled.

Step 4   Use the template to lay out and mark where each hole will be located.

Step 5   Set up a drill for drilling the hole in the concrete.

Step 6   Identify the type and size fastener needed.

Step 7   Select a drill bit to match the fastener chosen. The drill bit selected for a wedge-type fastener should be the same size as the fastener.

Step 8   Assemble the drill and drill bit.

Step 9   Drill the holes for the fasteners. The drill bit can be lubricated with water during the drilling process. The depth of the hole can exceed the length of the fastener. There is always a possibility of hitting rebar when drilling in concrete. If you hit rebar, use a drill bit that is specially made to drill through rebar.

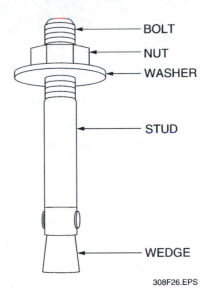

Figure 26 ◆ Wedge-type fastener.

Step 10   Clean out the holes when finished drilling.

Step 11   Lay out the fastener and the parts for each hole.

Step 12   Remove the nut and washer, and drive the fastener into the hole.

Step 13   Place the support or bracket over the fastener stud.

Step 14   Attach the washer and nut, and tighten the nut against the bracket. As the nut is tightened, the wedge at the end of the fastener expands against the inside of the hole, securing the fastener. *Figure 27* shows an installed wedge-type fastener; *Figure 28* shows types of anchors.

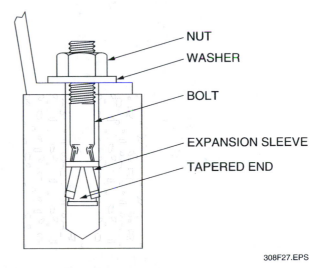

Figure 27 ◆ Installed wedge-type fastener.

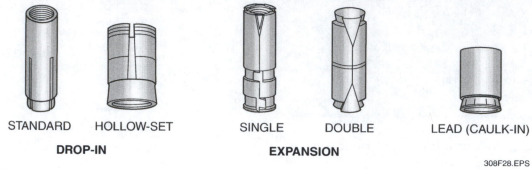

STANDARD   HOLLOW-SET   SINGLE   DOUBLE   LEAD (CAULK-IN)

DROP-IN   **EXPANSION**

308F28.EPS

Figure 28 ◆ Types of anchors.

5.4.2 Expansion Cases

An expansion case is a threaded insert that is placed into a drilled hole and expands to secure itself when a threaded rod is screwed into it. Expansion cases can be used with ¼-, ⅜-, and ½-inch threaded rods and are used for horizontal or floor use. *Figure 29* shows an expansion case.

Follow these steps to install an expansion case:

Step 1   Identify the type of hanger to be used.

Step 2   Locate the area where the support is to be fastened or attached.

Step 3   Make a layout template of the expansion case holes to be drilled.

Step 4   Use the template to lay out and mark where each hole will be located.

Step 5   Set up a drill for drilling the hole in the concrete.

Step 6   Identify the kind and size of expansion case needed.

Step 7   Select a drill bit to match the expansion case chosen. The drill bit selected for an

expansion case must be larger than the threaded rod being used. For a ¼-inch rod, use a ⁵⁄₁₆-inch drill bit; for a ⅜-inch rod, use a ¾-inch drill bit; and for a ½-inch rod, use a 1-inch drill bit.

Step 8   Assemble the drill and drill bit.

 WARNING!
Carefully inspect the drill and bit to make sure they are in good operating condition. If they are not, you could be injured by parts that chip and fly off during operation.

Step 9   Drill the holes for the fasteners. There is always a possibility of hitting rebar when drilling in concrete. If you hit rebar, use a drill bit that is specially made to drill through rebar.

 WARNING!
Be sure to wear ear protectors, safety glasses, gloves, and a safety belt if working at heights. Drilling in concrete generates noise, dust, and possible flying objects. Make sure others in the area also wear protective equipment. Maintain a firm grip on the drill and ensure that your safety belt is properly tied off because drilling in concrete produces a great amount of torque.

Step 10   Clean out the holes when finished drilling.

Step 11   Insert the expansion case into the hole.

Step 12   Screw the threaded rod into the expansion case. The sides of the expansion case will spread and grip the inside of the hole. The fastener is now ready to have a hanger attached to it.

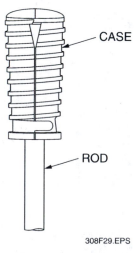

CASE

ROD

308F29.EPS

Figure 29 ◆ Expansion case.

5.5.0 Toggle Bolts

Toggle bolts are used to fasten a part to a hollow wall or panel. Since this is not a very strong attachment, it would only be used for light pipe. To install a toggle bolt, a hole large enough for the wings of the toggle bolt to slide through must be drilled. The wings are inserted into the wall where they snap open on the other side. The bolt is then tightened to complete the fastener. The connecting unit or attachment must be installed on the bolt before the bolt is installed. If you remove the bolt after it is in place, the wings will be lost inside the wall. *Figure 30* shows three types of toggle bolts.

Follow these steps to install a toggle bolt:

Step 1 Identify the type of hanger or support to be used.

Step 2 Locate the area where the support is to be fastened or attached.

Step 3 Make a layout template of the bolt holes to be drilled.

Step 4 Use the template to lay out and mark where each hole will be located.

Step 5 Set up a drill for drilling the hole.

Step 6 Identify the kind and size of toggle bolt needed.

Step 7 Select a drill bit to match the toggle bolt chosen. The drill bit selected should be slightly larger than the toggle bolt.

WARNING!

Be sure to wear ear protectors, safety glasses, and gloves because drilling in concrete generates noise, dust, and possible flying objects. Make sure others in the area also wear protective equipment.

Step 8 Assemble the drill and drill bit.

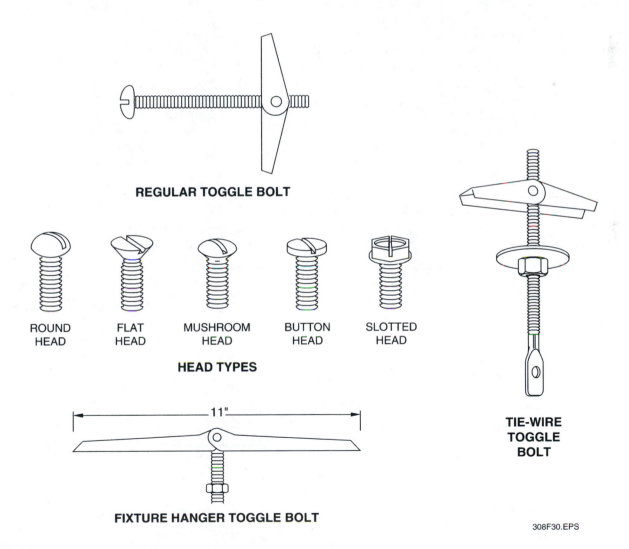

REGULAR TOGGLE BOLT

ROUND HEAD   FLAT HEAD   MUSHROOM HEAD   BUTTON HEAD   SLOTTED HEAD

HEAD TYPES

11"

FIXTURE HANGER TOGGLE BOLT

TIE-WIRE TOGGLE BOLT

308F30.EPS

Figure 30 ◆ Toggle bolts.

Step 9 Drill the holes for the toggle bolts.

Step 10 Unscrew the wings from the toggle bolt.

Step 11 Insert the toggle bolt into the support or bracket to be fastened to the wall.

Step 12 Screw the wings onto the toggle bolt.

Step 13 Insert the toggle bolt into the hole.

Step 14 Tighten the toggle bolt. *Figure 31* shows installing the toggle bolt.

6.0.0 ◆ FABRICATING BRACKETS

It is often necessary to fabricate angle iron brackets in the field. It is the pipefitter's responsibility to ensure that the brackets are fabricated before the time that the piping is erected. Most brackets are simply right triangles made of steel that are fastened to an existing structure and used to support the pipe. Fabricating and laying out brackets requires figuring the lengths of the sides of right triangles. Always take careful measurements to ensure that the bracket will fit, and be sure to account for any obstructions in the area in which the bracket will be placed. There are numerous types of brackets that can be fabricated in the field. The following sections explain how to fabricate 45-degree and 30- by 60-degree angle iron brackets that are commonly used in the field.

6.1.0 Fabricating 45-Degree Angle Iron Brackets

A 45-degree angle iron bracket can be fabricated in either a one-piece unit or a three-piece unit. *Figure 32* shows a 45-degree angle iron bracket.

Follow these steps to fabricate a 45-degree angle iron bracket:

Step 1 Check the piping specifications for the required size of the bracket.

Step 2 Obtain a length of angle iron and a cutting torch.

Step 3 Lay out the angle iron, using a soapstone. *Figure 33* shows the layout of the angle iron bracket.

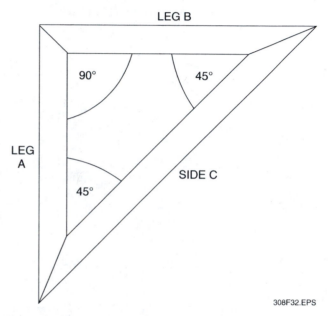

308F32.EPS

Figure 32 ◆ 45-degree angle iron bracket.

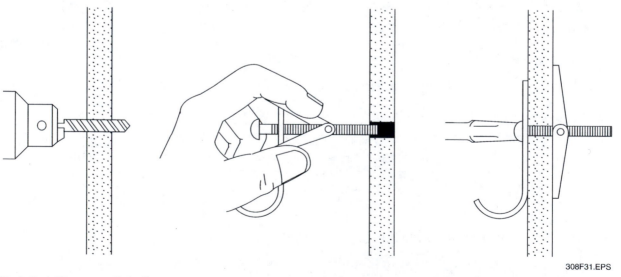

308F31.EPS

Figure 31 ◆ Installing a toggle bolt.

Step 4 Measure and mark the length of one leg of the bracket with a straight line on the angle iron. This is leg A.

Step 5 Measure and mark the second leg on the angle iron. This is leg B and will be the same length as leg A.

Step 6 Multiply the length of leg A times 1.414 to determine the length of the knee brace.

Step 7 Measure and mark the knee brace on the angle iron, and label the knee brace side C.

Step 8 Cut off the angle iron at the edge of side C.

Step 9 Lay out the notches between legs A and B and between leg B and side C according to the formulas in *Figure 33*.

Step 10 Lay out the angles on the ends of the angle iron, using the formula in *Figure 33*.

Step 11 Cut the notches out of the angle iron, using the cutting torch.

Step 12 Heat the angle iron between legs A and B until the metal is red-hot.

WARNING!

Do not breathe the fumes created when galvanized steel is heated. Heat galvanized steel in an open, well-ventilated area, and wear a respirator if it is required on your job site. If you are slightly affected, drink milk to ease the discomfort.

Step 13 Bend leg A up so that it forms a 90-degree angle with leg B.

Step 14 Heat the angle iron between leg B and side C until the metal is red-hot.

Step 15 Bend side C up until the edge the edge of leg A and the angle iron forms a right triangle.

Step 16 Have the welder weld the bracket together. This bracket can be welded or bolted to a beam.

Follow the same procedures to fabricate a 30- by 60-degree angle iron bracket. Use the formulas provided in *Figure 34*, which shows the layout of a 30- by 60-degree angle iron bracket.

A = B	C = A × 1.414	E = D × 2.414	G = D – THICKNESS
B = A	D = WIDTH OF ANGLE IRON	F = D × 2.414	OF METAL

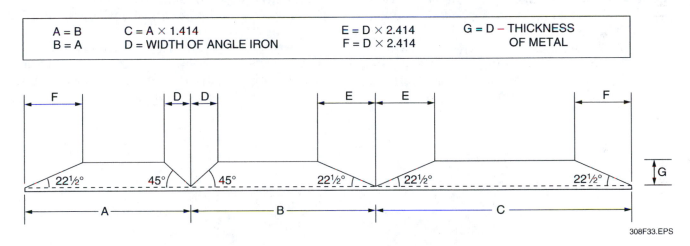

Figure 33 ◆ Layout of a 45-degree angle iron bracket.

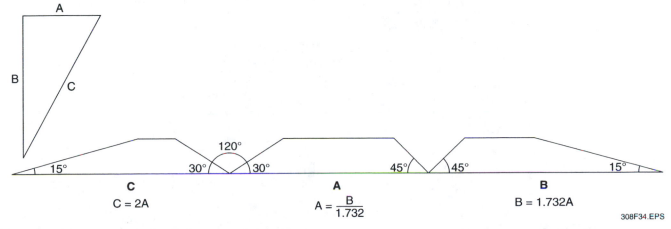

Figure 34 ◆ Layout of a 30- by 60-degree angle iron bracket.

7.0.0 ◆ SPRING CAN SUPPORTS

The proper selection and installation of spring can supports are critical for controlling the expansion and contraction of piping systems. Improperly controlled piping systems can cause structural failures to flanges, valves, piping, equipment, and supports and can also be a safety hazard. If a piping system containing hazardous materials fails, it could cause contamination of the environment, an explosion, or a fire.

To control and support a piping system, a pipefitter must understand the effects of expansion and contraction in piping systems and the importance of properly installing, adjusting, and maintaining support devices.

8.0.0 ◆ TYPES OF SPRING CAN SUPPORTS

Whenever piping moves vertically, it must be supported by a pipe hanger that will support the piping while allowing vertical movement. A spring-type hanger performs this function. All spring can supports have a load scale and a load indicator that show the load rating being imposed on the support. Spring can supports can be adjusted to a preset **cold load** setting before the pipeline is put in service and a **hot load** setting after the pipeline is in service. Spring-type hangers and supports can be classified as variable spring can supports or constant spring can supports.

8.1.0 Variable Spring Can Supports

A variable spring can support is used to support piping that moves vertically and that does not require constant support. The support of a variable spring can varies with the deflection of the spring. A variable spring can is used when the movement of the pipeline through thermal expansion and contraction does not exceed the allowable spring deflection and when the change in supporting force caused by the spring deflection is allowable. As the piping moves in a vertical direction, it causes the variable spring to deflect an equal distance. As the spring deflects, the supporting force it is exerting on the pipe changes by an amount equal to the vertical movement times the spring rate. The spring deflection rate is a constant, which is unique for each size of a particular support series. During the assembly of the variable spring unit, the spring is **precompressed** to a position that establishes a standard minimum load for each size. Precompressing the spring minimizes the **head room** required for the use of variable spring supports. It also acts as a preadjustment of the spring, which reduces the time and labor required during installation to set the hanger to the desired position. *Figure 35* shows a typical variable spring can support.

Variable spring can supports use several spring configurations, depending on the load and application. Spring configurations can differ within the standard types of variable spring can hangers.

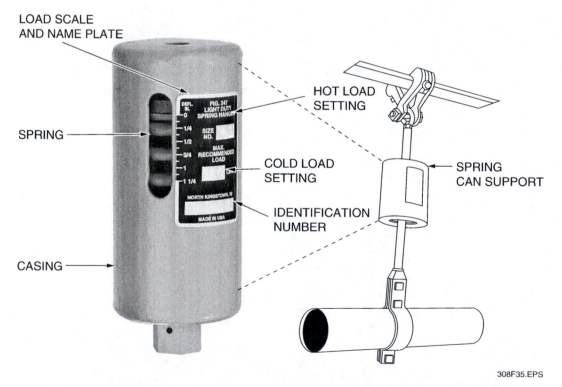

308F35.EPS

Figure 35 ◆ Variable spring can support.

The common spring configurations are:

- Standard
- Short
- Double
- Triple
- Quadruple

The standard spring configuration uses a single spring enclosed in a steel casing. The short spring configuration is used in confined areas where thermal movement of the piping is relatively small. Double, triple, and quadruple spring configurations use two, three, or four springs arranged in series enclosed inside a single, steel casing. Centering guides are located between the springs to keep them permanently aligned within the casing. *Figure 36* shows variable spring can spring configurations.

There are seven types of variable spring can supports that are identified by letters from Type A through Type G. Each type support, A through G, is manufactured in a variety of sizes ranging from size 000 to size 22 and are capable of supporting loads from 10 to 50,000 pounds. A support unit can weigh from approximately 5 to 1,500 pounds depending on the size and type. Each type of support is available in all of the above spring configurations. *Figure 37* shows the seven types of variable spring can supports.

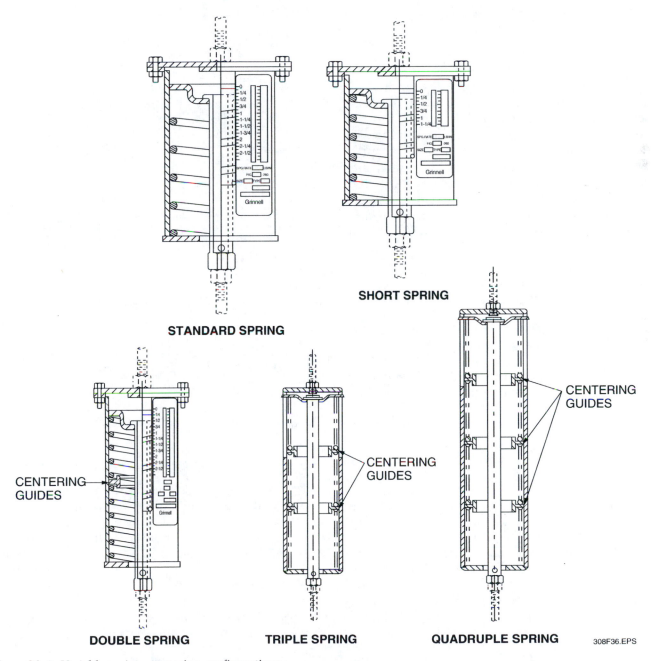

Figure 36 ◆ Variable spring can spring configurations.

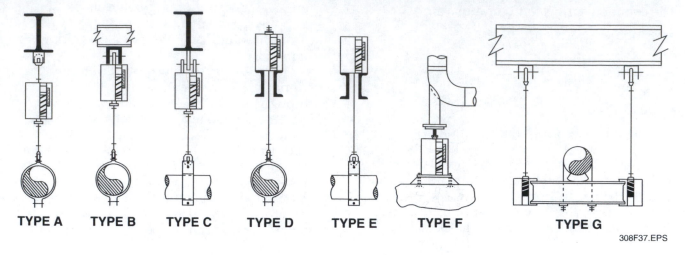

TYPE A   TYPE B   TYPE C   TYPE D   TYPE E   TYPE F   TYPE G

308F37.EPS

Figure 37 ◆ Types of variable spring can supports.

8.1.1 Type A Variable Spring Can Support

The Type A support is a basic spring support. It is designed to be attached to its supporting member by screwing a rod into a tapped hole in the top cap of the support. The upper jam nut should then be locked, securing the support. A Type A support unit is used where extensive head room is available and where it may be desirable to locate the spring at a specific elevation in the hanger assembly. It is also frequently installed flush against the underside of a pair of channels. *Figure 38* shows a Type A variable spring can support.

8.1.2 Type B and C Variable Spring Can Supports

The Type B and C supports are similar to Type A in basic design and construction. Type B has a single lug, and Type C has two lugs, so they can be attached to the building structure. The lugs permit the use of a clevis, a pair of angles, or an eyebolt for attachment where head room is limited. *Figure 39* shows Type B and C variable spring can supports.

8.1.3 Type D and E Variable Spring Can Supports

Type D and E variable spring can supports are designed to sit on top of a pair of supporting beams with the hanger rod extending down between the beams. The major difference between Type D and Type E supports is that Type D supports can be adjusted from the top of the casing, and Type E supports can be adjusted above or below the casing. *Figure 40* shows Type D and E variable spring can supports.

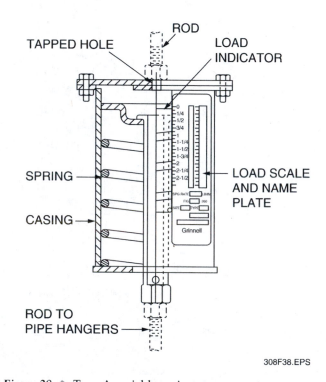

308F38.EPS

Figure 38 ◆ Type A variable spring can support.

8.1.4 Type F Variable Spring Can Support

The Type F support is used under a base elbow or piping that must be supported directly from the floor. A base plate with four bolt holes or slots is mounted to the casing of the support to allow the support to be bolted into place. The load flange on the top of the support can be attached to the pipe with a saddle or with a pipe roll. Be careful not to lose or misplace the load flange during support installation. *Figure 41* shows a Type F variable spring can support.

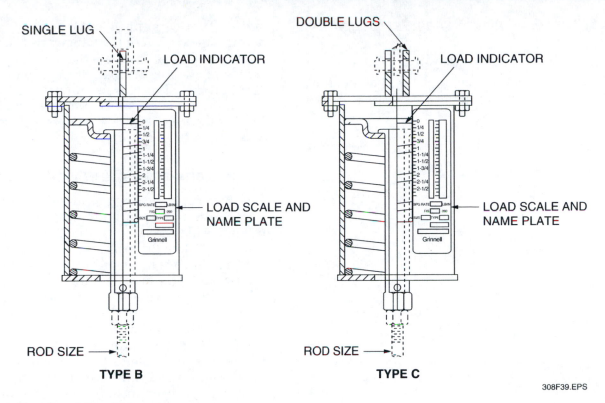

Figure 39 ◆ Type B and C variable spring can supports.

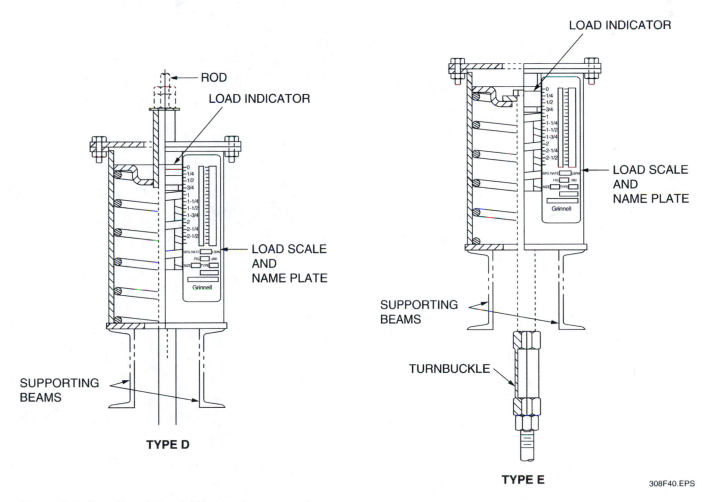

Figure 40 ◆ Type D and E variable spring can supports.

8.1.5 Type G Variable Spring Can Support

The Type G support unit is a complete trapeze assembly. The support consists of two standard spring units plus two back-to-back channel irons welded at each end to the support casing. When sizing Type G supports, you must take into account the weight of the hanger and, since there are two supports, calculate that each support will carry only half of the load. If the pipe is not centered on the channel iron, however, the support that the pipe is closer to will carry more of the load of the pipe. *Figure 42* shows a Type G variable spring can support.

8.2.0 Constant Spring Can Supports

Constant support hangers provide constant support to the piping throughout the full range of thermal expansion and contraction of the piping. This means that the load of the pipe is evenly distributed throughout the pipeline during any movement of the line. Constant supports are usually calibrated at the factory before shipment to the exact load specified by engineering personnel, and this load can range from 30 to over 80,000 pounds. Constant supports are generally used where it is desirable to prevent the pipe weight load from transferring to connected equipment. They are also used to support critical piping systems.

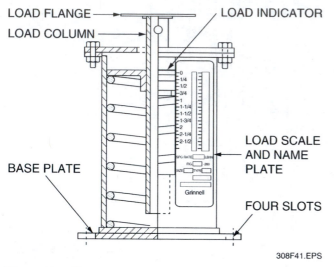

308F41.EPS

Figure 41 ◆ Type F variable spring can support.

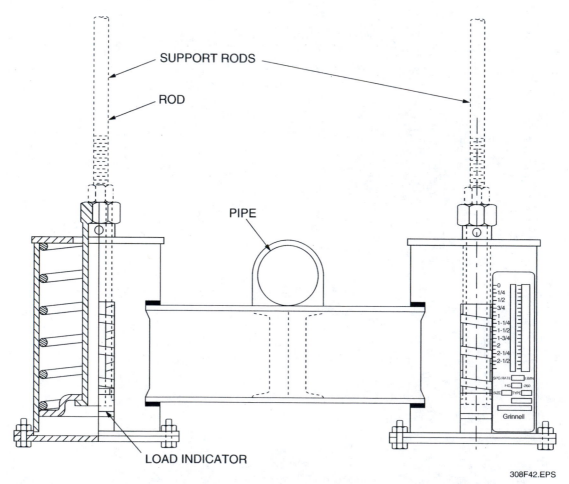

308F42.EPS

Figure 42 ◆ Type G variable spring can support.

Constant supports consist of a helical coil spring that works in conjunction with a bell crank lever to ensure constant support. The spring force times the distance to the lever pivot point is always equal to the pipe load times its distance to the lever pivot point. *Figure 43* shows constant supports.

Constant supports are available in vertical and horizontal configurations in Types A through G. The constant support types designate how they are attached to the existing structure and are identified in the same manner as the variable spring can supports. *Figure 44* shows types of vertical and horizontal constant supports.

8.2.1 Type A Constant Supports

Vertical Type A constant supports have a tapped hole in the top cap of the hanger. A threaded rod is screwed through the cap to attach the support to a beam. When using a Type A hanger, the support must be mounted flush with its supporting beam. A sight hole located in the top of the spring casing allows you to correctly thread the rod in the hole without screwing the rod too far into the support. Horizontal Type A constant supports have a threaded hole located at each end of the hanger frame. *Figure 45* shows Type A constant supports.

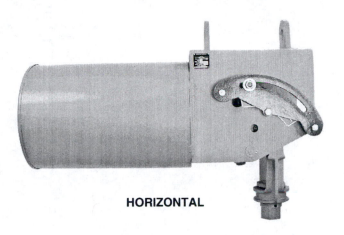

HORIZONTAL

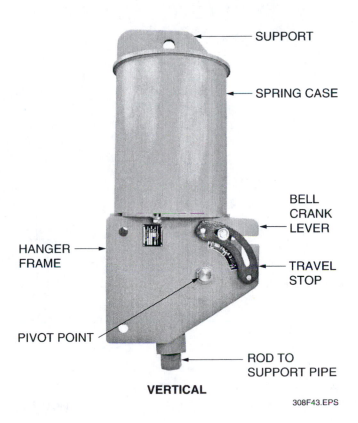

SUPPORT

SPRING CASE

BELL
CRANK
LEVER

TRAVEL
STOP

HANGER
FRAME

PIVOT POINT

ROD TO
SUPPORT PIPE

VERTICAL

308F43.EPS

Figure 43 ◆ Constant supports.

8.2.2 Type B and C Constant Supports

Vertical Type B constant supports have a single lug, and Type C have a pair of lugs welded to the top cap for attachment to the existing structure, using a welded beam clamp or eye rod attached to the existing structure. Type B horizontal supports have a single welded lug located at each end of the hanger frame, while horizontal Type C supports have a pair of welded lugs located at each end of the hanger frame. *Figure 46* shows Type B and C constant supports.

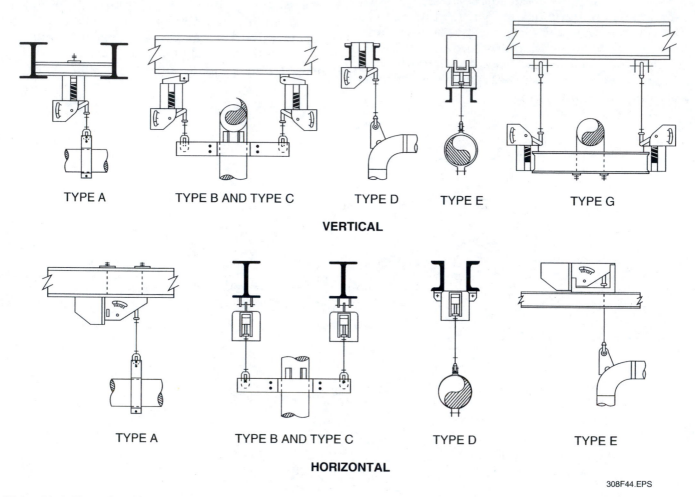

Figure 44 ◆ Vertical and horizontal constant supports.

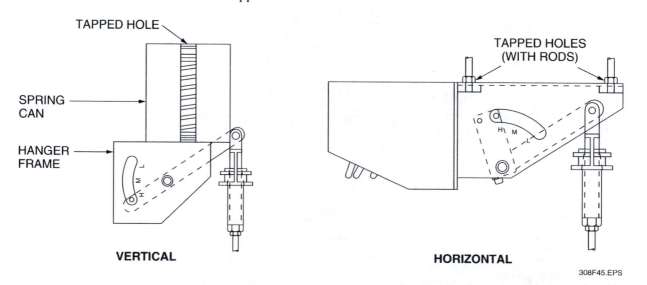

Figure 45 ◆ Type A constant supports.

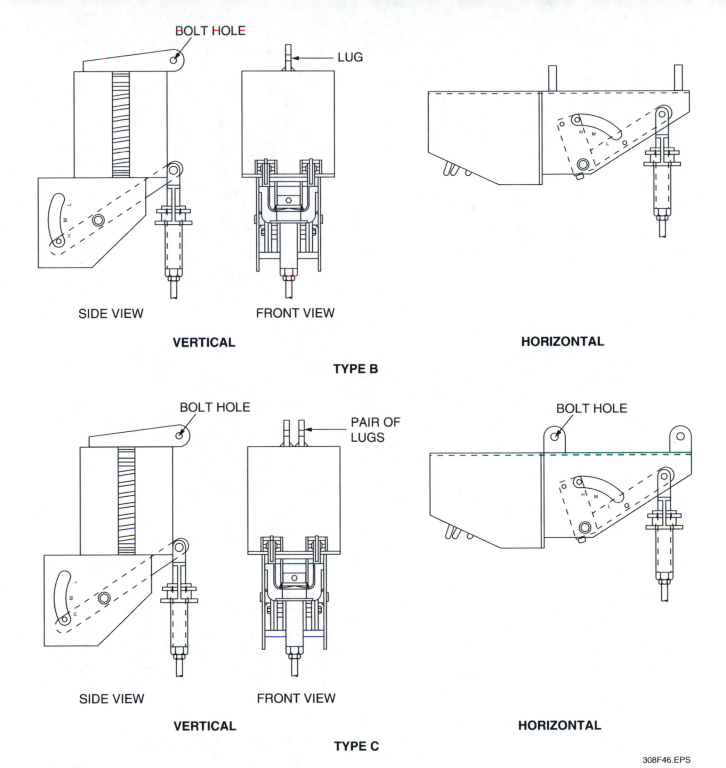

BOLT HOLE

LUG

SIDE VIEW   FRONT VIEW

VERTICAL

HORIZONTAL

TYPE B

BOLT HOLE

PAIR OF
LUGS

BOLT HOLE

SIDE VIEW   FRONT VIEW

VERTICAL

HORIZONTAL

TYPE C

308F46.EPS

Figure 46 ◆ Type B and C constant supports.

8.2.3 Type D Constant Supports

A vertical Type D constant support has two support ears welded to the sides near the top of the can to allow the support to rest on the tops of a pair of supporting beams, with most of the support hanging down between the beams. The thickness of the supporting beams from bottom to top is limited to the height of the support can. The supporting ears have bolt holes so that the support can be bolted in place. A horizontal Type D constant support is bolted to the bottoms of a pair of supporting beams through the bolt holes in the supporting ears. *Figure 47* shows Type D constant supports.

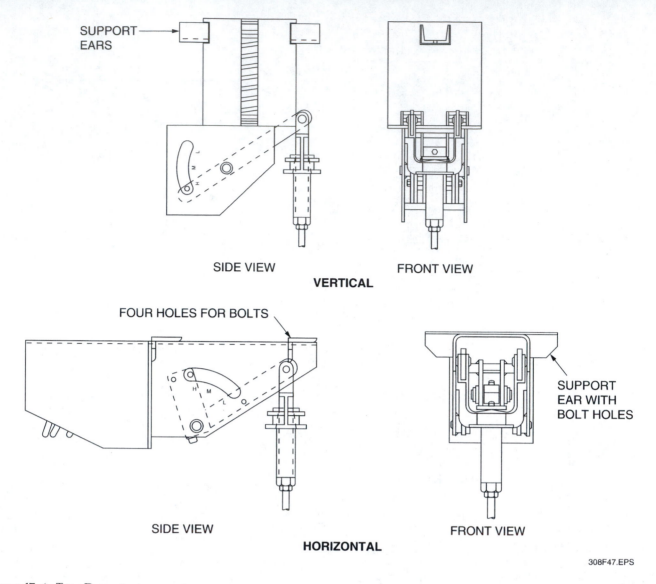

SUPPORT EARS

SIDE VIEW   FRONT VIEW

VERTICAL

FOUR HOLES FOR BOLTS

SUPPORT EAR WITH BOLT HOLES

SIDE VIEW   FRONT VIEW

HORIZONTAL

308F47.EPS

Figure 47 ◆ Type D constant supports.

8.2.4 Type E Constant Supports

Vertical and horizontal Type E constant supports have brackets on each side of the hanger frame to allow the support to rest on top of a pair of supporting beams and be bolted into place. If the hanger rod does not exceed the thickness of the beams from top to bottom, a rod coupling must be attached between the hanger and the pipeline. The entire Type E support rests above the top of the supporting beams. *Figure 48* shows Type E constant supports.

8.2.5 Type F Constant Supports

Type F constant supports are designed to support the pipeline from below the pipe. The pipe is supported by a saddle or shoe attached to the support lever and secured by a removable load pin. The Type F support has a base plate mounted to the bottom of the can to allow it to be bolted to the floor or to beams. *Figure 49* shows Type F constant supports.

8.2.6 Type G Constant Supports

Type G constant supports are a complete trapeze assembly. The support consists of two vertical constant supports with two back-to-back channel irons welded between the support casings. When sizing Type G hangers, you must take into account the weight of the hanger and, since there are two hangers, calculate that each hanger will carry only half of the load. If the pipe is not centered on the channel iron, however, the support that the pipe is closer to will carry more of the load of the pipe. *Figure 50* shows a Type G constant support.

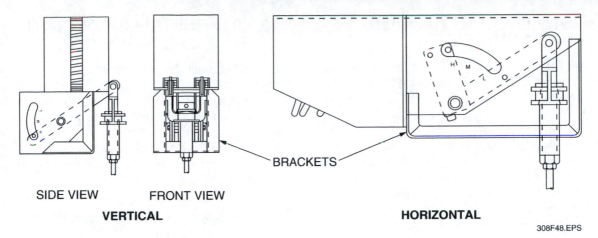

SIDE VIEW   FRONT VIEW

VERTICAL   **HORIZONTAL**

BRACKETS

308F48.EPS

Figure 48 ◆ Type E constant supports.

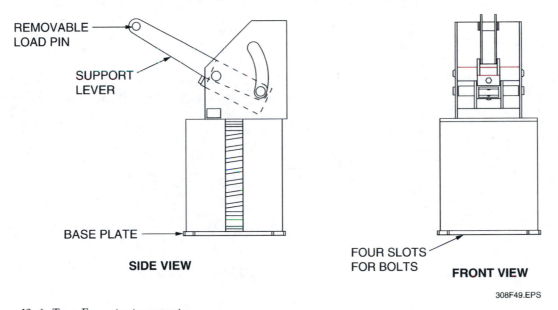

REMOVABLE LOAD PIN

SUPPORT LEVER

BASE PLATE

SIDE VIEW

FOUR SLOTS FOR BOLTS   **FRONT VIEW**

308F49.EPS

Figure 49 ◆ Type F constant supports.

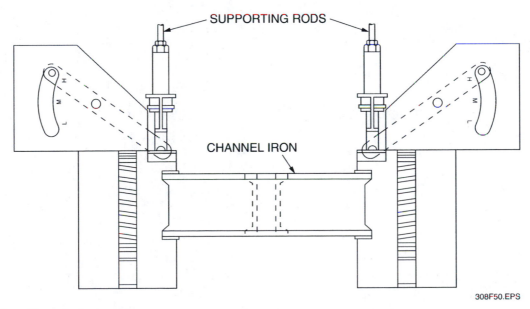

SUPPORTING RODS

CHANNEL IRON

308F50.EPS

Figure 50 ◆ Type G constant supports.

9.0.0 ◆ STORING AND HANDLING PROCEDURES

Special storing and handling procedures are required to protect supports from physical damage and **corrosion**. When a support is rusted or corroded, the spring mechanism will not move freely and accurate adjustments cannot be made. Physical damage to support rods, the spring can, or the mounting system will cause the support to malfunction. Supports must be stored indoors on pallets to protect them from the weather, and they should be stored in the shipping containers if possible until they are needed for installation. Always be careful to keep all the **components**, such as turnbuckles, clevises, and threaded rods, together. Proper methods for rigging and lifting the supports should be followed to prevent damaging the support.

Support units weighing 100 pounds or more can be ordered with special lifting lugs to help lift the assemblies into position for installation. *Figure 51* shows lifting lug locations.

10.0.0 ◆ INSTALLING SPRING CAN SUPPORTS

Installing variable spring can supports requires special rigging and handling skills to ensure that the support is installed safely. Accurate records are kept on each support, so it can be easily located when needed for replacement or installation. To install spring can supports, you must identify the location of the support, install the support, and then adjust the support.

10.1.0 Identifying Location of Supports

Pipefitters follow engineer drawings to properly locate variable spring can supports. *Figure 52* shows a sample detail sheet.

The detail sheet provides all of the information needed to install a spring can support. The material description lists all of the materials needed, including the exact type of spring can support and the cold and hot load settings for the support. Center

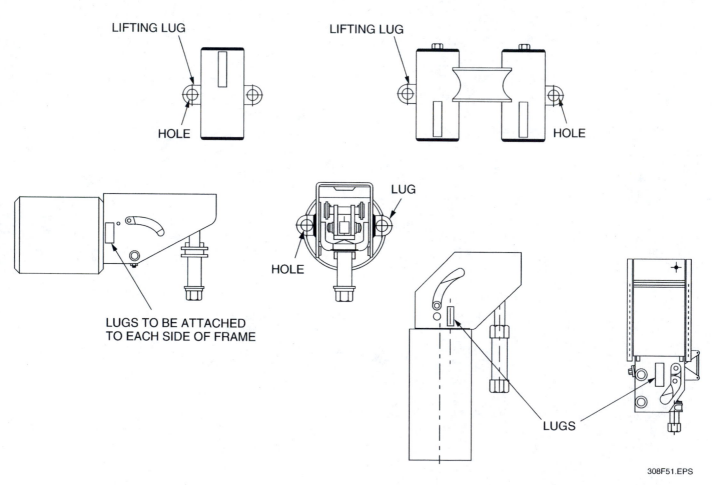

Figure 51 ◆ Lifting lug locations.

line pipe elevation and top of steel elevation are both given, so you can install the support at the correct height. A location plan is shown to the right of the support drawing to designate the exact location in the plant that the support is to be located. Most location plans use the grid system to designate the location of the support. By using all of the information shown on the detail sheet, you can install the spring can support in the proper location and make the correct adjustments to the support.

ITEM NO.	QTY.	FIGURE NO.	SIZE	MATERIAL DESCRIPTION	PRICE CODE	WEIGHT EACH	SHIP ORDER NO.
1	1	401	3/4	WELDED BEAM ATTACHMENT W/BOLT AND NUT			23
2	2	340	3/4	EYE NUT-WELDLESS-RIGHT HAND THREAD			23
3	4	380	3/4	HEX NUTS			23
4	1	300	3/4 " × 3'-10"	THREADED ROD-RIGHT HAND BOTH ENDS			23
5	1	568	11 TYPE = A	VARIABLE SPRING HL=1903 CL=1605 MVT.=0.875"DN	WTS		23
6	1	300	3/4 " × 4'-0"	THREADED ROD-RIGHT HAND BOTH ENDS			23
7	1	102	10"	DOUBLE BOLT PIPE CLAMP			23

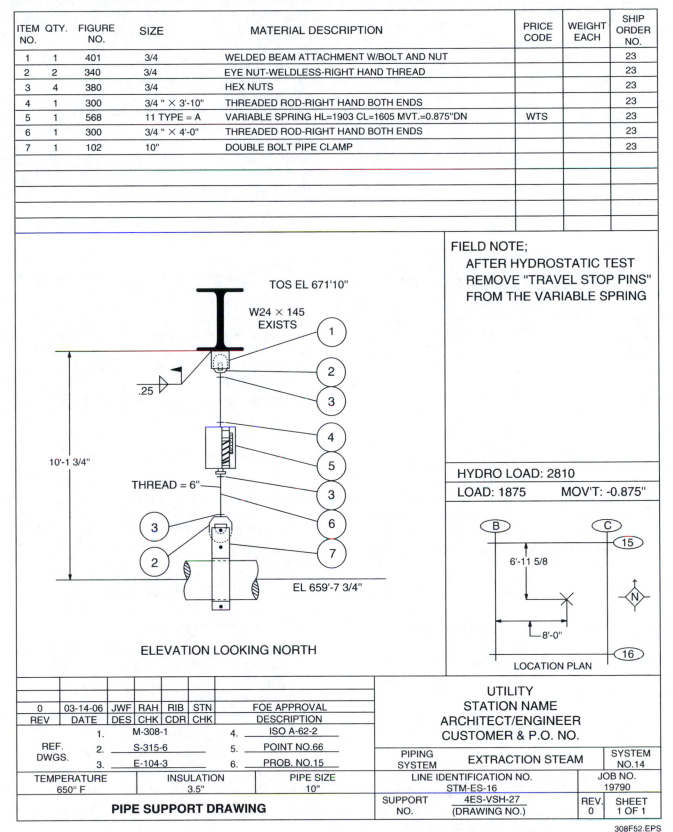

FIELD NOTE;
AFTER HYDROSTATIC TEST REMOVE "TRAVEL STOP PINS" FROM THE VARIABLE SPRING

TOS EL 671'10"

W24 × 145 EXISTS

.25

10'-1 3/4"

THREAD = 6"

EL 659'-7 3/4"

ELEVATION LOOKING NORTH

HYDRO LOAD: 2810

LOAD: 1875   MOV'T: -0.875"

6'-11 5/8

8'-0"

N

LOCATION PLAN

0	03-14-06	JWF	RAH	RIB	STN	FOE APPROVAL		UTILITY STATION NAME ARCHITECT/ENGINEER CUSTOMER & P.O. NO.
REV	DATE	DES	CHK	CDR	CHK	DESCRIPTION		

REF. DWGS.
1. M-308-1   4. ISO A-62-2
2. S-315-6   5. POINT NO.66
3. E-104-3   6. PROB. NO.15

TEMPERATURE 650° F	INSULATION 3.5"	PIPE SIZE 10"

PIPE SUPPORT DRAWING

PIPING SYSTEM	EXTRACTION STEAM	SYSTEM NO.14
LINE IDENTIFICATION NO. STM-ES-16		JOB NO. 19790
SUPPORT NO.	4ES-VSH-27 (DRAWING NO.)	REV. 0	SHEET 1 OF 1

308F52.EPS

Figure 52 ◆ Detail sheet.

10.2.0 Installing Supports

Follow these steps to install a variable spring can support:

Step 1 Clear all removable obstacles from the area in which you will be working; barricade the area; and position the rigging equipment and devices.

Step 2 Check the identification tag to verify that the support is correct for that location.

Step 3 Obtain the detail sheet for the hanger. The detail sheet gives the location, type of brackets, and welding specifications.

Step 4 Obtain all materials that are indicated on the detail sheet.

Step 5 Install the spring can according to the detail sheet instructions.

CAUTION

Do not remove the travel stops from the variable spring can support if the piping system is to be hydrostatically tested, or damage to the support will result.

10.3.0 Removing Travel Stops

When specified on the order, spring can supports are shipped to the job site with travel stops in place to prevent the spring from moving. These stops must stay in place on the hanger for temporary conditions of underload and overload that may occur during installation, hydrostatic testing, or chemical clean-out. The travel stop is installed at the factory with the support in the specified cold position. After installation, hydrostatic testing, and chemical clean-out of the pipeline, remove the travel stops from the support. A red caution tag including removal instructions is attached to travel stops to call attention to the device and remind the pipefitter to remove the travel stop before the pipeline is put into service. After removing the travel stop, keep it with the supports in case you have to reinstall it. Different types of travel stops are used with variable spring and constant supports.

WARNING!

Always preadjust the load on the spring can so that the travel stop can be removed easily. The spring can will have to be adjusted past the cold setting to release the stops.

10.3.1 Variable Spring Can Support Travel Stops

Travel stops for Grinnell variable spring can supports are attached to the spring through the opening on the front of the can. The travel stop consists of an up-limit stop piece and a down-limit stop piece that are held in place with metal bands that must be cut to remove the travel stops. Travel stops for Corner & Lada variable spring can supports consist of two down-limit bolts that extend through the top cap of the hanger and screw into the load indicator above the spring. Spacers attached to the down-limit bolts prevent upward movement of the spring. *Figure 53* shows variable spring can support travel stops. The stops are not to be removed until the run has been tested and the system is under load. Do not cut the bands until then, and be very careful that no one has his or her fingers in the can when the bands are cut or an injury may occur.

10.3.2 Constant Support Travel Stops

Travel stops are attached to each side of the hanger frame on constant supports. The travel stops consist of two plates with a series of holes and are attached to the hanger frame with two or more cap screws. The two plates hold the hanger in the cold position until the plates are removed by removing the cap screws. *Figure 54* shows a constant support travel stop.

10.4.0 Adjusting Spring Can Supports

After the piping or equipment is erected and after all testing that might temporarily change loading conditions is completed, the variable spring should be adjusted to its cold position. If travel stops have been furnished, they should be removed before the equipment is adjusted and put into operation. The procedure for adjusting a spring can support depends on the type of hanger. Use the following guidelines to adjust the various types of supports:

- *Types A, B, C and G variable spring can supports* – Turn the coupling on the lower hanger rod until the load position indicator is in the proper position.

- *Type D variable spring can supports* – Turn the load adjusting nut above the can until the load position indicator is in the proper position.

- *Type E variable spring can supports* – Turn the adjusting turnbuckle below the can until the load position indicator is in the proper position.

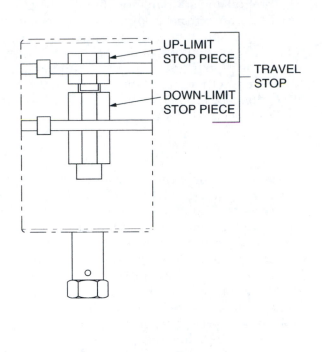

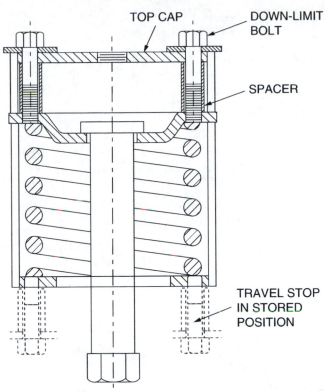

308F53.EPS

Figure 53 ◆ Variable spring can support travel stops.

- *Type F variable spring can supports* – Insert a bar into the holes in the **load column** and turn the column until the load position indicator is in the proper position.

- *All constant supports* – Turn the single-load adjustment bolt until the load position indicator is in the proper position.

After all spring can supports are adjusted to their cold positions, the piping system can be put into service. When the piping or equipment reaches its full operating conditions, a final adjustment to the hot position should be made to

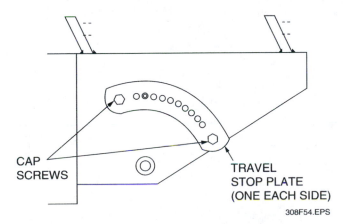

308F54.EPS

Figure 54 ◆ Constant support travel stops.

the spring can supports. The hot position is specified by the engineers, calibrated at the factory by the support manufacturer, and is indicated by a red button labeled with the letter H on the load scale. After adjusting the hanger to the hot position, no further adjustments are necessary.

11.0.0 ◆ MAINTAINING VARIABLE SPRING CAN SUPPORTS

Variable spring can supports require regularly scheduled maintenance inspections. During these inspections, the findings are documented and used for routine maintenance activities. The following maintenance checks are performed on supports:

- Check for rust and corrosion.
- Check for worn or broken parts.
- Check for missing identification tag.
- Check for missing parts.
- Check the load indicator settings against the specifications.

Supports with incorrect load settings must be readjusted to the correct load specification, and supports that are not functioning properly must be replaced with a new assembly. Correct or report problems in accordance with company policy.

11.1.0 Replacing Variable Spring Can Supports

When a support unit becomes worn or corroded and does not function properly, it must be replaced. Follow these steps to replace a variable spring can support.

Step 1 Clear all removable obstacles from the area in which you will be working, and position the rigging equipment and devices.

Step 2 Support the pipe, using a chain hoist or come-along of the proper load rating.

Step 3 Attach the rigging to the spring can lifting lugs or the mounting assembly.

Step 4 Remove the bolts and clamps that attach the support to the structure.

Step 5 Remove the spring can support.

Step 6 Verify that the new support is correct for that location.

Step 7 Attach the new support assembly securely to the building.

CAUTION

Never use another pipe run as a supporting member. Pipe support assemblies must be attached to the structure.

Step 8 Attach the lower hanger rod.

Step 9 Adjust the spring can support until the load indicator is positioned at the desired setting indicated on the load scale.

Step 10 Adjust the support unit to specification.

Vibration and sudden movement is common in pipelines, both from pumps and drivers in use, and from cavitation, water hammer, and sudden release from blowoffs being activated. This is a risk to the pipeline, the supports, and hangers. In some areas, other lateral or longitudinal hazards arise, due to extreme weather events or to earthquakes. Special hangers and braces exist for such hazards, including **snubbers** and **sway braces**.

Snubbers *(Figure 55)* are either hydraulic, pneumatic, or mechanical shock absorbers attached to the piping. Snubbers do not interfere with thermal expansion and contraction of pipe, because such movements take place slowly. Fast movements are slowed dramatically by the snubber.

Sway braces *(Figure 56)* are either a device like a spring can or a set of wire ropes attached to the pipe, preventing movement either laterally or along the axis of the pipe. The stress is communicated to the structure of the building, to which the sway brace or snubber is attached, with attachments similar to those used with hangers.

308F55.EPS

Figure 55 ◆ Snubber.

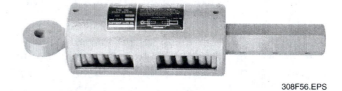

308F56.EPS

Figure 56 ◆ Sway braces.

1. Every pipe hanger and support has a maximum _____.
   a. length
   b. flow rate
   c. load rating
   d. height

2. The load rating safety factor for a hanger is _____.
   a. 2 to 1
   b. 3 to 1
   c. 4 to 1
   d. 5 to 1

3. Adjustable ring hangers can be adjusted vertically _____.
   a. 1 to 2 inches
   b. 3 to 4 inches
   c. 4 to 5 inches
   d. 5 to 6 inches

4. The height of an adjustable clevis is adjusted by loosening or tightening the top nut of the two nuts.
   a. True
   b. False

5. Double-bolt pipe clamps are available in _____.
   a. one size only
   b. aluminum and carbon steel only
   c. carbon steel and chrome molybdenum steel only
   d. two sizes only

6. Trapeze hangers are designed for _____.
   a. thermoplastic pipe only
   b. top loading only
   c. small pipe only
   d. very large pipe only

7. Pipe hanger connecting units attach the pipe hanger to the _____.
   a. floor
   b. roof
   c. siding
   d. supporting structure

8. The most basic type of connecting unit is the _____.
   a. angle iron
   b. clevis
   c. eyebolt
   d. beam clamp

9. Turnbuckles allow hangers to be adjusted to _____.
   a. different heights
   b. different angles
   c. different pipe sizes
   d. different types of pipes

10. Rod attachments are used to link _____.
    a. pipes together
    b. clevises
    c. pipe supports to hangers
    d. threaded rods together

11. Most commonly, beam clamps attach to the _____.
    a. roof of the building
    b. lower flange of beams
    c. upper flange of beams
    d. web of a beam

12. A C-clamp is placed over the _____ and tightened.
    a. edge of the roof
    b. flange of an I-beam
    c. pipe
    d. face of a fitting

13. If vertical adjustment is needed, weld the welded beam attachment _____.
    a. to the side of a vertical beam
    b. to the pipe
    c. to the top flange
    d. upside down

14. A pipe support is a device that is normally attached _____.
    a. to the top of the pipe
    b. under piping
    c. to the side of the pipe
    d. to a spring can

15. When riser clamps are used to support insulated pipes, the insulation is _____.
    a. left off where the riser clamps go
    b. compressed by the clamp
    c. put on over the clamp also
    d. removed from the entire spool

16. Some pipe clamps are attached to the walls directly.
    a. True
    b. False

17. Pipe supports can be fabricated in the field as long as they _____.
    a. are made of metal
    b. meet design specifications
    c. are not adjustable
    d. are not welded

18. Hanger extension rods are screwed into concrete inserts after the _____.
    a. concrete is mixed
    b. concrete has set
    c. insert is pulled out of the concrete
    d. rod is dipped in the concrete

19. An expansion case expands to secure itself when a threaded rod is _____.
    a. screwed into it
    b. taken out of it
    c. driven into it with a hammer
    d. cut off it

20. Toggle bolts are used to fasten a part to a _____.
    a. beam web
    b. solid concrete wall
    c. concrete floor
    d. hollow wall

21. Spring cans are set to a _____ setting before the pipeline is put into service.
    a. hot load
    b. high load
    c. cold load
    d. no-load

22. A variable spring can support is used to support piping that moves vertically and _____.
    a. requires constant support
    b. horizontally
    c. does not require constant support
    d. needs to bounce

23. Spring deflection rate is a constant for each size of a support series.
    a. True
    b. False

24. The short spring can is used where _____.
    a. movement is relatively small
    b. movement doesn't occur
    c. no can is needed
    d. movement is lateral only

25. A Type A spring can is attached at the top by _____.
    a. one lug
    b. two lugs
    c. a clevis
    d. a threaded rod

26. The difference between Type B and Type C variable spring cans is the _____.
    a. length of the threaded rod
    b. length of the can
    c. number of lugs at the top
    d. number of lugs at the bottom

27. Type D and E spring cans are _____.
    a. attached to the bottom of the beam flanges
    b. mounted on top of beams
    c. held with threaded rods at the top
    d. underneath the pipe

28. A Type F spring can is _____.
    a. attached to the top of a beam
    b. supported on the floor
    c. not used for piping
    d. supported on the side of a beam

29. In constant supports the spring force times the distance to lever pivot point is equal to the _____ times the distance to the pivot point.
    a. weight of the spring
    b. distance from the beam
    c. pipe load
    d. number of the springs

30. Travel stops are removed _____.
    a. as soon as you receive the can
    b. after installation and testing of the run
    c. before testing
    d. before installation of the pipe

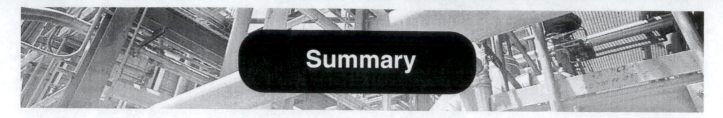

Summary

Pipe hangers and supports do what their names imply. They support the pipeline in service from either above or below the pipe. Using the proper types of hangers and supports controls the movement caused by thermal expansion and the movement of fluid through the pipe. Correct pipe support is as important as correct alignment and joining. The most important thing to remember when installing pipe hangers and supports is that the pipeline must run straight without sagging.

Spring can supports are used on pipelines that are subjected to movement through thermal expansion and contraction. They support the pipeline in service from either above or below the pipe. Pipefitters install and maintain spring can supports, but it is the engineer's job to specify the type of support used in a particular system. When handling large supports, safe rigging practices must be followed to prevent serious injury to workers and damage to equipment. When handling spring can supports, you must be safety conscious at all times.

Notes

Trade Terms Introduced in This Module

Cold load: A piping system before it is filled with product.

Component: A single part in a system.

Corrosion: The result of wearing away slowly, especially by chemical action or abrasive wear through friction.

Fabricate: Putting together component parts to form an assembly.

Head room: The space between an object and the ceiling.

Hot load: The actual load of the pipeline in service.

Load column: Part of a support assembly that is used for adjustment purposes.

Load rating: The usable weight capacity that a piece of hardware can support.

Pipe riser: A vertical pipe extending from an opening in a floor or ceiling.

Precompress: To place pressure on a spring before it is installed in a spring can.

Run: A length of pipe that is made up of more than one piece of pipe.

Snubber: A hydraulic device that slows sudden up and down movement in pipe systems to prevent damage to supports or piping.

Structural member: A supporting part of a building or structure.

Sway brace: Either a spring or hydraulic shock absorber, or a set of cables, designed to prevent dramatic side-to-side or longitudinal movement.

Resources & Acknowledgments

Additional Resources

This module is intended to be a thorough resource for task training. The following reference works are suggested for further study. These are optional materials for continued education rather than for task training.

Cooper B-Line provides information on pipe supports at http://www.cooperbline.com/product/PDFLibrary/PipeHangers/index.htm.

Anvil International provides information on mechanical supports and hangers at http://www.anvilintl.com/ps_hanger/afs/15060_100 32214_01_SD_13656.pdf.

NIBCO provides information on hangers, including Tolco products, at http://www.nibco.com/cms.

Piping Technology and Products provides information on spring cans, hangers, and supports at http://www.pipingtech.com/products/pipe_supports.htm.

Figure Credits

Anvil International, 308F01, 308F02 (photo), 308F03 (photo), 308F06, 308F07 (photo), 308F08 (photo), 308F09 (photo), 308F10, 308F11, 308F12, 308F13 (photo), 308F15, 308F16 (photo), 308F17 (photo), 308F18, 308F22 (photo), 308F35 (photo), 308F36, 308F38–308F43, 308F56

Basic-PSA, Inc./www.basicpsa.com, 308F55

CONTREN® LEARNING SERIES — USER UPDATE

NCCER makes every effort to keep these textbooks up-to-date and free of technical errors. We appreciate your help in this process. If you have an idea for improving this textbook, or if you find an error, a typographical mistake, or an inaccuracy in NCCER's Contren® textbooks, please write us, using this form or a photocopy. Be sure to include the exact module number, page number, a detailed description, and the correction, if applicable. Your input will be brought to the attention of the Technical Review Committee. Thank you for your assistance.

Instructors – If you found that additional materials were necessary in order to teach this module effectively, please let us know so that we may include them in the Equipment/Materials list in the Annotated Instructor's Guide.

Write: Product Development and Revision
National Center for Construction Education and Research
3600 NW 43rd St, Bldg G, Gainesville, FL 32606

Fax: 352-334-0932

E-mail: curriculum@nccer.org

Craft _____ Module Name _____

Copyright Date _____ Module Number _____ Page Number(s) _____

Description _____

_____

_____

_____

(Optional) Correction _____

_____

_____

(Optional) Your Name and Address _____

_____

_____

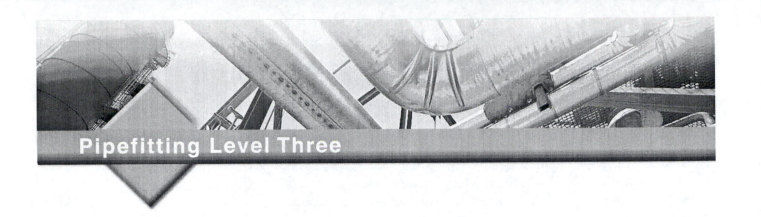

Pipefitting Level Three

08309-07

Testing Piping Systems and Equipment

08309-07

Testing Piping Systems and Equipment

Topics to be presented in this module include:

Overview

Part of the process of fitting pipe is testing the assemblies. Any work must be tested to prevent leakage and to prevent possible catastrophic breakdown. The method of testing may be service or flow testing, or hydrostatic pressure testing, raising the pressure to one and one-half times the anticipated operating pressure. The other kind of testing is NDE, nondestructive evaluation, examining all the welds both inside and out. This includes both tools such as the borescope and videoscope, and specific tests for welds. You will learn about several methods of examining welds, including dye-penetrant tests, radiological examinations, acoustic testing, and magnetic particle and magnetic eddy tests.

Objectives

When you have completed this module, you will be able to do the following:

1. Perform pretest requirements.
2. Perform service and flow tests.
3. Perform head pressure tests.
4. Perform hydrostatic tests.
5. Explain how to perform steam blow tests.
6. Explain nondestructive examinations (NDE).

Trade Terms

Acceptable weld	Porosity
Antisyphon	Root opening
Concavity	Root penetration
Crack	Snubber
Crater	Steam target
Defects	Test boundaries
Design pressure	Test medium
Discontinuity	Testing
Head pressure	Toe crack
Hydrostatic test	Toe of the weld
Inclusions	Underbead crack
NDE	Undercut
Operating pressure	Weld root
Pig	Weldment

Required Trainee Materials

1. Pencil and paper
2. Appropriate personal protective equipment

Prerequisites

Before you begin this module, it is recommended that you successfully complete *Core Curriculum*; *Pipefitting Level One*; *Pipefitting Level Two*; and *Pipefitting Level Three*, Modules 08301-07 through 08308-07.

This course map shows all of the modules in the third level of the *Pipefitting* curriculum. The suggested training order begins at the bottom and proceeds up. Skill levels increase as you advance on the course map. The local Training Program Sponsor may adjust the training order.

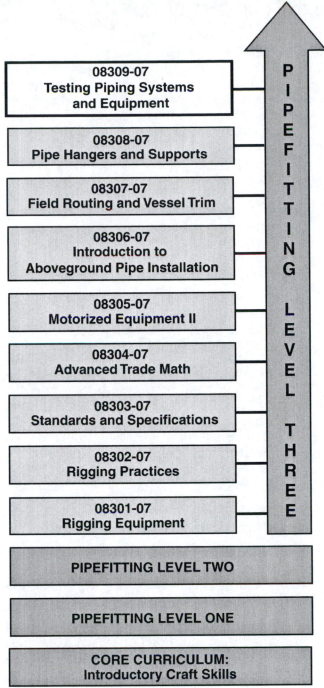

309CMAP.EPS

1.0.0 ◆ INTRODUCTION

During the erection of a piping system, the welds and fittings are examined by various nondestructive and visual examination procedures. This module examines the basics of nondestructive examination (**NDE**) procedures, including weld inspection, ultrasonic, electromagnetic, and radiographic **testing**. Holiday testing, a technique used to examine coatings for possible leaks, is described.

After a piping system is erected, most piping specifications require that the system be flushed and tested before it is put into service to ensure the quality and integrity of the joints. Some pipelines require flushing before testing, and others require flushing after testing. There are many types of tests used for piping systems. This module explains how to perform service, flow, **head pressure**, hydrostatic, and steam blow tests.

2.0.0 ◆ PERFORMING VISUAL INSPECTIONS

Weld joints are visually inspected before the welding is started, as welding proceeds, and after the welding is completed to ensure that a quality weld is produced and that all code requirements are satisfied. If at any point a weld fails to pass inspection, it must be repaired or redone to the satisfaction of the inspection criteria.

2.1.0 Performing Prefit-Up Inspections or Hold Points

A hold point is an area that must be inspected before proceeding with the work activities, such as code welding. Perform a prefit-up inspection following these procedures.

- Ensure that the base metal is clean and free of all foreign material. Foreign materials such as chemicals, paint, grease, dirt, and scale can combine with the weld metal and unfavorably affect the weld.
- Check the bevel angle on the pipe to ensure that it is within tolerance.
- Check the weld area to ensure that it is free of damage, such as nicks and gouges.

2.2.0 Performing Visual Inspections of Fit-Ups

Perform the following procedures to visually inspect the fit-up.

Step 1 Check the fit-up for misalignment or mismatch, using a Hi-Lo gauge. *Figure 1* shows measuring a mismatch, using a Hi-Lo gauge.

Step 2 Check the fit-up for the proper root gap, using a taper gauge. *Figure 2* shows measuring a root gap, using a taper gauge.

Step 3 Inspect the tack welds for any weld **defects**. Tack welds can exhibit defects such as **cracks**, **crater** cracks, **inclusions**, poor fusion, **undercut**, inadequate penetration, and burn-through. The starting and stopping ends of the tack weld should be ground so that they can be incorporated into the root pass without developing defects.

2.3.0 Inspecting Root Passes

Each root pass of a weld must be thoroughly inspected to identify any defects. All defects in the root pass must be repaired before the welding is resumed because defects, especially cracks, in the root pass can crack out or propagate through the entire finished weld. The weld should be inspected on the inside and outside if possible.

Codes and standards define the quality requirements necessary to achieve the integrity and reliability of the **weldment**. These quality require-

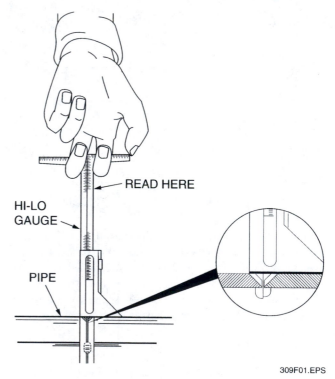

309F01.EPS

Figure 1 ◆ Measuring mismatch, using a Hi-Lo gauge.

ments help ensure that welded joints are capable of serving their intended function for the intended life of the weldment. Weld discontinuities can prevent a weld from meeting the minimum quality requirements.

The American Welding Society (AWS) defines a **discontinuity** as an interruption of the typical structure of a weldment, such as a lack of homogeneity in the mechanical, metallurgical, or physical characteristics of the material or weldment. A discontinuity is not necessarily a defect. A defect found during inspection will require the weld to be rejected. A single excessive discontinuity or a combination of discontinuities can make the weldment defective (unable to meet minimum quality requirements). However, a weld can have one or more discontinuities and still be an **acceptable weld**.

The pipefitter should be able to identify discontinuities and understand the effect they have on weld integrity. Some can be seen during visual inspection. Those that are internal to the weldment can only be detected through other testing methods. The most common weld discontinuities are:

- **Porosity**
- Inclusions
- Cracks
- Incomplete joint penetration
- Incomplete fusion
- Undercuts
- Arc strikes
- Spatter
- Unacceptable weld profiles

2.4.0 Porosity

Porosity is the presence of voids or empty spots in the weld metal. It is the result of gas pockets being trapped in the weld as it is being made. As the molten metal hardens, the gas pockets form voids. Unless the gas pockets work up to the surface of the weld before it hardens, porosity cannot be detected through visual inspection.

Porosity can be grouped into four major types:

- *Uniformly scattered porosity* – May be located throughout single-pass welds or throughout several passes in multiple-pass welds (*Figure 3B*).
- *Clustered porosity* – A localized grouping of pores that results from improperly starting or stopping the welding.
- *Linear porosity* – May be aligned along a weld interface, the root of a weld, or a boundary between weld beads (*Figure 3A*).
- *Piping porosity* – Normally extends from the root of the weld toward the face. These elongated gas pores are also called wormholes. When they do not extend to the surface, the porosity cannot be visually detected (*Figure 3C*).

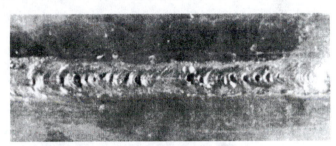

(A) LINEAR POROSITY

(B) SCATTERED SURFACE POROSITY

(C) PIPING POROSITY   309F03.EPS

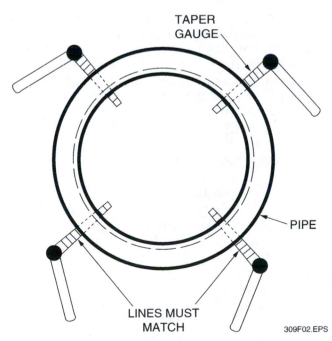

TAPER GAUGE

PIPE

LINES MUST MATCH   309F02.EPS

Figure 2 ◆ Measuring root gap, using taper gauge.

Figure 3 ◆ Examples of porosity.

Most porosity is caused by improper welding techniques or contamination. Improper welding techniques may cause inadequate shielding gas to be formed. As a result, parts of the weld are left unprotected. Oxygen in the air or moisture in the flux or on the base metal that dissolves in the weld pool can become trapped and produce porosity.

The intense heat of the weld can decompose paint, dirt, oil, or other contaminants, producing hydrogen. This gas can become trapped in the solidifying weld pool and produce porosity.

2.5.0 Inclusions

Inclusions are foreign matter trapped in the weld metal (*Figure 4*), between weld beads, or between the weld metal and the base metal. Inclusions are sometimes jagged and irregularly shaped. Sometimes they form in a continuous line. This concentrates stresses in one area and reduces the structural integrity (strength) of the weld.

Inclusions generally result from faulty welding techniques, improper access to the joint for welding, or both. A typical example of an inclusion is slag, which normally forms over a deposited weld. If the electrode is not manipulated correctly, the force of the arc will cause some of the slag particles to be blown into the molten pool. If the pool solidifies before the inclusions can float to the top, they become lodged in the metal, producing a discontinuity. Sharp notches in joint boundaries or between weld passes also can result in slag entrapment.

SURFACE SLAG INCLUSIONS

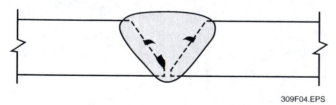

309F04.EPS

Figure 4 ◆ Examples of nonmetallic inclusions.

Inclusions are more likely to occur in out-of-position welding because the tendency is to keep the molten pool small and allow it to solidify rapidly to prevent it from sagging.

With proper welding techniques and the correct electrode used with the proper setting, inclusions can be avoided or kept to a minimum. Other remedies include:

* Positioning the work to maintain slag control
* Changing the electrode to improve control of molten slag
* Thoroughly removing slag between weld passes
* Grinding the weld surface if it is rough and likely to entrap slag
* Removing heavy mill scale or rust on weld preparations
* Avoiding the use of electrodes with damaged coverings

2.6.0 Cracks

Cracks are narrow breaks that occur in the weld metal, in the base metal, or in the crater formed at the end of a weld bead (*Figure 5*). They are caused when localized stresses exceed the ultimate strength of the metal. Cracks are generally located near other weld or base metal discontinuities.

2.6.1 Weld Metal Cracks

Three basic types of cracks can occur in weld metal: transverse, longitudinal, and crater. As seen in *Figure 5*, weld metal cracks are named to correspond with their location and direction.

Transverse cracks run across the face of the weld and may extend into the base metal. They are more common in joints that have a high degree of restraint.

Longitudinal cracks are usually located in the center of the weld deposit. They may be the continuation of crater cracks or cracks in the first layer of welding. Cracking of the first pass is likely to occur if the bead is thin. If this cracking is not eliminated before the other layers are deposited, the crack will progress through the entire weld deposit.

Crater cracks have a tendency to form in the crater whenever the welding operation is interrupted. These cracks usually proceed to the edge of the crater and may be the starting point for longitudinal weld cracks. Crater cracks can be minimized or prevented by filling craters to a slightly convex shape prior to breaking the welding arc.

Figure 6 shows examples of various types of weld metal cracks.

Weld metal cracking can usually be reduced by taking one or more of the following actions:

- Improving the contour or composition of the weld deposit by changing the electrode manipulation or electrical conditions
- Increasing the thickness of the deposit and providing more weld metal to resist the stresses by decreasing the travel speed
- Reducing thermal stress by preheating
- Using low-hydrogen electrodes
- Balancing shrinkage stress by sequencing welds
- Avoiding rapid cooling conditions

2.6.2 Base Metal Cracks

Base metal cracking usually occurs within the heat-affected zone of the metal being welded. The possibility of cracking increases when working with hardenable materials. These cracks usually occur along the edges of the weld and through the heat-affected zone into the base metal. Types of base metal cracking include **underbead cracking** and **toe cracking**. Hot cracks occur while the weld is solidifying. They can be caused by insufficient ductility at high temperature. Cold cracks occur after the weld has solidified. They are often caused by improper welding technique.

Underbead cracks are limited mainly to steel. They are usually found at regular intervals under the weld metal and usually do not extend to the surface. Because of this, they cannot be detected by visual methods of inspection.

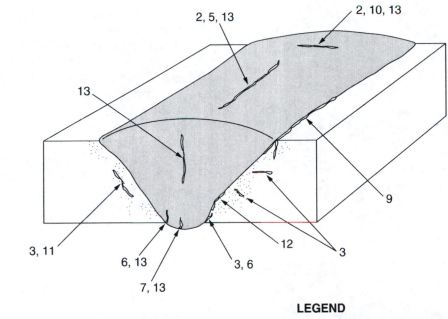

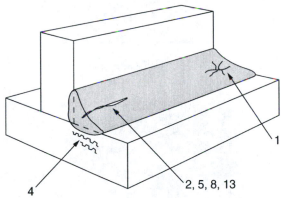

LEGEND

1. CRATER CRACK
2. FACE CRACK
3. HEAT-AFFECTED ZONE CRACK
4. LAMELLAR TEAR
5. LONGITUDINAL CRACK
6. ROOT CRACK
7. ROOT SURFACE CRACK
8. THROAT CRACK
9. TOE CRACK
10. TRANSVERSE CRACK
11. UNDERBEAD CRACK
12. WELD INTERFACE CRACK
13. WELD METAL CRACK

309F05.EPS

Figure 5 ◆ Types of cracks.

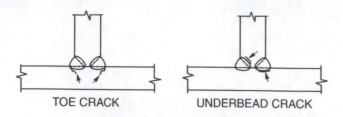

TOE CRACK   UNDERBEAD CRACK

TOE CRACK

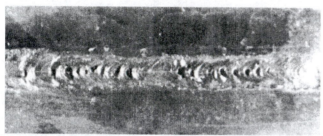

LONGITUDINAL CRACK AND LINEAR POROSITY

CRATER CRACK

LONGITUDINAL CRACK OUT OF CRATER CRACK

FILLET WELD THROAT CRACK

309F06.EPS

Figure 6 ◆ Types of weld metal cracks.

Toe cracks are generally the result of strains caused by thermal shrinkage acting on a heat-affected zone that has been embrittled. They sometimes occur when the base metal cannot accommodate the shrinkage strains that are imposed by welding.

Base metal cracking can usually be reduced or eliminated by one of the following means:

- Controlling the cooling rate by preheating
- Controlling heat input
- Using the correct electrode
- Controlling welding materials

2.7.0 Incomplete Joint Penetration

Incomplete joint penetration (*Figure 7*) occurs when the filler metal fails to penetrate and fuse with an area of the weld joint. This incomplete penetration will cause weld failure if the weld is subjected to tension or bending stresses.

Insufficient heat at the root of the joint is a frequent cause of incomplete joint penetration. If the metal being joined first reaches the melting point at the surfaces above the root of the joint, molten metal may bridge the gap between these surfaces and screen off the heat source before the metal at the root melts.

Improper joint design is another leading cause of incomplete joint penetration. If the joint is not prepared or fitted accurately, an excessively thick root face or an insufficient root gap may cause incomplete penetration. Incomplete joint penetration is likely to occur under the following conditions:

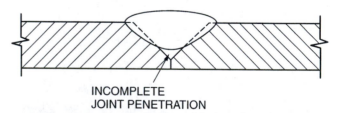

INCOMPLETE
JOINT PENETRATION

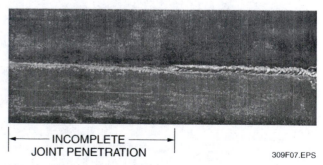

|← INCOMPLETE →|
JOINT PENETRATION

309F07.EPS

Figure 7 ◆ Incomplete joint penetration.

- If the root face dimension is too big even though the **root opening** is adequate
- If the root opening is too small
- If the included angle of a V-groove is too small

Figure 8 shows correct and incorrect joint designs.

Even if the welding heat is correct and the joint design is adequate, incomplete penetration can result from poor control of the welding arc. Examples of poor control include:

- Using an electrode that is too large
- Traveling too fast
- Using a welding current that is too low

Incomplete penetration is always undesirable in welds, especially in single-groove welds if the root of the weld is subject either to tension or bending stresses. It can lead directly to weld failure or can cause a crack to start at the unfused area.

2.8.0 Incomplete Fusion

Many pipefitters confuse incomplete joint penetration with incomplete fusion. It is possible to have good penetration without complete root fusion. Incomplete fusion is the failure of a welding process to fuse, or join together, layers of weld metal or weld metal and base metal.

Incomplete fusion may occur at any point in a groove or fillet weld, including the root of the weld. Often the weld metal simply rolls over onto the plate surface. This is generally referred to as overlap. In many cases, the weld has good fusion at the root and at the plate surface, but because of poor technique and insufficient heat, the **toe of the weld** does not fuse. *Figure 9* shows incomplete fusion and overlap. Causes for incomplete fusion include:

- Insufficient heat as a result of low welding current, high travel speeds, or an arc gap that is too close

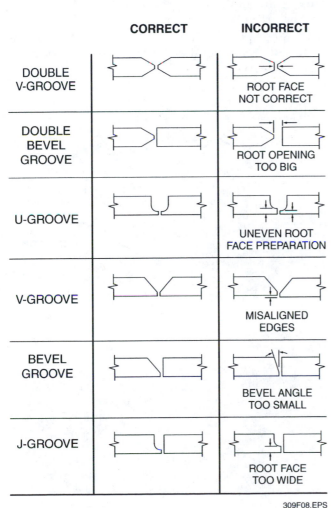

Figure 8 ◆ Correct and incorrect joint designs.

309F08.EPS

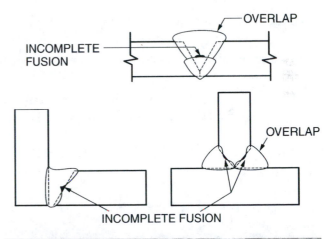

INCOMPLETE FUSION AT WELD FACE

INCOMPLETE FUSION BETWEEN INDIVIDUAL WELD BEADS

309F09.EPS

Figure 9 ◆ Incomplete fusion and overlap.

- Wrong size or type of electrode
- Failure to remove oxides or slag from groove faces or previously deposited beads
- Improper joint design
- Inadequate gas shielding

Incomplete fusion discontinuities affect weld joint integrity in much the same way as porosity and slag inclusion.

2.9.0 Undercut

Undercut is a groove melted into the base metal beside the weld. It is the result of the arc removing more metal from the joint face than is replaced by weld metal. On multilayer welds, it may also occur at the point where a layer meets the wall of a groove. *Figure 10* shows undercut and overlap.

Undercutting is usually due to improper electrode manipulation. Other causes of undercutting include:

- Using a current adjustment that is too high
- Having an arc gap that is too long
- Failing to fill up the crater completely with weld metal

Most welds have some undercut that can be found upon careful examination. When it is con-trolled within the limits of the specifications and does not create a sharp or deep notch, undercut is usually not considered a weld defect. However, when it exceeds the limits, undercutting can be a serious defect because it reduces the strength of the joint.

2.10.0 Arc Strikes

Arc strikes are small, localized points where surface melting occurs away from the joint. These spots may be caused by accidentally striking the arc in the wrong place or by faulty ground connections.

Striking an arc on base metal that will not be fused into the weld metal should be avoided. Arc strikes can cause hardness zones in the base metal and can become the starting point for cracking. *Figure 11* shows an example of arc strikes.

2.11.0 Spatter

Spatter (*Figure 12*) is made up of very fine particles of metal on the plate surface adjoining the weld area. It is usually caused by high current, a long arc, an irregular and unstable arc, or improper shielding. Spatter makes a poor appearance on the weld and base metal and can make it difficult to inspect the weld.

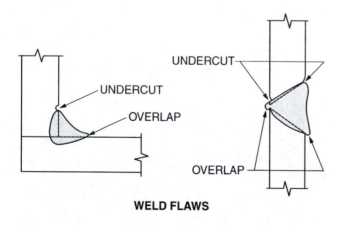

WELD FLAWS

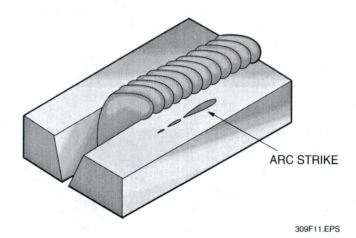

309F11.EPS

Figure 11 ◆ Arc strikes.

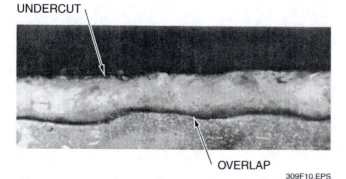

309F10.EPS

Figure 10 ◆ Undercut and overlap.

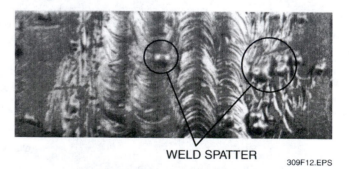

309F12.EPS

Figure 12 ◆ Weld spatter.

2.12.0 Acceptable and Unacceptable Weld Profiles

The profile of a finished weld can affect the performance of the joint under load as much as other discontinuities affect it. This applies to the profile of a single-pass and to a layer of a multiple-pass weld. An unacceptable profile for a single-pass or multiple-pass weld could lead to the formation of discontinuities such as incomplete fusion or slag inclusions as the other layers are deposited. *Figure 13* shows acceptable and unacceptable weld profiles for both fillet and groove welds.

The ideal fillet has a uniform concave or convex face, although a slightly nonuniform face is acceptable. The convexity of a fillet weld or individual surface bead will be approximately 0.07 times the actual face width or the width of the individual surface bead, plus $\frac{1}{16}$ inch.

Butt welds should be made with slight reinforcement (not exceeding $\frac{1}{8}$ inch) and a gradual transition to the base metal at each toe. Butt welds should not have excess reinforcement, insufficient throat, excessive undercut, or overlap. If a butt weld has any of these defects, it should be repaired. The bead width should not exceed the groove width by more than $\frac{1}{8}$ inch. The tie-in of weld to base metal should be smooth.

2.13.0 Inspecting Completed Welds Using Borescopes

Borescopes are another form of visual inspection and are used to detect surface flaws inside pipe, valves, and rotating equipment in areas that are inaccessible. *Figure 14* shows three typical types of borescopes.

The flexible borescopes are most commonly used by pipefitters. Two of the most common types of flexible borescopes are flexible fiberscopes and videoscopes with CCD probes.

2.13.1 Flexible Fiberscopes

A fiberscope consists of a light guide bundle, an image guide bundle, an objective lens, interchangeable viewing heads, and remote controls to control the rotation of the tip. Fiberscopes are available in diameters from 1.4 to 5 inches and come in lengths up to 40 feet. The interchangeable viewing heads provide various directions and fields of view on a single fiberscope. Fields of view are typically 40 to 60 degrees; special applications can range from 10 to 120 degrees. Most fiberscopes provide adjustable focusing of the objective lens.

Before actually operating a flexible fiberscope on the job, you must attend specialized training. The following is a general procedure for operating fiberscopes to give you an understanding of the equipment.

Follow these steps to operate a flexible fiberscope:

Step 1 Obtain from the supervisor or field engineer the procedure for the system to be inspected.

Step 2 Ensure that the system has been vented and drained.

Step 3 Review the procedure to determine what equipment must be removed so that the inspection can be performed.

Step 4 Remove the equipment listed in the procedure.

CAUTION

Do not allow any debris or tools to fall into the system.

Step 5 Route an electrical cord overhead to the location where the flexible fiberscope is to be used.

Step 6 Connect the flexible fiberscope to the electrical cord.

Step 7 Turn on the flexible fiberscope, and check its operation.

Step 8 Insert the flexible fiberscope into the system to be inspected.

Step 9 Move the flexible fiberscope along the system to the area to be inspected.

Step 10 Rotate the flexible fiberscope end while monitoring the eyepiece lens to identify any defects.

Step 11 Document any deficiencies, and report them to the supervisor.

Step 12 Communicate with the supervisor and field engineer or quality control personnel to determine if repairs are needed.

Step 13 Remove the flexible fiberscope from the system.

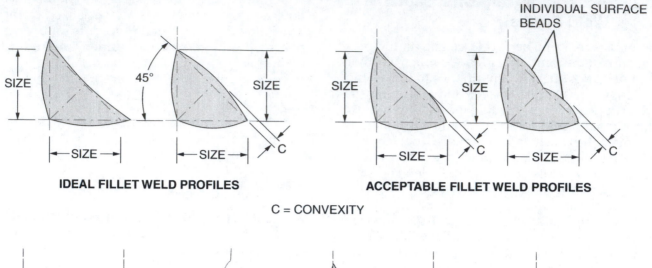

IDEAL FILLET WELD PROFILES

ACCEPTABLE FILLET WELD PROFILES

INDIVIDUAL SURFACE BEADS

C = CONVEXITY

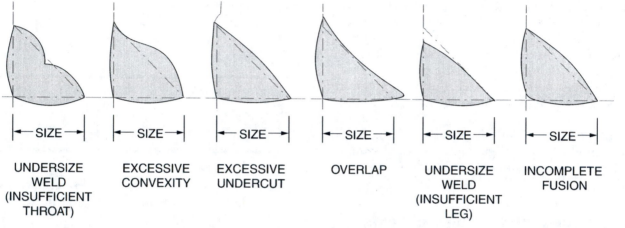

UNDERSIZE WELD (INSUFFICIENT THROAT)

EXCESSIVE CONVEXITY

EXCESSIVE UNDERCUT

OVERLAP

UNDERSIZE WELD (INSUFFICIENT LEG)

INCOMPLETE FUSION

UNACCEPTABLE FILLET WELD PROFILES

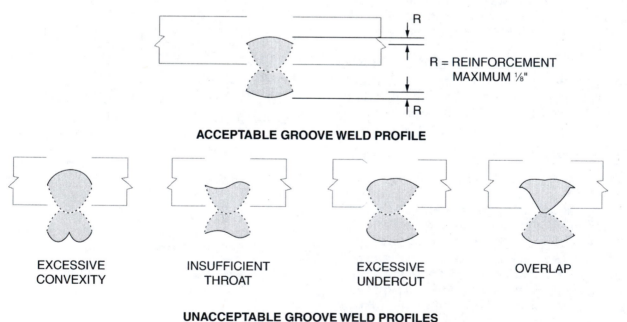

R = REINFORCEMENT MAXIMUM ⅛"

ACCEPTABLE GROOVE WELD PROFILE

EXCESSIVE CONVEXITY

INSUFFICIENT THROAT

EXCESSIVE UNDERCUT

OVERLAP

UNACCEPTABLE GROOVE WELD PROFILES

309F13.EPS

Figure 13 ◆ Acceptable and unacceptable weld profiles.

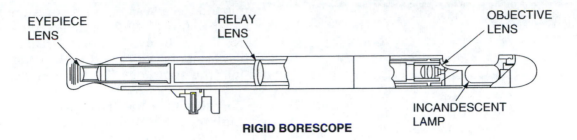

RIGID BORESCOPE

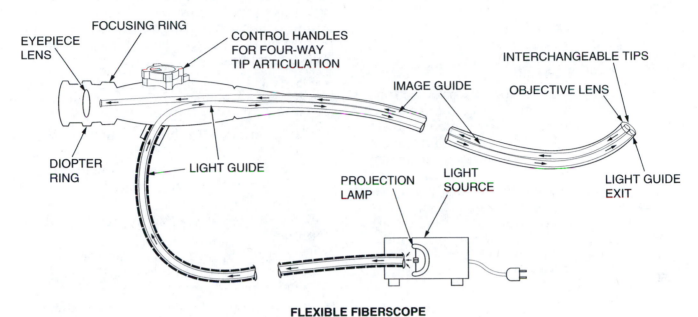

FLEXIBLE FIBERSCOPE

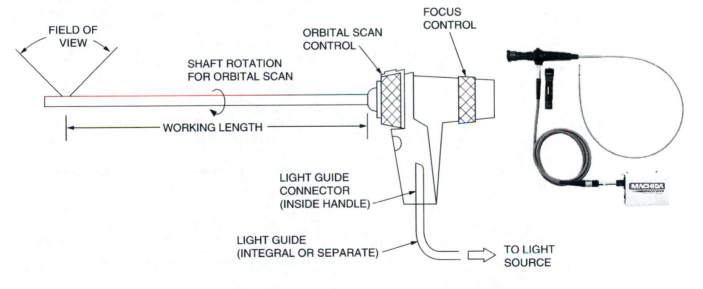

RIGID BORESCOPE WITH LIGHT

309F14.EPS

Figure 14 ◆ Borescopes.

NOTE

Perform and document any required repairs.

Step 14 Inspect the system to verify that no debris or foreign materials have fallen in.

Step 15 Replace any equipment that was removed to allow access.

2.13.2 Videoscopes with CCD Probes

Videoscopes with CCD probes transmit color or black-and-white images to a video monitor. The end of the electronic videoscope contains a CCD chip, which consists of thousands of light-sensitive elements in a pattern of rows and columns. Videoscopes with CCD probes produce images similar to borescopes, but they normally produce a higher resolution. The most significant difference is that videoscopes with CCD probes have a longer working length than fiberscopes. Videoscopes with CCD probes are operated in the same manner as flexible fiberscopes. *Figure 15* shows a typical videoscope.

3.0.0 ◆ NONDESTRUCTIVE EXAMINATION AND EVALUATION

Nondestructive examination methods are any tests that are used to test equipment and components without damaging them. Nondestructive

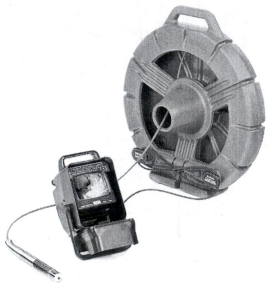

309F15.EPS

Figure 15 ◆ Videoscopes.

examination is a system of one or more types of tests that quantify defects or detect future failures. The results can be used to find the real cause of the change or abnormality and correct the problem, not just the symptom. *Table 1* lists the types of NDE.

Safety should be the first concern when preparing equipment for NDE testing. Most equipment to be tested must be cleaned using a grinder with a flapper wheel, wire brush, or 14-inch half-round bastard file. The following procedures should be performed when preparing for NDE testing activities:

- Have the safety engineer check the area to ensure adequate ventilation when using chemicals to perform NDE testing.
- Inspect grinders for broken, missing, or loose parts.
- Check the grinding wheel to ensure that the rpm rating is higher than the grinder it is to be used on.
- Barricade the work area to prevent through-traffic.
- Take special precautions when preparing to perform X-ray testing to prevent overexposure to radiation.
- Inspect the scaffolding and work platform for safe access and egress.

As a pipefitter, you will not be expected to perform NDE testing unless you are prepared to do so through specialized training and certification. Your main duty is to prepare the weld for inspection and assist the technician who is performing the test. Follow these steps to prepare a weld for inspection:

Step 1 Chip away all slag from the weld, using a chipping hammer.

CAUTION

Use light chipping strokes to remove the slag from the weld. Do not damage the weld or base metal with the hammer.

Step 2 Remove any weld spatter from the weld and base metal, using a bastard file.

CAUTION

Use light pressure on the grinder when removing weld spatter to prevent grinding away any weld metal or base metal.

Step 3 Clean any grit or scale from the weld and surrounding area thoroughly, using a grinder with a flapper wheel. The type NDE test to be performed determines how clean the weld and surrounding area must be.

3.1.0 Visual Inspection

In visual inspection, the surface of the weld and the base metal are observed for visible imperfections. Certain tools and gauges may be used during the inspection. Visual inspection is the examination method most commonly used by welders and inspectors. It is the fastest and most inexpensive method for examining a weld. However, it is limited to what can be detected by the naked eye or through a magnifying glass.

Properly done before, during, and after welding, visual inspection can detect more than 75 percent of discontinuities before they are found by more expensive and time-consuming nondestructive examination methods.

Table 1 Types of Nondestructive Examination

NDE	Type of Reading	Advantages	Limitations	Examples of Use
Liquid penetrant	Detects surface openings due to cracks, porosity, seams, or folds.	Is inexpensive, easy to use, readily portable, and sensitive to small surface flaws.	Flaw must be open to the surface. Is not useful for porous materials.	Turbine blades for surface cracks or porosity.
Magnetic particles	Detects leakage of magnetic flux caused by surface or near-surface cracks, voids, inclusions, and material or geometry changes.	Is inexpensive and sensitive to both surface and near-surface flaws.	Is limited to ferromagnetic material; surface preparation and postinspection demagnetization may be required.	Railroad wheels for cracks. Detection of weld defects.
Ultrasonics	Detects changes in acoustic impedance caused by cracks, nonbonds, inclusions, or interfaces.	Can penetrate thick materials; excellent for crack detection; can be automated.	Normally requires coupling either by contact to surface or immersion in a fluid. Orientation, detection, or interpretation of the defect can present problems.	Adhesive assemblies for bond integrity. Detection of cracks.
Radiography	Detects changes in material density from voids, inclusions, material variations, and placement of internal parts.	Can be used to inspect a wide range of materials and thicknesses; is versatile; film provides a record of inspection.	Radiation safety requires precautions, is expensive, and detection of cracks can be difficult.	Pipeline welds for penetration, inclusions, and voids. Verification of parts in assemblies.
Visual-optical	Detects surface characteristics, such as finish, scratches, cracks, or color, and strain in transparent materials.	Is often convenient and can be automated.	Can be applied only to surfaces, through surface openings, or to transparent material.	Paper, wood, or metal for surface finish and uniformity.
Eddy currents	Detects changes in electrical conductivity or magnetic permeability caused by material variations, cracks, voids, or inclusions.	Is readily automated, and the cost is moderate.	Is limited to electrically conducting materials and has limited penetration depth. Interpretation of defect signals can be difficult.	Heat exchanger tubes for wall thinning and cracks. Verification of material heat treatment.

309T01.EPS

Prior to welding, the base metal should be examined for conditions that may cause weld defects. Dimensions, including edge preparation, should also be confirmed by measurements. If problems or potential problems are found, corrections should be made before proceeding any further.

After the parts are assembled for welding, the weld joint should be visually checked for a proper root opening and any other aspects that might affect the quality of the weld. The following should be visually examined:

• Proper cleaning
• Joint preparation and dimensions
• Clearance dimensions for backing strips, rings, or consumable inserts
• Alignment and fit-up of the pieces being welded
• Welding procedures and machine settings
• Specified preheat temperature (if applicable)
• Tack weld quality

During the welding process, visual inspection is the primary method for controlling quality. Some of the aspects that should be visually examined include:

• Quality of the root pass and the succeeding weld layers
• Sequence of weld passes
• Interpass cleaning
• Root preparation prior to welding a second side
• Conformance to the applicable procedure

After the weld is completed, the weld surface should be thoroughly cleaned. A thorough visual examination may disclose weld surface defects such as cracks, shrinkage cavities, undercuts, incomplete penetration, nonfusion, overlap, and crater deficiencies before they are discovered using other nondestructive inspection methods.

An important aspect of visual examination is checking the dimensional accuracy of the weld after it is completed. Dimensional accuracy is determined by conventional measuring gauges. The purpose of using the gauges is to determine if the weld is within allowable limits as defined by the applicable codes and specifications.

Some of the more common welding gauges are:

• Undercut gauge
• Butt weld reinforcement gauge
• Fillet weld blade gauge set

3.1.1 Undercut Gauge

An undercut gauge is used to measure the amount of undercut on the base metal. Typically, codes allow for undercut to be no more than 0.010" deep when the weld is transverse to the primary stress in the part that is undercut. Several types of gauges can be used to check the amount of undercut. These gauges have a pointed end that is pushed into the undercut. The back side of the gauge indicates the measurement in either inches or millimeters. Two types of undercut gauges currently used are the bridge cam gauge and the V-WAC gauge (*Figure 16*). These gauges can be used for measuring undercut and for many other measurements.

3.1.2 Butt Weld Reinforcement Gauge

The butt weld reinforcement gauge has a sliding pointer calibrated to several different scales that are used to measure the size of a fillet weld or the reinforcement of a butt weld. To use the gauge for a fillet weld, position it as shown in *Figure 17* and slide the pointer to contact the base metal or weld metal. Be sure to read the correct scale for the measurement being taken. The other end of the gauge is used for butt welds.

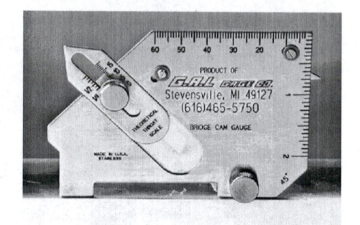

BRIDGE CAM GAUGE

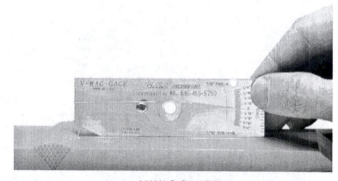

V-WAC GAUGE

309F16.EPS

Figure 16 ◆ Undercut gauges.

3.1.3 Fillet Weld Blade Gauge Set

The fillet weld blade gauge set has seven individual blade gauges for measuring convex and concave fillet welds. The individual gauges are held together by a screw secured with a knurled nut. The seven individual blade gauges can measure eleven concave and convex fillet weld sizes: ⅛", 3⁄16", ¼", 5⁄16", ⅜", 7⁄16", ½", ⅝", ¾", ⅞", and 1" and their metric equivalents. The same blade size cannot be used for measuring both concave and convex fillet welds.

To use the fillet weld blade gauge set, identify the type of fillet weld to be measured (concave or convex) and the size. Select the appropriate blade and position it. Be sure the gauge blade is flush to the base metal with the tip touching the vertical member.

Figure 18 shows an application of a fillet weld blade gauge.

3.2.0 Liquid Penetrant Inspection

Liquid penetrant inspection (PT) is a nondestructive method for locating defects that are open to the surface. It cannot detect internal defects. The technique is based on the ability of a penetrating liquid, which is usually red in color, to wet the surface opening of a discontinuity and to be drawn into it. A liquid or dry powder developer, which is usually white in color, is then applied over the metal. If the flaw is significant, red penetrant bleeds through the white developer to indicate a discontinuity or defect.

The dye, cleaner, and developer are available in spray cans for convenience. Some solvents used in the cleaners and developers contain high amounts of chlorine, a known health hazard, to make the liquids nonflammable.

WARNING!
Refer to the material safety data sheet (MSDS) for hazards associated with the liquid penetrant solvent.

The most common defects found using this process are surface cracks. Most cracks exhibit an irregular shape. The width of the bleed-out (the red dye bleeding through the white developer) is a relative measure of the depth of a crack.

Surface porosity, metallic oxides, and slag will also hold penetrant and cause bleed-out. These indications are usually more circular and have less width than a crack. *Figure 19* shows liquid penetrant materials and an example of the results of liquid penetrant inspection.

Figure 20 shows the sequence of a liquid penetration test.

Liquid penetrant inspection involves wetting the surface of a test object so that the liquid flows over that surface to form a continuous and uniform coating and migrates into cracks and surface defects. The liquid is then washed off the surface, and a developer is placed on the surface. As the developer draws the penetrant out of the defects, it is stained by the penetrant and highlights the cracks and surface irregularities to form a visual image of the flaw.

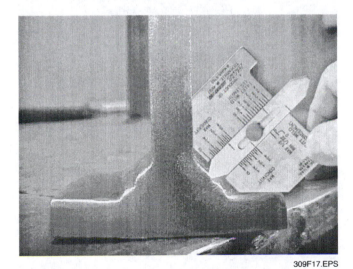

309F17.EPS

Figure 17 ◆ Automatic weld size gauge (AWS).

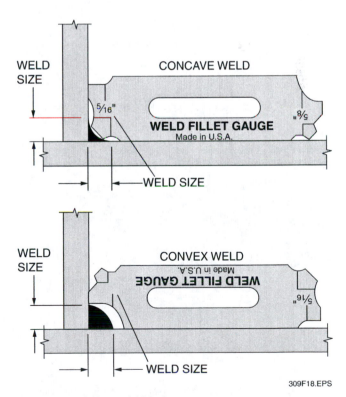

309F18.EPS

Figure 18 ◆ Fillet weld blade gauge.

With liquid dye penetrants, two types of indications are revealed: true indications and false indications. True indications are caused by the penetrant bleeding out from actual discontinuities in the metal. The standard defects indicated are stress or fatigue cracks, underbead cracks, pits, and porosities. Large stress cracks are indicated by wide lines that become apparent quickly after the developer is applied. Underbead cracks are undersurface cracks that bleed through the surface. Underbead cracks are indicated by a line of dots that appear a few minutes after the developer is applied. Porosity is indicated by dots that come to the surface almost immediately.

The major reasons for false indications are failure to apply the penetrant correctly or rough, irregular surfaces of the test metal. The procedure for liquid dye penetrant testing must be carefully performed in order to achieve the proper results.

Follow these steps to inspect welds, using a liquid dye penetrant:

Step 1 Remove any dirt, grease, scale, acids, chromates, or other contaminants from the surface of the metal, using a grinder with a flapper wheel, a wire brush, or a bastard file.

Step 2 Apply the liquid penetrant to the test area. The liquid dye penetrant can be applied by dipping the part being tested into the penetrant or by brushing or spraying the penetrant onto the surface being tested.

Step 3 Allow the liquid dye penetrant to sit for the required waiting period. The length of the required waiting period depends on the type of penetrant used and the type of metal being tested. Follow the penetrant manufacturer's recommendations.

Step 4 Remove the excess penetrant, using the recommended solution for the type of penetrant being used. Do not allow the excess penetrant to dry. If the penetrant is allowed to dry, the metal must be recleaned and the penetrant reapplied. Be careful not to remove the penetrant from the possible defects in the metal.

Step 5 Apply the developer to the test area.

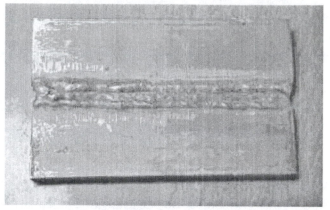

(1) WELD READY FOR PENETRANT TEST

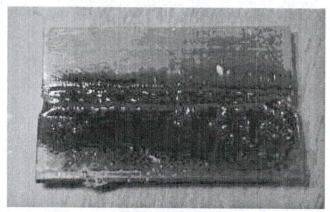

(2) PENETRANT APPLIED

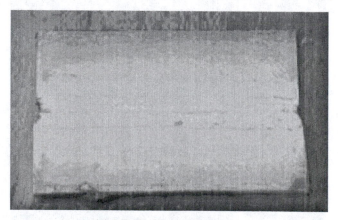

(3) DEVELOPER ON WELD

309F20.EPS

Figure 20 ◆ Liquid penetrant test.

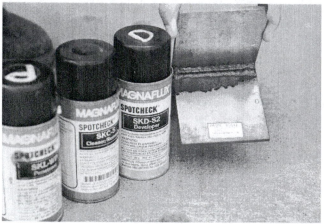

309F19.EPS

Figure 19 ◆ Liquid penetrant materials and inspection example.

NOTE

The two major types of developers are the wet developer and the dry developer. The wet developer can be either solvent-based or water-based. Both are applied by a pressurized spray. The dry developer performs the same functions as the wet developer but is applied in a powder form. The wet developer can be applied to the surface while it is still damp to the touch, but the dry developer can only be applied to a dry surface.

Step 6   Inspect the test area for defects.

Step 7   Clean the test area thoroughly, using solvent.

3.2.1 Advantages and Disadvantages of Liquid Penetrant Inspection

The advantages of liquid penetrant inspection are that it can find small defects not visible to the naked eye, it can be used on most types of metals, it is inexpensive, and it is fairly easy to use and interpret. It is most useful to examine welds that are susceptible to surface cracks. Except for visual inspection, it is perhaps the most commonly used nondestructive examination method for surface inspection of nonmagnetic parts.

The disadvantages of liquid penetrant inspection are that it takes more time to use than visual inspection, and it can only find surface defects. The presence of weld bead ripples and other irregularities can also hinder the interpretation of indications. Because chemicals are used, care must be taken when performing the inspection. When testing the rough, irregular surfaces produced by welding, the presence of nonrelevant indications may also make interpretation difficult.

3.3.0 Magnetic Particle Test

When a magnetic field is generated in and around a part made of a ferromagnetic material, such as steel or iron alloys, and the lines of magnetic flux are intersected by a defect such as a crack, magnetic poles are generated on either side of the defect. Magnetic particle inspection works only for materials that can be magnetized, such as steel and nickel alloys.

In a magnetic particle test, the sample is magnetized by an electrical current, and magnetic particles are applied over the surface. These particles can be applied either dry or in a liquid carrier such as oil or water. The flaws on or just under the surface of the sample modify the magnetic field so

that the particles outline the defect in a pattern and highlight the shape and location of the flaw.

Maximum sensitivity is obtained when the part is oriented perpendicular to the magnetic field and the strength of the field saturates the entire part. The magnetic particle equipment uses either AC or DC current and can be applied to the part with coils, probes, or flexible conductors that can wrap around a test piece.

Due to the variety of shapes of samples, the inductor for the test must come in a variety of sizes and shapes. Some sample parts, such as gears, balls, and ring-shaped parts, can be placed in the magnetic field without contact from a probe. Other samples use coils, clamps, central conductors, or other forms of direct contact with the sample. A welded joint would use a yoke to induce a magnetic field around the weld seam and show any discontinuities as extra particles scattered outside the weld seam.

Magnetic particle inspection is used to test welds for such defects as surface cracks, nonfusion, porosity, and slag inclusions. It can also be used to inspect plate edges for surface imperfections prior to welding. Defects can be detected only at or near the surface of the weld. Defects much deeper than this are not likely to be found. Certain discontinuities exhibit characteristic powder patterns that can be identified by a skilled inspector.

For magnetic particle examination, the part to be inspected must be ferromagnetic (made of steel or a steel alloy), smooth, clean, dry, and free from oil, water, and excess slag. The part is then magnetized by using an electric current to set up a magnetic field within the material. The magnetized surface is covered with a thin layer of magnetic powder. If there is a defect, the powder is held to the surface at the defect because of the powerful magnetic field. *Figure 21* shows magnetic particle examination.

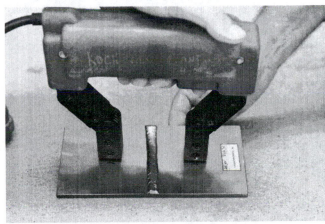

309F21.EPS

Figure 21 ◆ Magnetic particle examination.

When this examination method is used, there is normally a code or standard that governs both the method and the acceptance/rejection criteria of indications.

Follow these steps to assist a technician inspecting welds using magnetic particles:

Step 1   Inspect the magnetic yoke to ensure that the electrical cord and plug are not damaged.

Step 2   Clean the part to be inspected thoroughly to remove any dust, dirt, or oil.

Step 3   Place the yoke on the part to be inspected. *Figure 22* shows electromagnetic yokes.

Step 4   Turn on the yoke.

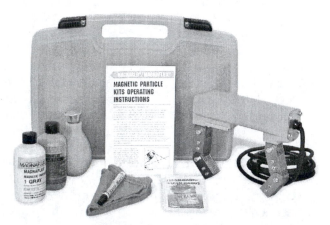

MAGNETIZING YOKE KIT

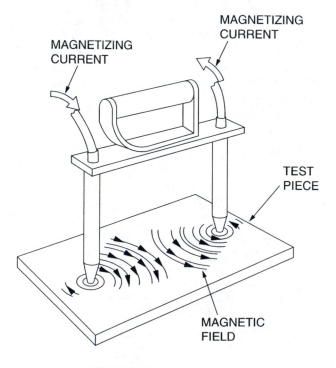

DOUBLE-PROD CONTACTS

309F22.EPS

Figure 22 ◆ Electromagnetic yokes.

Step 5   Dust the area between the probes of the yoke with dry ferromagnetic particles.

Step 6   Inspect the area between the probes for irregularities in the ferromagnetic particles, which indicate defects in the metal or weld.

Step 7   Rotate the yoke on the part 90 degrees, or one quarter turn.

Step 8   Repeat Steps 5 and 6.

Step 9   Perform Steps 5 through 8 until all areas of the inspection are covered.

Step 10   Clean the part to remove the ferromagnetic particles.

Step 11   Clean the yoke thoroughly.

Step 12   Store the yoke in its proper place.

3.4.1 Advantages and Disadvantages of Magnetic Particle Inspection

The advantages of magnetic particle inspection are that it can find small defects not visible to the naked eye, and it is faster than liquid penetrant inspection.

The disadvantages of magnetic particle inspection are that the materials must be capable of being magnetized, the inspector must be skilled in interpreting indications, rough surfaces can interfere with the results, the method requires an electrical power source, and it cannot find internal discontinuities located deep in the weld.

3.5.0 Ultrasonic Test

Ultrasonic inspection (UT) is a relatively low-cost nondestructive examination method that uses soundwave vibrations to find surface and subsurface defects in the weld material. Ultrasonic waves are passed through the material being tested and are reflected back by any density change caused by a defect. The reflected signal is shown on the screen display of the instrument.

The term ultrasonic indicates that these frequencies are above those heard by the human ear. Ultrasonic devices operate very much like depth sounders or fish finders.

Ultrasonic examination can be used to detect and locate cracks, laminations, shrinkage cavities, pores, slag inclusions, incomplete fusion, and incomplete joint penetration as well as other discontinuities in the weld. A qualified inspector can interpret the signal on a screen to determine the approximate position, depth, and size of the discontinuity. *Figure 23* shows a portable ultrasonic device.

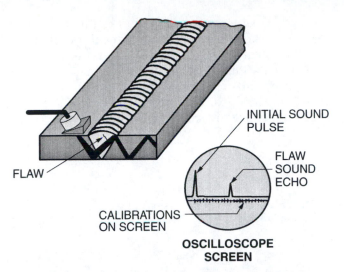

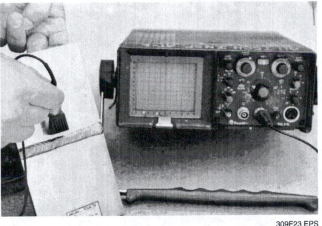

INITIAL SOUND PULSE

FLAW SOUND ECHO

CALIBRATIONS ON SCREEN

FLAW

OSCILLOSCOPE SCREEN

309F23.EPS

Figure 23 ◆ Portable ultrasonic device.

Ultrasonic instruments measure thickness in materials; amplified stethoscopes listen for turbulence in pipes to detect internal leaks and malfunctions. Other ultrasonic test instruments send focused, amplified sound waves into solid materials and compare them with the return wave to find flaws below the surface, such as corrosion, bad welds, or casting flaws. The frequency range for ultrasonic monitoring is from 20,000 Hz to 100 kHz.

In many ways, a beam of ultrasound is similar to a beam of light. Both are waves and travel in a constant velocity in a consistent medium. The velocity depends on the medium, not on the properties of the wave form. Like a beam of light, ultrasonic waves are bent, reflected from surfaces, refracted when they cross boundaries of a different surface, and diffracted at edges or around obstacles. This beam, when reflected back, creates a signature which is interpreted as an image by the ultrasonic monitor.

Ultrasonic beams can penetrate deep and with high sensitivity and accuracy. The output is readily digitized, and operation of the monitor is entirely nonhazardous. Some of the different types of ultrasonic techniques are pulse-echo, pulse-transmission, ultrasonic attenuation, continuous wave resonance, and ultrasonic spectroscopy. Ultrasonic monitors are very portable but require experienced technicians to monitor equipment. Reading an ultrasonic flaw detector requires some practice because the screen image shows the peaks of back reflection. *Figure 24* shows an ultrasonic flaw detector and an angle beam transducer.

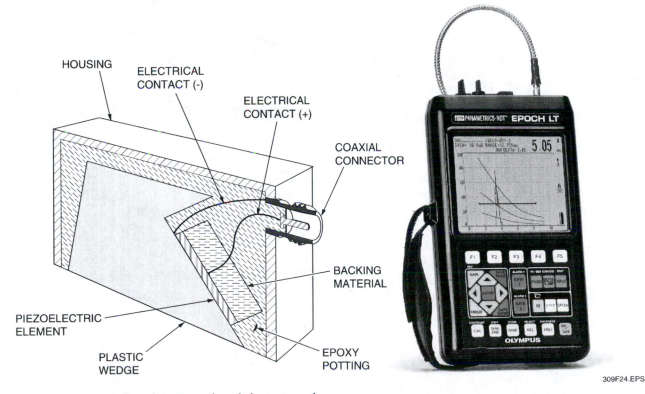

HOUSING

ELECTRICAL CONTACT (-)

ELECTRICAL CONTACT (+)

COAXIAL CONNECTOR

BACKING MATERIAL

PIEZOELECTRIC ELEMENT

PLASTIC WEDGE

EPOXY POTTING

309F24.EPS

Figure 24 ◆ Ultrasonic flaw detector and angle beam transducer.

The following steps are presented to give you a better understanding of how ultrasonic tests are performed. Follow these steps to assist a technician performing an ultrasonic test.

Step 1 Remove any slag, rust, or paint from the area around the test piece, using a wire brush or a half-round bastard file.

Step 2 Spread the approved couplant on an angle beam test block.

Step 3 Place the transducer on the test block to determine the skip distance. By using a test block, the skip distance can easily be measured. Once the skip distance is known, the region over which the transducer should be moved to scan the test piece can be determined.

Step 4 Adjust the ultrasonic test machine to the approved number of pulses per minute. Contact the supervisor or field engineer to obtain the pulses per minute if the procedure does not specify.

Step 5 Spread the approved couplant on the test plate.

Step 6 Place the transducer on the test plate.

Step 7 Move the transducer back and forth between one-half and one skip distance from the weld to be tested to define the location, depth, and size of a flaw. *Figure 25* shows moving the transducer back and forth.

Step 8 Move the transducer in an arc to detect a gas hole, slag inclusion, or surface crack. *Figure 26* shows moving the transducer in an arc.

Step 9 Mark on the test piece the location of any discontinuities.

Step 10 Document the test results.

Step 11 Repair any unacceptable discontinuities.

3.5.1 Advantages and Disadvantages of Ultrasonic Inspection

The advantages of ultrasonic inspection are that it can find defects throughout the material being examined, it can be used to check materials that cannot be radiographed, it is nonhazardous to personnel and equipment, and it can detect even small defects from one side of the material.

The disadvantages of ultrasonic inspection are that it requires a high degree of skill to properly interpret the patterns, and that very small or thin weldments are difficult to inspect.

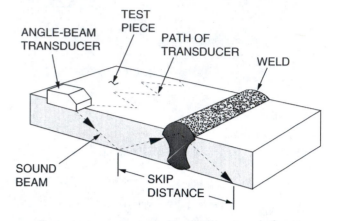

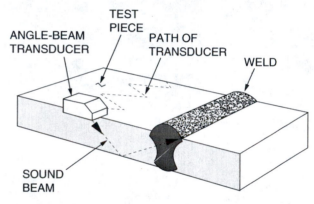

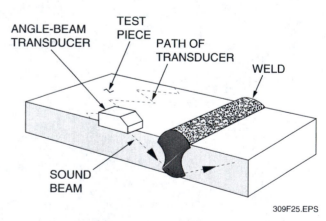

309F25.EPS

Figure 25 ◆ Moving transducer back and forth.

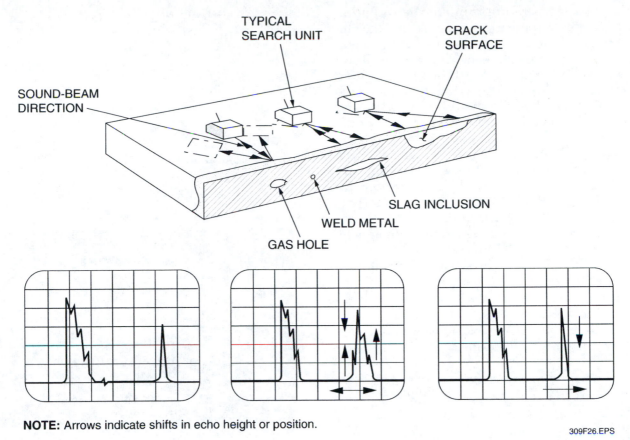

NOTE: Arrows indicate shifts in echo height or position.

Figure 26 ◆ Moving transducer in arc.

3.6.0 Radiography

Radiography is a nondestructive examination method that uses a beam of X-ray radiation that penetrates the weldment. X-rays are very sensitive and can be used to inspect the density of almost any ferrous, nonferrous, organic, or inorganic material. Some of the radiation energy is absorbed, and the intensity of the beam is reduced. The variations in the beam intensity are recorded on film. These variations are seen as differences in shading that are typical of the types and sizes of any discontinuities present.

X-rays can detect both surface and subsurface discontinuities. Because of the extremely powerful nature of gamma rays, much radiographic work is now done with radium or Cobalt 60 sources. The principal advantages of gamma ray inspection are the low cost and portability of the source, although special precautions must be taken when working with X-ray materials to protect yourself from overexposure to radiation.

When a joint is radiographed, the radiation source is placed on one side of the weld and the film on the other. The joint is then exposed to the radiation source. The radiation penetrates the metal and produces an image on the film. The film is called a radiograph and provides a permanent record of the weld quality. Radiography should only be used and interpreted by trained, qualified personnel. *Figure 27* shows a radiography examination.

Radiographic inspection can produce a visible image of weld discontinuities, both surface and subsurface, when they are different in density from the base metal and different in thickness parallel to the radiation. Surface discontinuities are better identified by visual, penetrant, or magnetic particle examination.

The X-ray method is used to detect internal defects and to check for proper alignment of assembled parts. X-rays are also used extensively to examine welds, check wall thickness, and examine the quality of casting and forging. *Figure 28* shows an X-ray, using gamma rays.

WARNING!

Only certified technicians are permitted to perform x-ray activities. **Never** cross a radiation barrier.

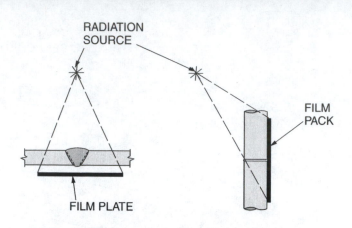

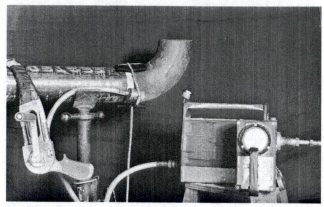

309F27.EPS

Figure 27 ◆ Radiography examination.

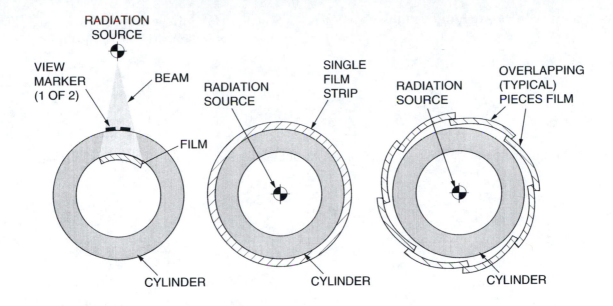

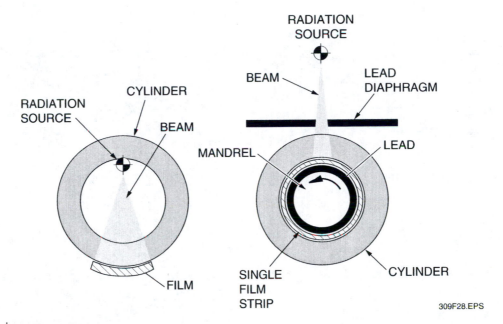

Figure 28 ◆ X-ray, using gamma rays.

The following procedure is presented to give you an understanding of how to X-ray welds. Follow these steps to assist a technician X-raying welds:

Step 1 Obtain the x-ray procedure, and review it to ensure that the correct weld identification is listed. The X-ray procedure must be signed and have a detailed explanation of the surface area to be tested.

Step 2 Clean the weld and surface area, using a wire brush or half-round bastard file to remove any scale, rust, paint, arc strikes, or gouges.

Step 3 Mark off the area on which to place the X-ray film, using a permanent-type marker.

Refer to the procedure to ensure that the X-ray film is placed in the correct position.

Step 4 Rope off the area using radiation tape and appropriate signs. The area must be roped off on the sides, above, and below the target area.

Step 5 Place the approved radiation source at the test area. The radiation source must be located at the approved distance from the weld to be tested.

Step 6 Place the target on the back side of the weld to be tested.

Step 7 Shoot the weld, using the radiation source.

Step 8   Remove the x-ray film from the weld.

Step 9   Have the x-ray film processed. *Figure 29* shows weld discontinuities.

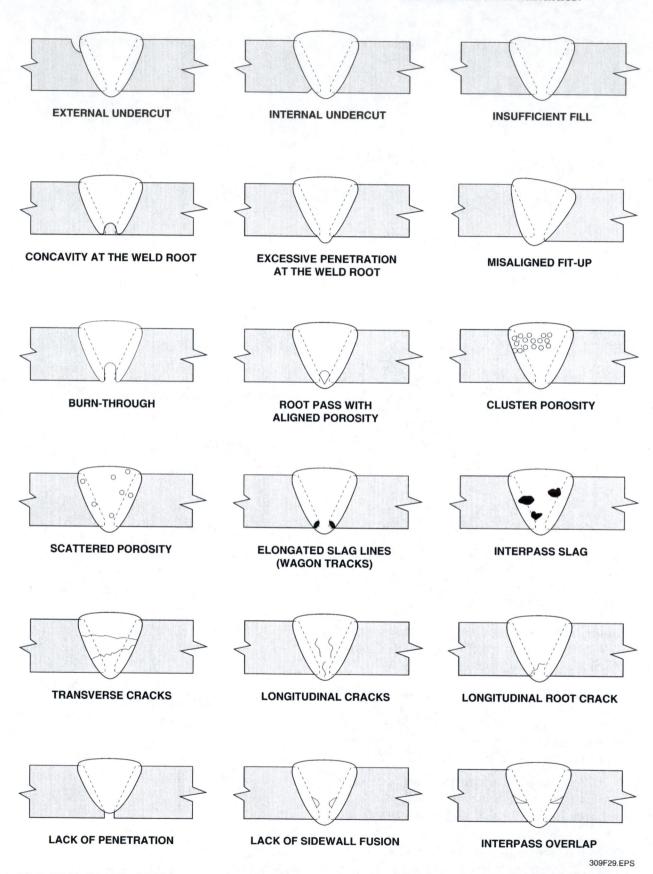

EXTERNAL UNDERCUT

INTERNAL UNDERCUT

INSUFFICIENT FILL

CONCAVITY AT THE WELD ROOT

EXCESSIVE PENETRATION AT THE WELD ROOT

MISALIGNED FIT-UP

BURN-THROUGH

ROOT PASS WITH ALIGNED POROSITY

CLUSTER POROSITY

SCATTERED POROSITY

ELONGATED SLAG LINES (WAGON TRACKS)

INTERPASS SLAG

TRANSVERSE CRACKS

LONGITUDINAL CRACKS

LONGITUDINAL ROOT CRACK

LACK OF PENETRATION

LACK OF SIDEWALL FUSION

INTERPASS OVERLAP

309F29.EPS

Figure 29 ◆ Weld discontinuities.

3.6.1 Reading Surface Discontinuities

The following are types of surface discontinuities that can be detected by X-rays:

- *External undercut* – External undercut is a gouging out of the piece to be welded, alongside the edge of the top or external surface of the weld. The film shows an irregular, darker density along the edge of the weld image.

- *Internal undercut* – Internal, or root, undercut is a gouging out of the piece to be welded, alongside the edge of the bottom or internal surface of the weld. The film shows an irregular, darker density near the center of the width of the weld and along the edge of the root pass.

- *Insufficient fill* – Longitudinal grooves indicate insufficient fill. The film shows a weld density darker than the density of the pieces being welded and extending across the full width of the weld image.

- **Concavity** *at the weld root* – The film shows an elongated, irregular, darker density with fuzzy edges in the center of the width of the weld image.

- *Excessive penetration at the weld root* – The film shows a lighter density in the center of the width of the weld image with a straight, darker density in the center of the width of the weld image, either extended along the weld or in isolated, circular drops.

- *Weld with offset or mismatch with lack of penetration of the pieces to be welded with insufficient filling of the bottom of the weld* – The film shows an abrupt density change across the width of the weld image with a straight, darker density line at the center of the width of the weld image along the edge of the density change.

- *Misaligned fit-up* – Weld with offset or a mismatch of the pieces to be welded. The film shows an abrupt change in film density across the width of the weld image.

- *Burn-through* – A burn-through is a severe depression or a crater hole at the bottom of the weld. The film shows a localized, darker density with fuzzy edges in the center of the width of the weld image, which could appear wider than the root pass image.

3.6.2 Reading Subsurface Discontinuities

The following are types of subsurface discontinuities that can be detected by X-rays:

- *Root pass with aligned porosity* – Involves rounded and elongated voids in the bottom of the weld along the center line of the weld. The film shows rounded and elongated darker density spots which may be connected in a straight line in the center of the weld image.

- *Cluster porosity* – Involves rounded or slightly elongated voids grouped together. The film shows rounded or slightly elongated darker density spots in clusters randomly spaced.

- *Scattered porosity* – Involves rounded voids random in size and location. The film shows rounded spots of darker densities random in size and location.

- *Elongated slag lines (wagon tracks)* – Impurities that solidify on the surface after welding and were not removed between passes. The film shows elongated, parallel, or single darker density lines irregular in width and slightly winding in the lengthwise direction.

- *Interpass slag inclusions* – Usually nonmetallic impurities that solidified on the weld surface and were not removed between weld passes. The film shows an irregularly shaped darker density spot usually slightly elongated and randomly spaced.

- *Transverse cracks* – A fracture in the weld metal running across the weld. The film shows feathery, twisting lines of darker density running across the width of the weld image.

- *Longitudinal cracks* – A fracture in the weld metal running lengthwise in the welding direction. The film shows feathery, twisting lines of darker density running lengthwise along the weld at any location in the width of the weld image.

- *Longitudinal root cracks* – A fracture in the weld metal at the edge of the root pass. The film shows feathery, twisting lines of darker density along the edge of the image of the root pass. The twisting feature helps to distinguish the root crack from incomplete **root penetration**.

- *Lack of penetration* – Occurs when the edges of the pieces have not been welded together, usually at the bottom of a single V-groove weld. The film shows a darker density band with very straight, parallel edges in the center of the weld image.

- *Lack of sidewall fusion* – Involves elongated voids between the weld beads and the joint surfaces to be welded. The film shows an elongated, parallel, or single darker density lines, sometimes with darker density spots along the lack of fusion lines, which are straight in the lengthwise direction of the weld, not winding like elongated slag lines. One edge of the lack of fusion line may be very straight as with the lack of penetration. The lack of sidewall fusion images will not be in the center of the weld image.

- *Interpass overlap* – Involves lack of fusion areas along the top surface and edge of lower passes. The film shows small spots of darker densities, some with slightly elongated tails aligned in the welding direction and not in the center of the weld image.

- *Tungsten inclusions* – Random bits of tungsten fused into but not melted into the weld metal. The film shows irregularly shaped, low-density spots randomly located in the weld image

3.6.3 Advantages and Disadvantages of Radiographic Inspection

The advantages of radiographic inspection are that the film gives a permanent record of the weld quality, the entire thickness can be examined, and it can be used on all types of metals.

The disadvantages of radiographic inspection are that it is a slow and expensive method for inspecting welds, some joints are inaccessible to radiography, and radiation of any type is very hazardous to humans. Cracks can frequently be missed if they are very small or are not aligned with the radiation beam.

3.7.0 Electromagnetic (Eddy Current) Inspection

Like magnetic particle testing, electromagnetic, or eddy current, inspection (ET) uses electromagnetic energy to detect defects in the joint. An alternating current (AC) coil, which produces a magnetic field, is placed on or around the part being tested. After being calibrated, the coil is moved over the part to be inspected. The coil produces a current in the metal through induction. The induced current is called an eddy current. If a discontinuity is present in the test part, it will interrupt the flow of the eddy currents. This change can be observed on the oscilloscope.

Eddy currents only detect discontinuities near the surface of the part. This method is suitable for both ferrous and nonferrous materials and is used in testing welded tubing and pipe. It can determine the physical characteristics of a material and the wall thickness in tubing. It can check for porosity, pinholes, slag inclusions, internal and external cracks, and nonfusion.

3.7.1 Advantage and Disadvantage of Eddy Current Inspection

The advantage of eddy current inspection is that it can detect surface and near-surface weld defects. It is particularly useful in inspecting circular parts like pipes and tubing.

The disadvantage of eddy current inspection is that eddy currents decrease with depth, so defects farther from the surface may go undetected. The accuracy of the examination depends in large part on the calibration of the instrument and the qualification of the inspector.

3.8.0 Leak Testing

Leak testing is used to determine the ability of a pipe or vessel to contain a gas or liquid under pressure. Testing methods vary depending on the application of the weldment. In some cases, the vessel is pressurized and tested by immersing it in water or applying a soap bubble solution to the weld (*Figure 30*). A leak is indicated by the presence of bubbles. An open tank can be tested using water that contains fluorescein, which can be detected by ultraviolet light.

A method called the vacuum box test is used to test a vessel where only one side of the weld is accessible. The base of a storage tank is an example. The vacuum box is a transparent box with a soft rubber seal. A vacuum pump is used to extract all the air from the box. In the helium spectrometer leak test, helium is used as a tracer gas inside the vacuum box. Because of the small size of helium atoms, they can pass through an opening so small that it might not be detectable by other test methods. Sensitive instruments are used to detect the presence of helium.

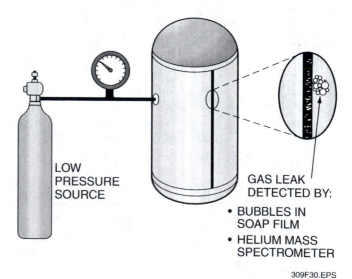

LOW PRESSURE SOURCE

GAS LEAK DETECTED BY:
- BUBBLES IN SOAP FILM
- HELIUM MASS SPECTROMETER

309F30.EPS

Figure 30 ◆ Leak testing.

3.9.0 Conductivity Tests

Pipefitters are sometimes called upon to perform conductivity tests, or holiday tests, on an underground piping system. Conductivity testing ensures that there is no unprotected piping that is not covered with coating. Conductivity tests use amperage to test the conductivity of a material. The holiday testers have variable amperage ratings, and the amperage rating used must comply with the specifications of the piping system. *Figure 31* shows a holiday tester.

Follow these steps to perform a conductivity test on a piping system.

Step 1   Check out a holiday tester from the tool room. If permits are required, procure them first.

Step 2   Press the battery test push button on the holiday tester to ensure that the battery is fully charged.

WARNING!
The pipe that is being tested must be grounded since the holiday tester can produce a spark.

Step 3   Drop the copper ground wire onto the ground near the piping system to be tested.

WARNING!
Do not touch the copper ground wire once the holiday tester has been turned on.

Step 4   Start up the holiday tester.

Step 5   Brush the wand past the ground wire to test the holiday tester. Brushing the wand past the ground wire causes the wand to produce an arc-striking sound and causes sparks to jump from the wand to the ground wire. This indicates that the holiday tester is operating properly.

Step 6   Wrap the wand spring around the pipe, and brush the wand along the entire length of the piping system, paying special attention to the hand-wrapped areas around the welds.

NOTE
On large-diameter pipe, two pipefitters are normally needed, one on each side of the pipe. Special attention should be given to tees and branches to avoid missing any bare spots. Tees and branches must be tested using a wand with a brass wool pad on the end. These devices can also be used to test the insides of glass-lined vessels and flanges.

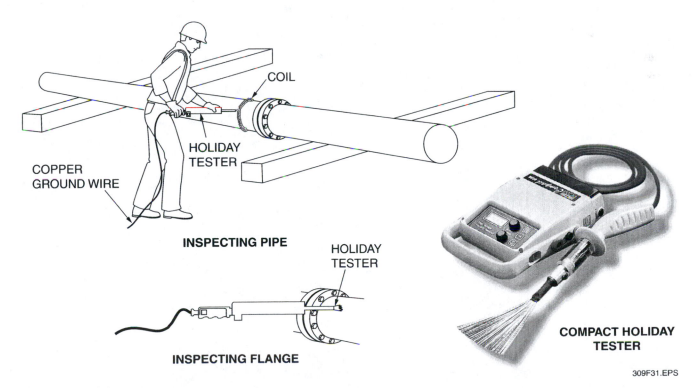

Figure 31 ◆ Holiday tester.

Step 7 Patch any bare spots, using an approved material.

Step 8 Return the holiday tester to the tool room after the test is complete.

4.0.0 ◆ PRETEST REQUIREMENTS

Before testing any component, spool, or piping system, there are checks, verifications, and inspections that must be made. The pretest requirements include identifying **test boundaries**, installing pressure gauges, preparing the system for testing, and cleaning the system. The requirements of your project will determine your actual duties. The procedures in this module are for training purposes only and may not correspond with the procedures used on your project.

4.1.0 Identifying Test Boundaries

One of the most important pretest requirements is defining the test boundaries. The test boundaries are defined on a test report that has been prepared by the appropriate engineering personnel. The test report indicates the associated drawings, codes, classification of the test items, type of test to be performed, testing medium, required time, and test pressure. *Figure 32* shows a sample test report.

By using the drawings identified in the test report, you can determine a number of details that are necessary to perform a test, including the following:

HYDROSTATIC AND PNEUMATIC TEST REPORT

ASSOCIATED TRAVELERS

REFERENCE DRAWINGS

PROCEDURE NO. _____ REV. _____

TEST BOUNDARIES _____

CODE _____

CLASS _____

DESIGN PRESSURE _____ psig

REQUIRED TEST PARAMETERS

TIME _____ psig:MIN. _____ MAX. _____

TEST TYPE

☐ HYDROSTATIC

☐ PNEUMATIC

☐ OTHER _____

TEST MEDIA

☐ DEMIN. WATER   ☐ FRESH WATER   ☐ OTHER _____

ENGINEER _____ DATE _____

PREREQUISITES VERIF. Q.C. ENGINEER _____ DATE _____

TIME TEST STARTED _____ GAGE READING _____ TIME AT PRESSURE _____

TIME TEST COMPLETED _____ GAGE READING _____ GAGE NO. USED _____

GAGE RANGE _____ | CALIBRATION DUE DATE _____

TEMP. OF TEST MEDIA (as required)

☐ VERIFIED SATISFACTORY

☐ MEASURED _____ °F

TEST ACCEPTED

☐ YES   IF NO, REASON _____

☐ NO   _____

DISPOSITION ACTION _____

TEST CONDUCTED BY _____ DATE _____ | Q.C. INSPECTOR _____ DATE _____

REMARKS _____

AUTHORIZED INSPECTOR _____ DATE _____

309F32.EPS

Figure 32 ◆ Test report.

- System or component material type
- Nominal pipe size of system
- System or component location
- System **design pressure**
- Welds and weld locations
- Valves, flanges, and instrumentation devices

- Temporary attachments
- Hanger and support locations
- High and low points of the system
- Equipment locations
- Classification of the system

4.2.0 Installing Pressure Gauges

Pressure gauges must be installed on the system or component to be tested according to project specifications. The pressure gauge should be connected directly to the high point of the system or component. If this gauge is not visible to the operator, an additional gauge should be placed so that it is visible at all times during the test. Pressure gauges used during testing must be indicating pressure gauges, either digital or analog, with dials that have a graduated scale range of approximately two times the intended maximum test pressure. The gauge must have a scale range of 50 percent more than the intended maximum test pressure. When performing **hydrostatic testing**, an **antisyphon** or **snubber** must be used between the pressure gauge and the system being tested to prevent fluctuation of the test gauge. If an antisyphon or snubber is not used, the positive displacement action of the test pump will cause the gauge needle to move erratically, and the gauge will give false readings. Antisyphons and snubbers smooth and dampen the pressure impulses, thereby prolonging the life and accuracy of the pressure gauges. *Figure 33* shows typical pressure gauges.

Each pressure gauge must have a current calibration sticker verifying that the gauge has been calibrated against a standard dead weight tester or a calibrated master gauge. As a minimum requirement, each gauge must be calibrated every 6 months. On some projects, test gauges are calibrated just before and immediately after pressure testing.

4.3.0 Preparing System For Testing

Before testing a piping system, you must prepare the system. Job-specific pretest inspections are listed on the job specifications and on the piping drawings.

Several areas and components of the piping system must be inspected, including the base material, alignment, joints, valves, instruments, and supports. If any of the pretest checks do not pass visual inspection, tag the part and write down why that part failed the inspection. Corrections and repairs must be made before testing the system. The field inspections may be written up as an inspection punchlist. *Figure 34* shows a sample field inspection punchlist.

This section provides a general checklist of the items that must be inspected. On your job site, you may or may not be responsible for many of the items in the following list. Always check with your immediate supervisor to determine your specific responsibilities. Most companies perform testing when going from shutdown status to operational status. Turnarounds from summer to winter production may trigger the same process also. The process is the same for either circumstance as for initial construction of a pipe system. The system must be tested thoroughly to find any flaws before the system goes into production.

DIFFERENTIAL PRESSURE GAUGE

DIGITAL PRESSURE GAUGE

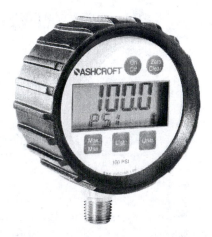

BATTERY POWERED DIGITAL PRESSURE GAUGE

309F33.EPS

Figure 33 ◆ Pressure gauges.

Field Installation Punchlist
Aboveground Piping Before
Hydrotest ☐
Pneumatic ☐

Contract No.:	Page: of
Owner:	
Location:	

Contractor:	Area:	Line No.:

Reference Drawing:	Description:

Items to Check	Installation O.K. (signature & date)	Items to Check	Installation O.K. (signature & date)
Line size		Stem orientation correct	
Material correct		Valves in open position	
Flange rating correct		Field support installed	
Welding complete		Sufficient supports	
Stress relieving complete		Anchors installed	
Installation straight and plumb		Guides installed	
Line slopes per drawing		Proper shoes installed and welded	
Branches located correctly		Enough clearance for insulation	
Branches reinforced		Spring hangers installed	
Weep holes in reinforcing pad		Pins not removed before hydrotest	
High-point vent in		Radiography complete	
Low-point drains in		Hardness checked after stress relieving	
Reducer located correctly		Joints and welds bare and not painted	
Reducer type correct		PSV's isolated by blinds	
Blinds installed		Flow restrictions removed	
Cold spring per drawing		Category M sensitive leak test	
Clearance for expansion		Hydrotest	
Gasket type and material correct			
Bolt size and material correct			
Valve No. correct			
Bypass installed			

Remarks: _____
_____
_____
_____

Contractor's Representative	Owner's Inspector	
_____	_____	_____

309F34.EPS

Figure 34 ◆ Field inspection punchlist.

One critical position in the testing process is the checkout coordinator. The checkout coordinator supports testing procedures. The coordinator controls tagout access, providing a central location for the information on lockouts. While testing a particular section, that section is locked out and tagged out by the group controlling the testing procedure. At completion, the coordinator controls turnover of the system from group lockout and tagging to the individual normally responsible for that part of the system. This makes it possible to control the process in a way that minimizes the chance of accidental loss of lockout.

Perform a pretest field inspection of a piping system. Ensure that:

- The piping system matches the ISOs, orthographic drawings, and P&IDs.
- The piping system components match the specifications list.
- All joints fit properly and are not angled.
- There is no damage or side pressure in the pipe.
- The system is properly sloped if required.
- Threaded joints are tightened securely.
- All flange bolts are installed and tightened securely.
- Ensure that the proper gaskets are installed.
- All expansion joints are secured or protected.
- Pressure-relief valves are blinded, gagged, or removed and that the lines are capped.
- Flow valves that will not be tested are locked in the open position or have been removed.
- There is an air vent at the high point of the system.
- System equipment not to be tested is isolated, locked out, and tagged out to prevent accidental involvement.
- One temporary pressure-relief valve is set at 110 percent of the test pressure if required by job specifications.
- The low point of the system is set up for draining after the test is complete.

- All instruments and instrument valves are removed.
- A pressure gauge with a range of one and one-half times the test pressure is installed.
- All hangers and supports are in the proper location.
- Temporary supports are installed if necessary.
- All valves and slip blinds are properly tagged.
- The system is capable of supporting the weight of the **test medium**.
- All NDE test requirements have been satisfied.
- Prealignment of all rotating equipment has been performed.
- The draining of the test medium complies with EPA standards and that all proper precautions are taken.
- All documentation is complete and properly signed.

4.4.0 Cleaning System

After the line or system has passed the field inspection, it must be cleaned thoroughly before testing. The purpose of cleaning a piping system is to remove all foreign material that may have accumulated in the system during installation, to detect any leaks in the system, and to detect any problems with the hangers and supports within the system. Before cleaning any system, all in-line devices, such as soft-seated control valves, magnetic flowmeters, probes, and orifice plates, must be removed from the system and replaced with temporary spools.

Cleaning a piping system can be done by pigging the system, by hydroblasting, or by chemical cleaning. Water is also normally used to clean pipelines, but certain applications might require other methods. For example, air systems are usually blown clean with air, and oil piping systems are flushed with oil.

4.4.1 Pigging

Pigging a piping system refers to sending an object, usually cylindrical or spherical, through a pipeline. The **pig** is either pulled with a cord, or pushed by air, gas, or pressurized liquid. Pigs are used to test the shape of the pipe (geometry pigs) for buckling or bending, to clean the pipeline, to test for obstructions (inline inspection pigs), or to separate two media in the pipe (batching pigs) in cases where one medium is replacing another.

The most commonly used pig is a polly pig, a pig made of solid polyurethane. The pig may be covered with abrasives, brushes, or metal pins to clean material from the pipe walls. Pigs are always slightly larger in diameter than the pipeline the pig is to run in, to ensure that the pig seals well in the line and is forced through the line by the launching pressure. Polly pigs are most commonly bullet-shaped *(Figure 35)* but sometimes spherical.

Polly pigs are either coated with harder polyurethane, or soft all the way to the surface. Similarly shaped pigs are solid cast of various materials, including neoprene, nitrile, or polyurethane, either as sealing or batching pigs, or as cleaning pigs, with exterior abrasive bands.

The other common type of pig is the steel mandrel pig *(Figure 36)*. This type is built around a steel mandrel, frequently articulated, with either discs or cups attached to maintain the seal with the pipe walls, and with various cleaning or other functional apparatus mounted around the mandrel. The steel mandrel type is both more expensive and more adaptable than the polly pig, as well as being more enduring. The discs or cups, as well as the tools attached, are usually replaceable.

One type of spherical pig is inflatable, usually with liquid. Spherical pigs are used in contexts where there are large numbers of turns and changes of direction to be navigated. The sphere is made of polyurethane, Viton®, neoprene, or nitrile, and has valves that allow it to be filled. The inflatable sphere should not be inflated with air, because air's compressibility may cause problems.

If a pig becomes immobilized by an obstruction, a detector pig can be sent to locate the obstruction and pig, allowing the workers to find and clear the obstruction. Some mandrel pigs are capable of producing a strong stream of water on demand, to flush such an obstacle out of the way.

Pig launchers may either be connected directly in line with the pipeline to be cleaned *(Figure 37)*, or may be connected to a wye line and launched into a segment of line. Pig receivers also may be in line or connected to a branch, with the pipeline cut off by a valve past the wye to the receiver.

309F35.EPS

Figure 35 ◆ Polly pigs.

309F36.EPS

Figure 36 ◆ Steel mandrel pigs.

4.4.2 Hydroblasting

It is becoming more and more common for pipelines to be cleaned and cleared with hydro-blasting apparatus. This is a technique involving pressures as high as 36,000 pounds per square inch (psi). The water is applied through a very small nozzle, for short runs with a hand-fed flex lance *(Figure 38)* and for longer runs with a line mole *(Figure 39)*—a set of nozzles on a fitting arranged to push themselves down the pipe while cutting the dirt and obstructions out of the way. This is an extremely powerful and efficient way of cleaning pipe, but a dangerous one. The same high-pressure water can injure or kill people as easily as it clears pipe obstructions. Do not attempt this unless you have been trained and certified to use hydroblasting equipment.

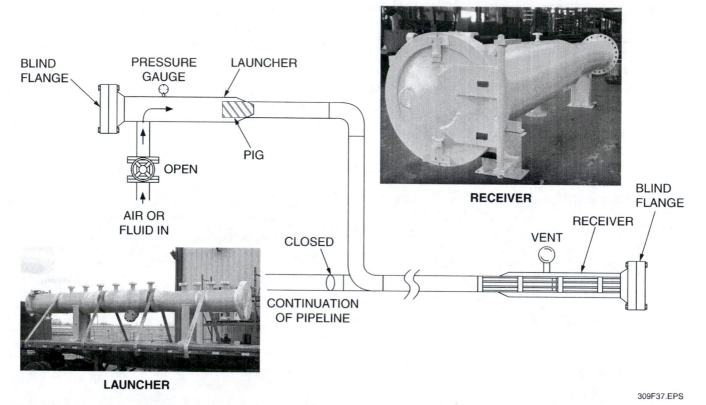

309F37.EPS

Figure 37 ◆ Pigging components.

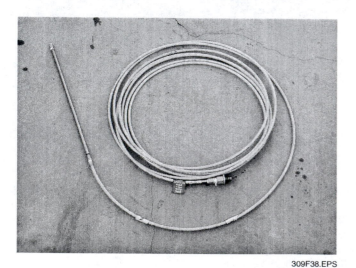

309F38.EPS

Figure 38 ◆ Hand-fed flex lance.

309F39.EPS

Figure 39 ◆ Line mole.

4.4.3 Chemical Cleaning

Depending on the piping codes and job specifications, it may be necessary to add specified chemicals to the system to clean it during the flushing procedure. Adding chemicals to clean the system is also known as pickle and passivate. Chemical cleaning is a process used to clean many piping systems before they are tested and put into service. Chemical cleaning involves circulating a strong detergent and heated water solution through the system for a specified length of time as recommended by the chemical manufacturer. The chemical manufacturer also provides written procedures for the types of chemicals to be used.

After the detergent and water solution is drained, a mild acid solution is circulated through the system, and lab samples of the solution are taken every 15 minutes. When the desired pH level is obtained, the acid wash is neutralized with soda ash and is then drained and flushed from the system. The system is then filled with demineralized water, oil, nitrogen, or whatever medium the specifications call for.

Chemical cleaning can be extremely hazardous; therefore, proper eye protection, rubber gloves, and protective clothing should be worn and the system must be tagged during all chemical cleanings. After the chemical cleaning is complete, the chemicals must be disposed of according to EPA standards and regulations.

5.0.0 ◆ PERFORMING SERVICE AND FLOW TESTS

Service and flow tests are the simplest tests to perform on a system. These tests are usually performed on utility or low-pressure piping systems, such as water, air, or nitrogen systems. When performing these tests on a system that contains a liquid, close the last valve in the system and open the high-point vent valve. Turn on the service, and allow the air to bleed from the system before performing the test. Close the high-point vent valve after all of the air is removed from the system. To perform a service test, close off the last valve in the system; turn on the service to that system at normal operating pressure, and check for leaks with the system closed. To perform a flow test, open the last valve in the system and check for leaks with the system in operation. The test medium used in service and flow tests is the medium that will be used in the system under normal operating conditions.

6.0.0 ◆ HEAD PRESSURE TESTING

Head pressure tests are water tests performed on low-pressure or gravity-flow systems, including sanitary systems, reinforced concrete pipe storm drains, chemical sewers, process systems, and gravity drain systems. Head pressure is defined as the pressure generated by the total weight of a column of water. The head pressure of water columns is the same, regardless of the volume of the column, as long as the columns are the same height. Head pressure is always 0.4335 psi per foot of water and is measured at the lowest point of the water column. Another formula that is used to determine the amount of head pressure is 1 psi equals 27.72 inches of water column. Therefore, if the test report requires 5 psi of pressure, multiply 5 times 27.72, and you will see that you must have 138.6 inches, or 11 feet 5½ inches of water column. The pressure applied by the water in the water column applies equal pressure in all directions in the piping system. *Figure 40* shows head pressure.

In order to have enough water pressure for the test, you normally need approximately 10 feet of vertical pipe above the highest joint being tested. The test report gives the exact pressure that is required for the test. The main vent stack in most drain pipe installations is high enough to provide the necessary pressure when filled with water. To test the head pressure of a horizontal pipe with no vertical vents, you may have to temporarily install 10 feet of vertical pipe. This temporary pipe is known as a standpipe. The standpipe is the pipe where you monitor the water level. The standpipe can be joined to the system with a rubber compression joint or a no-hub joint. If the standpipe has to be erected in an open area, erect a scaffold and tie off the standpipe to the scaffold. After erecting the standpipe, isolate the components to be tested.

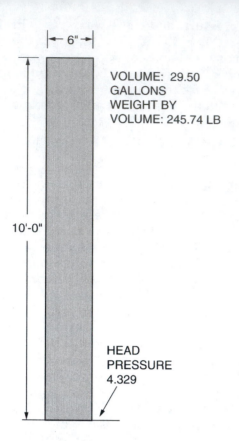

VOLUME: 29.50
GALLONS
WEIGHT BY
VOLUME: 245.74 LB

6"

10'-0"

HEAD
PRESSURE
4.329

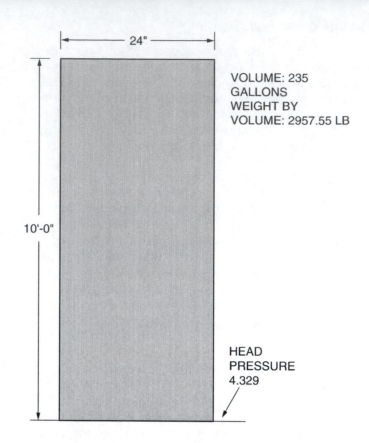

VOLUME: 235
GALLONS
WEIGHT BY
VOLUME: 2957.55 LB

24"

10'-0"

HEAD
PRESSURE
4.329

HEAD PRESSURE (psi) = HEIGHT (ft) × 0.4329

HEAD PRESSURE (psi) = 10 ft × 0.4329

4.329 psi = 10 ft OF WATER COLUMN

1 psi HEAD PRESSURE = 27.72 in. WATER COLUMN

4.329 psi = 4.329 × 27.72 in.

4.329 psi = 119.9998 in. WATER COLUMN

4.329 psi = 10 ft. OF WATER COLUMN

309F40.EPS

Figure 40 ◆ Head pressure.

6.1.0 Isolating Components to be Tested

Test plugs can be used to seal off openings and other sections of the pipeline that are not being tested. When ordering test plugs, you must specify the inside diameter of the pipe to be tested to ensure that the test plug will be the proper size. Test plugs should always be properly secured in the pipeline so that they do not blow out of the pipe and cause injury. Most test plugs are either pneumatic or mechanical.

6.1.1 Pneumatic Test Plugs

Pneumatic test plugs, commonly referred to as plumber's plugs, are hollow neoprene plugs that are inserted into the pipe and filled with air, causing the plug to expand and fit tightly inside the pipe. They are used only for low-pressure systems and are normally inflated using a hand pump. Pneumatic test plugs can be used for ABS, cast iron, carbon steel, clay, concrete, copper, or PVC pipe. The test plug manufacturer provides maximum back pressure ratings for all plugs. *Figure 41* shows applications of pneumatic test plugs.

Most pneumatic test plugs have a retainer ring that is larger than the intended pipe diameter to prevent the plug from being pulled down the pipe when the air is released from the plug. A water removal plate can be attached to the test plug to prevent water from flooding out the opening when the plug is removed. *Figure 42* shows a pneumatic test plug with a water removal plate.

6.1.2 Mechanical Test Plugs

A mechanical test plug consists of a large wing nut, a steel hub plate, a neoprene sealing cup, and a coned pressure plate. As the wing nut is turned

| CLOSET BENDS | STRAIGHT TEE | SANITARY TEE | FLOOR DRAIN |

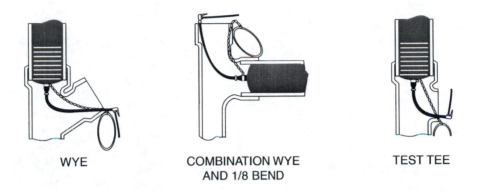

| WYE | COMBINATION WYE AND 1/8 BEND | TEST TEE |

309F41.EPS

Figure 41 ◆ Applications of pneumatic test plugs.

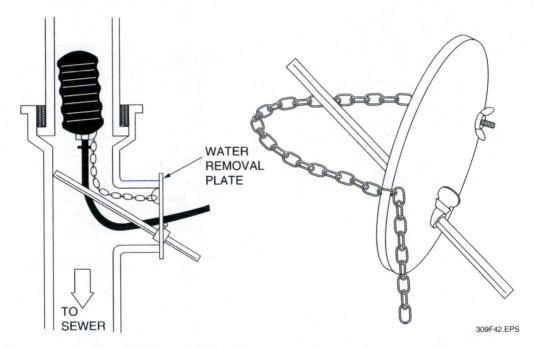

WATER REMOVAL PLATE

TO SEWER

309F42.EPS

Figure 42 ◆ Pneumatic test plug with water removal plate.

clockwise, the steel hub plate and the coned pressure plate are drawn together, causing the rubber sealing cup to expand and form a tight seal inside the pipe. Many mechanical plugs have a pipe with external threads through the plug that the wing nut is tightened on. This pipe allows the system being tested to be filled through the test plug.

Always check with the manufacturer of the test plug to find out the installation instructions and pressure ratings of the plugs. Most manufacturers give pressure ratings in air psig, water psig, and maximum feet of water. *Figure 43* shows mechanical test plugs.

6.2.0 Performing Head Pressure Test

Before performing a head pressure test, check the specifications to determine if any allowable leakage of the pipeline is permitted depending on the type of piping being tested. There is normally an allowable leakage with concrete-lined piping due

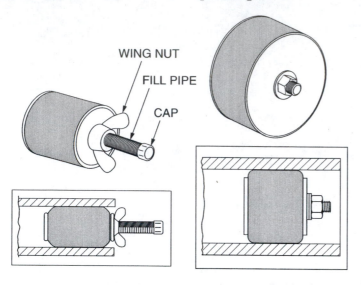

MECHANICAL TEST PLUGS

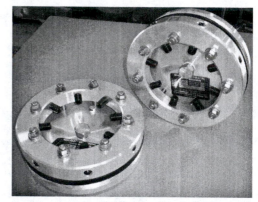

HIGH PRESSURE

MEDIUM PRESSURE

309F43.EPS

Figure 43 ◆ Mechanical test plugs.

to the concrete absorbing the liquid in the pipeline. Follow these steps to perform a head pressure test.

Step 1   Ensure that all pretest requirements have been completed and that the system is ready to be tested. Erect a standpipe, if necessary.

Step 2   Plug all openings except the highest vent opening or the standpipe to isolate the section or components to be tested. When testing a drain pipe in a high-rise structure, only test 5 floors or 50 vertical feet at a time. If you test more than this, the pressure on the lower pipes will be too great.

Step 3   Fill the standpipe to the top with water or another medium as specified by the test report.

Step 4   Allow the water to stay in the system for as long as the test report specifies.

Step 5   Check the water level in the standpipe. If the water level has not dropped, there are no leaks in the system. If the water level has dropped, inspect the system to locate the leak.

Step 6   Remove the plugs, and allow the water to drain from the system.

7.0.0 ◆ HYDROSTATIC TESTING

Hydrostatic testing is performed by filling a piping system with water and applying pressure to the system to check for leaks. When pressure is applied to water in a closed container, the water applies equal pressure in all directions. The test report specifies the test pressure and how long it must be maintained. If it does not, the test pressure must be at least one and one half times the maximum design pressure of the installation.

CAUTION

When you are dealing with piping systems which have been in existence for some period of time, you cannot always rely on ratings given for new systems. Against the chance of corrosion loss of pipe wall, or other weakening, do not test the existing section at one and one half the rating. Test at the rated pressure only.

Hydrostatic testing at high pressures is very dangerous. If a portion of the system fails during hydrostatic testing, serious injury and even death can result; therefore, the test must be performed only under close supervision by qualified personnel. If you are not involved in the test, stay out of the area where the test is being conducted. If you are required to be in the area, pay close attention to the instructions given to you by your supervisor. Do not attempt to make repairs to the system while it is under pressure. If repairs are necessary, stop the test, release the pressure, and drain the system.

Most hydrostatic tests use clean water as a test medium, but always verify this with the test report or engineering specifications. The temperature of the test medium should not be lower than the outside surrounding temperature because this can cause a brittle fracture to the piping system. It also causes the piping system to sweat, making it difficult to detect water leaks. Most job sites have a minimum temperature at which systems can be tested. When testing a piping system in extremely cold weather, some method must be used to heat the system being tested and the test medium, or you should use glycol as a test medium because it will not freeze. Engineering specifications govern the minimum temperatures at which a system can be tested, and this may vary from job to job.

The major tasks involved in performing hydrostatic tests include performing the pretest requirements, preparing the pump, sealing the system to be tested, and performing the test.

7.1.0 Performing Pretest Requirements

The pretest requirements presented earlier in this module must be completed before performing the hydrostatic test. The line must be adequately restrained to prevent movement; all pressure-relief and control valves must be removed, and all in-line valves locked and tagged open.

7.2.0 Preparing Pump

A hydrostatic test pump is used to fill the system with the test medium and apply the pressure to the system. Hydrostatic test pumps can be manual, electric, pneumatic, or gasoline-engine. The pumps are connected directly to the piping system to be tested. It is best to fill the system at the low point so that the air can be more easily removed from the high-point vents. *Figure 44* shows hydrostatic test pumps.

The test pump is normally connected to the piping system by screwed connections coming out of the discharge side of the pump. This screw pipe is connected to a wire-reinforced, high-pressure hose that is attached to a test tree going into the system. A test tree is a job-fabricated connection normally consisting of a main disconnect valve, a low-point

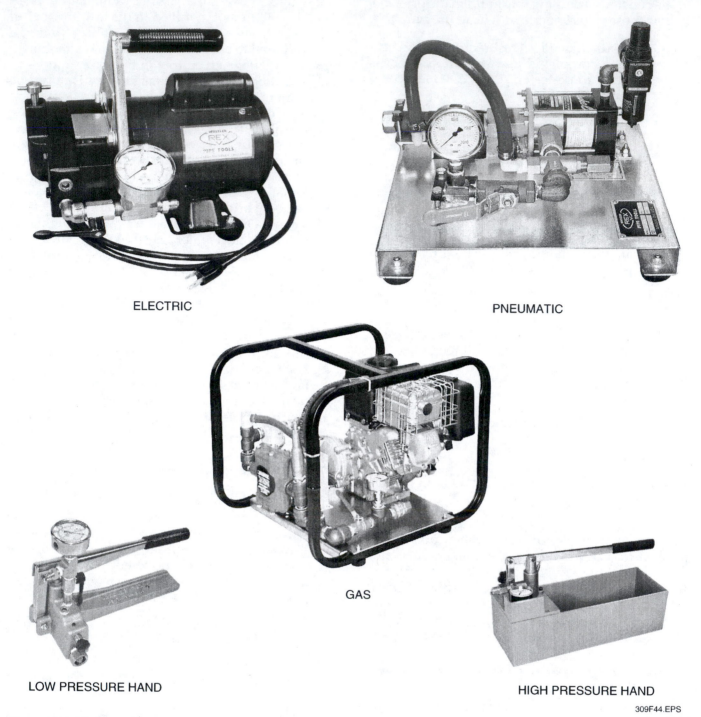

ELECTRIC

PNEUMATIC

GAS

LOW PRESSURE HAND

HIGH PRESSURE HAND

309F44.EPS

Figure 44 ◆ Hydrostatic test pumps.

drain valve, a high-point vent valve, and a valve attached to a pressure gauge. The pressure ratings of all connecting devices must be higher than the rated maximum pressure output of the pump.

Hydrostatic testing can be performed on piping systems or spool pieces in the shop or on piping systems in the field. *Figure 45* shows a test pump attached to a piping system in the shop.

Figure 46 shows a test pump attached to a piping system in the field.

7.3.0 Sealing System

Bleeder valves, also called high-point vent valves, are installed at the highest point in the piping system to vent all of the air from the system while it is being filled with water. If vent valves are not installed, air pockets may form and create dangerous situations and cause false pressure readings. After the high-point vent valves are installed, the section of the pipe or system to be tested must be isolated or sealed off from the rest of the system.

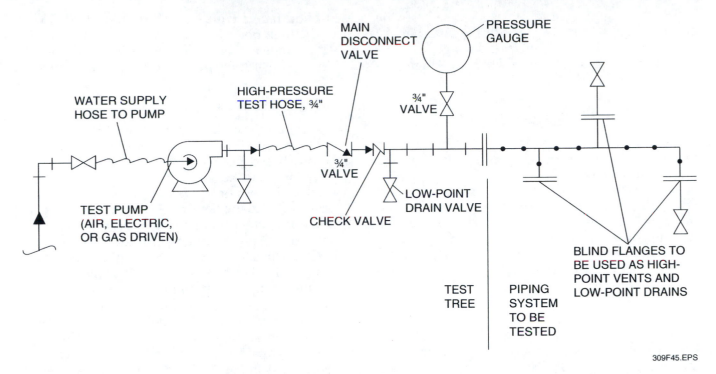

Figure 45 ◆ Test pump attached to piping system in shop.

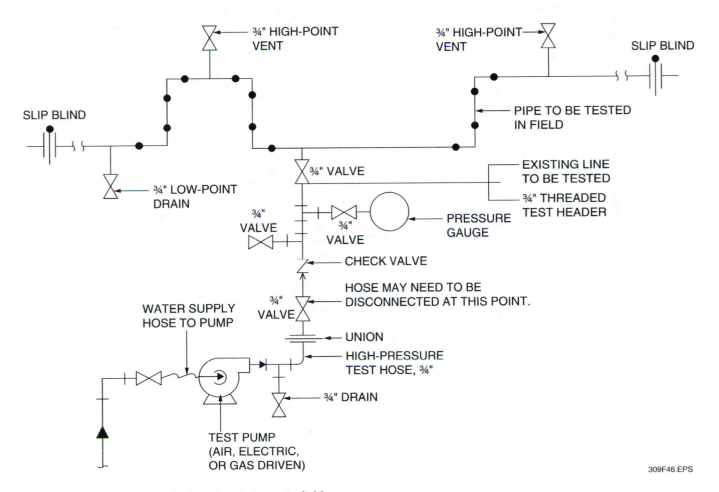

Figure 46 ◆ Test pump attached to piping system in field.

Three methods normally used to seal off the ends of pipe for a hydrostatic test are mechanical test plugs, welded caps, and blind flanges. If a section of pipe needs to be sealed off at a flange connection, a job-fabricated slip blind can be used.

7.3.1 Mechanical Test Plugs

The mechanical test plugs used for hydrostatic testing are similar to the mechanical plugs described earlier in this module. They can be used to seal off the ends of pipe for hydrostatic testing when the test does not require extremely high pressures. Always refer to the plug specifications to determine the maximum allowable pressure the plug can withstand. Clean the inside of the pipe and verify that the plug is the proper size before installing the test plug.

7.3.2 Welded Caps

When welding caps to the ends of pipes to be tested, follow the same welding procedures that are used when installing the piping system. This means that if special welding codes are followed for alloy pipe steel, all requirements concerning preheating and stress-relieving of the weld at the cap must be followed. When the test is complete, remove the cap, using a torch, or mechanically cut through the weld.

7.3.3 Blind Flanges

A blind flange is used to seal off a flange at the end of a pipe. The blind flange must have the same pressure rating as all other flanges in the system.

A blind flange with a bleeder valve is often used at the high point in the system to serve as a vent. This flange can be easily removed at the completion of the test. *Figure 47* shows a blind flange.

7.3.4 Slip Blinds

A slip blind is a piece of plate steel that has been cut to match a flange in the system. Slip blinds can be purchased from suppliers or fabricated on the job. The pancake flange must be made of solid steel and meet or exceed engineering specifications for all flanges in the system. The slip blind must be of the proper thickness for the test pressure. If it is not thick enough, the slip blind will bend during the test and you will not be able to remove it from the system. A T-handle is usually cut out with the slip blind and extends from the top to make the slip blind easy to remove.

The slip blind is placed between two flanges in the system and held with the flange bolts of the two flanges. This provides a temporary seal from sections of the system that are not being tested. After the test is complete, remove the bolts from the top half of the flanges and remove the slip blind. *Figure 48* shows a slip blind.

309F47.EPS

Figure 47 ◆ Blind flange.

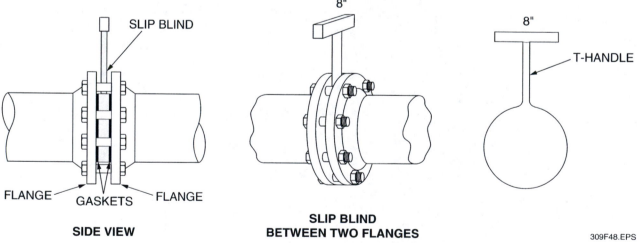

309F48.EPS

Figure 48 ◆ Slip blind.

7.4.0 Performing Hydrostatic Test

After all pretest requirements have been met and the system is sealed, the hydrostatic test can be performed. Follow these steps to perform a hydrostatic test.

Step 1 Ensure that all pretest requirements have been completed and that the proper documentation has been filled out.

Step 2 Barricade the area around the system to be tested.

Step 3 Open the bleeder valve at all high-point vents.

Step 4 Ensure that all plugs, caps, and blind flanges are in place.

Step 5 Begin to fill the system with water or the required test medium, allowing the air to escape from all high-point vent valves.

 WARNING!
Be sure to keep your face away from the valve when opening it.

Step 6 Monitor the water and air coming from the high-point vent valve, and leave the valve open until the water flows smoothly from the valve.

Step 7 Close the high-point vent valve.

Step 8 Attach the test pump to the system according to the project specifications.

Step 9 Allow the test pressure to build up slowly to 50 percent of the required test pressure. Monitor the pressure of the system at the pressure gauge attached to the test tree. The person operating the pump must be able to see the pressure gauge at all times.

 WARNING!
Never leave an operating hydrostatic test pump unattended.

Step 10 Increase the test pressure in increments of 10 percent of the test pressure until the final test pressure is reached.

 CAUTION
If the needle on the pressure gauge fluctuates erratically, this is a good indication that there is still air pocketed in the piping system.

Step 11 Close the valve between the test pump and the piping system.

Step 12 Disconnect the test pump from the system.

Step 13 Allow the system to remain at full test pressure for the amount of time specified in the test report.

Step 14 Inspect the entire system for leaks. If a welded or flanged joint leaks, the test is rejected, and the system must be drained, repaired, and retested. If there are no leaks, the system can be drained after the proper personnel have approved the test.

Step 15 Crack the valve at the test tree slowly to relieve the pressure from the system.

 WARNING!
Stand to the back side of the valve when releasing pressure from the system.

Step 16 Open the high-point vent valve to allow air back into the system.

Step 17 Open the drain valve at the test tree completely to allow the system to drain.

8.0.0 ◆ PNEUMATIC TESTING

Performing a hydrostatic test is often not practical; therefore, you will be required to perform a pneumatic test on a system or spool. A pneumatic test is required when the following conditions exist:

- When components or systems are designed or supported in a manner in which they cannot be safely filled with water due to the weight of the water in the pipe.
- When the test items cannot come in contact with the water.
- When a preliminary test is needed to locate major leaks.
- When a system has been previously hydrostatically tested and a retest of a part of the system is necessary.
- When it is required by the job specifications.

Pneumatic testing is especially used in the context of systems normally run below 100 psi, or in carbon dioxide lines. In storm drain or sewer systems, 5 psig should be the maximum pressure used.

Pneumatic testing uses either air or compressed gas as a test medium. If air is used as the test medium, the air must be clean air that has been filtered to remove unwanted elements, oil, and moisture. The clean air is applied to the system being tested by use of an air compressor to build pressure within the system. On those projects that use compressed gas as a test medium, an inert gas, such as nitrogen, must be used.

The test report specifies the test pressure and how long it must be maintained. If the test pressure is not specified in the test report, the pressure should be obtained from the field engineers or other appropriate personnel. The requirements before testing and for the actual pneumatic test procedure are essentially the same as those for hydrostatic testing. Use the following list to review the requirements for hydrostatic and pneumatic testing:

- All supporting documentation must be checked and verified.
- The test boundaries must be clearly defined.
- Instrumentation gauges must be removed and plugged.
- The entire system must be properly supported by hangers and supports.
- Interfacing systems and equipment not being tested must be blocked off or disconnected.
- Relief valves must be isolated or removed.
- Temporary relief valves should be in place to protect from overpressurizing the system.
- Expansion joints must be restrained, isolated, or protected.
- Calibrated pressure gauges must be installed as necessary.
- Mechanical or weld joints must be accessible for inspection.
- Flanged connections must be taped with a small pin hole pricked in the tape to observe leakage.
- The test gauge range must not be less than one and one half times the test pressure.
- Initial pressure must not exceed one half of test pressure.
- Pressure must be increased in 10 percent increments of the test pressure until the required pressure is reached.
- The quality control inspector must check the system for unacceptable leakage.
- Connections are checked by a bubble test using soapy water, a 40:1 mix of water to dish detergent.

- Documentation of the pressure test must be complete, legible, and signed by qualified personnel.
- The system must be restored to its original condition by removing items such as plugs, temporary hangers, and temporary valves after it has been tested and accepted.

9.0.0 ◆ EQUIPMENT TESTING

Equipment such as heat exchangers, vessels, and tanks must also be tested by pipefitters. The equipment manufacturer provides specific procedures that must be precisely followed to test their equipment. This section lists some general guidelines that must be followed when testing equipment. These guidelines are as follows:

- Barricade the area around the tank or piece of equipment being tested.
- Do not enter the testing area unless you have been authorized to do so.
- Close all valves at or below test elevation.
- Determine the volume of water needed to fill the tank and the length of time it will take to fill the tank. This can be done by measuring the amount of water in the tank after 1 hour of filling and calculating how long it will take to completely fill the tank based on the size of the tank.
- Ensure that the equipment being tested and the ground or platform on which the equipment is placed can withstand the weight of the water or other test medium.

NOTE

When filling a large vessel on the ground for the first time, the weight of the water or other test medium in the vessel will cause the vessel to settle. Some groundwater will seep from underneath the tank.

- Ensure that the high-point vent is a valve of appropriate size for the tank being tested.
- Ensure that the high-point vent is open when filling and draining the equipment.
- Fill the equipment in stages. Do not fill the equipment all at once.
- Ensure that there is a valve located at the supply source for the test medium and at the piece of equipment.
- Determine how and where to drain the equipment according to EPA standards.
- Thoroughly clean the area after the test.

10.0.0 ◆ STEAM BLOW TEST

Before a steam line is put into service, a steam blow must be performed to clean the inside of the line. Before a steam blow is performed, all pretest requirements and inspections must be completed, and the local authorities must be notified that you are conducting a steam blow. A steam blow is a very dangerous procedure and requires close attention to detail. The steam blow must be engineered and performed according to plant and design specifications. Because the pipe is brought to operating temperature during the steam blow, thermal expansion of the pipeline must be considered. A temporary vent line is attached to the end of the steam line to vent the steam outside the building or to the atmosphere. This line, which can be 10 or 15 feet long or more, must be anchored securely and must direct the hot steam to a safe area. This area must be roped off to ensure the safety of others. A pre-engineered muffler or silencer (*Figure 49*) that meets federal standards for allowable decibel ranges in the area must be attached to the end of the line.

10.1.0 Steam Targets

Many piping specifications require that **steam targets** be used to detect the presence of foreign material in the pipeline. A steam target is a piece of soft metal, such as copper, brass, or aluminum, that is inserted into the vent line or between the vent line and the muffler. Any foreign material in the steam line pits and damages the target during a steam blow. The steam blow is repeated until there are no imperfections or a limited number of pits on the steam target. The number of pits allowed depends on the codes used or the plant specifications. *Figure 50* shows steam targets.

Various types of steam targets are used on the job site. One type of steam target is a carbon steel ring cut about the same size as a gasket, with a piece of soft metal bolted across the middle of the ring. The ring is inserted between two flanges in the steam line. If the target shows imperfections after a steam blow, rotate the target 180 degrees, and blow the line again. This allows you to get two tests out of each steam target. The carbon steel

309F49.EPS

Figure 49 ◆ Steam silencer.

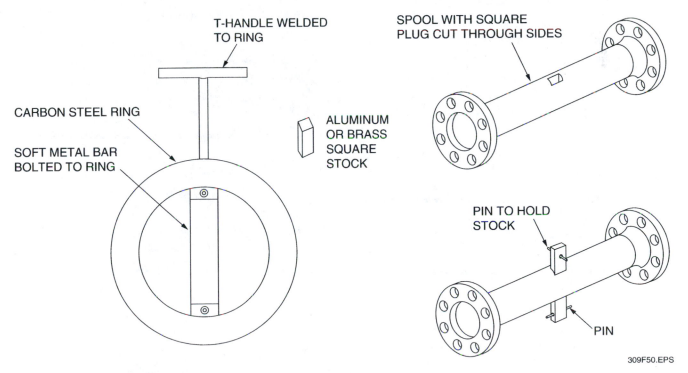

309F50.EPS

Figure 50 ◆ Steam targets.

ring can be used over and over again. A T-handle is welded to the ring to make it easy to remove.

Another type of steam target consists simply of a piece of aluminum or brass square bar stock. A special spool is fabricated to be inserted into the steam line. The spool has a square plug cut into it so that the bar stock can be inserted through the pipe. After each steam blow, this target can be rotated 90 degrees so that you can get four steam blows out of each target.

If a steam target is not required by the plant specifications, the steam line is usually blown for a specified length of time as determined in the test report. To reduce the amount of expansion and contraction of the line, do not allow the line to cool completely between blows. If a target is used, the pipeline is blown until an acceptable target is obtained.

10.2.0 Performing Steam Blow Test

A steam blow test is extremely dangerous because of the high temperature and pressure of the steam being used. The high pressure of the steam generates a great amount of noise; therefore, ear protection is required. The area must be completely barricaded, and anyone not involved with the test must stay clear of the area. Radio communication between everyone involved in the test is mandatory to effectively monitor the test and prevent damage and injury. Follow these steps to perform a steam blow test.

Step 1 Ensure that all pretest requirements have been completed and that the proper documentation has been filled out.

Step 2 Barricade the area around the system to be tested.

Step 3 Ensure that all valves have been locked open and that all control valves have been removed.

Step 4 Install a temporary vent line to the end of the steam line.

 NOTE
Install a steam target into the vent line if required.

Step 5 Barricade the area that the steam will be blown into.

Step 6 Crack the bypass valve to preheat the steam line.

 CAUTION
The line will expand; therefore, all hangers and supports must be closely monitored during preheating.

Step 7 Inspect all supports and hangers, and adjust those that have moved due to thermal expansion.

 WARNING!
The steam line is very hot. Wear heat-resistant gloves and use extreme caution when adjusting the hangers and supports.

Step 8 Allow the line to preheat to the operating temperature.

Step 9 Open the main supply valve completely.

Step 10 Blow the steam line for the specified length of time as determined by the test procedure.

Step 11 Close the main supply valve.

 NOTE
If a target is used, inspect it for imperfections.

Step 12 Drain the drip legs between each steam blow to purge the lines of condensate and prevent water hammer.

Step 13 Repeat the steam blow until an acceptable target is obtained.

Step 14 Allow the system to cool completely.

1. Foreign matter trapped in the weld metal is called a(n) _____.
   a. intrusion
   b. inclusion
   c. porosity
   d. bubble

2. Three basic types of cracks can occur in weld metal: transverse, longtitudinal, and _____.
   a. horizontal
   b. crater
   c. inclusion
   d. vertical

3. Arc strikes are small, localized points where _____ occurs away from the weld.
   a. grinding
   b. cutting
   c. surface melting
   d. spatter

4. The borescope most commonly used by pipefitters is the _____.
   a. rigid borescope
   b. right-angle borescope
   c. diopter
   d. flexible borescope

5. Liquid penetrant tests detect _____.
   a. weld undercut
   b. surface openings
   c. inclusions
   d. slag pockets

6. Magnetic particle tests are sensitive to _____.
   a. surface and near-surface flaws
   b. surface flaws only
   c. root problems
   d. paint problems

7. Both light beams and ultrasound beams are _____.
   a. liquid
   b. solid
   c. waves
   d. particles

8. Ultrasonic examination can be used to detect _____.
   a. surface cracks
   b. shrinkage cavities
   c. splatter
   d. arc strikes

9. Conductivity tests use amperage to find flaws in _____ of pipe.
   a. welds
   b. flanges
   c. fittings
   d. coatings

10. Pretest requirements include identifying _____.
    a. weld types
    b. test boundaries
    c. flange types
    d. leaks

11. Applicable codes are indicated on the _____.
    a. lift plan
    b. test boundaries
    c. test report
    d. test medium

12. The pressure gauge in a test should be connected directly to the _____ of the system or component.
    a. high point
    b. low point
    c. test boundary
    d. end

13. As a minimum requirement, each pressure gauge must be calibrated every _____.
    a. time it is used
    b. two weeks
    c. three months
    d. six months

14. Field pretest inspections may be written up as a(n) _____.
    a. inspection punchlist
    b. test report
    c. lift plan
    d. load chart

15. As part of the pretest field inspection of a piping system _____.
    a. run the system up to pressure
    b. take out the gaskets
    c. ensure that all equipment not to be tested is locked and tagged out
    d. open all the pressure relief valves

16. Cleaning a system can be done by _____.
    a. hand scrubbing
    b. torches
    c. pigging
    d. reamers

17. Using chemicals to clean pipe systems is sometimes called _____.
    a. pigging out
    b. pickle and passivate
    c. flushing
    d. flow testing

18. To perform a service test, close the last valve in the system and turn on the service to that system at _____.
    a. one half design pressure
    b. one and one half times rated pressure
    c. three and a half times design pressure
    d. normal operating pressure

19. To perform a flow test, check for leaks with the _____.
    a. last valve closed
    b. system shut down
    c. supply valve closed
    d. system in operation

20. Head pressure tests are performed on _____.
    a. high-pressure systems
    b. nuclear reactor coolants
    c. low-pressure systems
    d. hydroblasters

21. The head of one foot of water is equal to _____ psi.
    a. 0.4335
    b. 5
    c. 11
    d. 27.72

22. When testing a drain pipe in a high-rise structure, only test _____ feet at a time.
    a. 50
    b. 100
    c. 200
    d. 300

23. If design pressure for a new installation is 200 psi, the hydrostatic test pressure is _____ psi.
    a. 100
    b. 200
    c. 250
    d. 300

24. In a hydrostatic test it is best to fill the system from the _____.
    a. highest point
    b. middle
    c. lowest point
    d. end

25. In a pneumatic pressure test, connections are checked for leaks with a _____.
    a. pneumatic plug
    b. mechanical plug
    c. mix of soap and water
    d. hygrometer

Summary

Testing piping systems can be extremely dangerous, depending on the test you are performing. You must use great caution and stay alert at all times when performing any test. Good communication, alertness, and caution will keep you and others free from injury when working around high-pressure test stations. The tests presented in this module serve as insurance that a piping system will perform as expected. Never overlook any details, and you will be able to perform valid tests on piping systems and components.

As you start NDE testing activities, you must call on all of your previously learned skills. NDE testing requires you to be proficient in preparing welds, valves, piping, and vessels for the appropriate NDE test. Remember to always put safety first and be aware of your working surroundings when performing NDE testing activities. Your safety and the safety of your co-workers depend on the way you perform your job.

Notes

Trade Terms Introduced in This Module

Acceptable weld: A weld that meets all the requirements and acceptance criteria prescribed by the welding specifications.

Antisyphon: A protective device used to prevent erratic movement of a pressure gauge on the pipeline being tested.

Concavity: The maximum distance from the face of a concave fillet weld perpendicular to a line joining the toes of the weld.

Crack: A fracture-type discontinuity characterized by a sharp tip and high ratio of length and width to the opening displacement.

Crater: A depression at the termination point of a weld bead.

Defects: Discontinuities that render a part or product unable to meet minimum applicable acceptance standards or specifications.

Design pressure: The maximum specified pressure that all the components of a piping system are designed to withstand.

Discontinuity: An interruption or gap in the typical structure of a weldment, such as a lack of consistency in the mechanical, metallurgical, or physical characteristics of the base material or weldment.

Head pressure: The pressure that water applies to a system from above the system. The head pressure for water is 0.4335 psi per foot regardless of the volume of the cylinder. To find total head pressure at a given point, multiply the height of the water above that point (in feet) times 0.4335.

Hydrostatic test: A test performed by filling a piping system with water and then applying pressure to the water to check for leaks.

Inclusions: Particles of impurities that are retained in the weld metal after the metal solidifies and can occur anywhere in the weld bead.

NDE: Nondestructive examination of piping and related components.

Operating pressure: The pressure at which a system is intended to operate. The operating pressure is lower than the design pressure of the system.

Pig: A plug that is sent through a pipeline to remove internal debris, such as scale, rust, and welding rods. Types of pigs include steel mandrel, polly pig, and detector.

Porosity: Cavity-type discontinuities formed by gas being entrapped in the weld during solidification.

Root opening: The separation or gap between the two members to be joined by welding. The root opening between two pipes being butt welded is normally $\frac{1}{16}$ to $\frac{1}{8}$ inch.

Root penetration: The depth that a weld extends into the root of a joint measured on the center line of the root cross section.

Snubber: A protective device that uses weights or hydraulics to dampen the shock placed on a pressure gauge on the pipeline being tested.

Steam target: A piece of soft metal that is inserted into a steam line during a steam blow test and used to detect the amount of debris contained within the line.

Test boundaries: An ISO or group of ISOs that define the pipelines, spools, or systems to be tested.

Test medium: The fluid used to test a piping system. Most of the time, this is distilled or process water, but always check the job specifications to determine the test medium to be used.

Testing: Verifying that an item can meet specified requirements by subjecting the item to a set of physical, chemical, environmental, or operating conditions.

Toe crack: A crack in the base metal occurring at the toe of the weld.

Toe of the weld: The intersection between the face of the weld and the base metal.

Underbead crack: A crack in the heat-affected zone of the base metal that generally does not extend to the surface of the base metal.

Undercut: A groove melted into the base metal adjacent to the toe or root of a weld and left unfilled by weld metal.

Weld root: The points at which the back of the weld intersects the base metal surfaces.

Weldment: An assembly in which the component parts are joined by welding.

Resources & Acknowledgments

Additional Resources

This module is intended to be a thorough resource for task training. The following reference works are suggested for further study. These are optional materials for continued education rather than for task training.

The United Kingdom's largest testing and calibration laboratory for pipe fittings and materials has a website at http://www.wrcnsf.com/plastic_pipe.htm.

Pipeline Maintenance. National Center for Construction Education and Research. Upper Saddle River, NJ: Prentice Hall.

Ashtead Technology Rentals describes many types of testing equipment at http://www.ashtead-technology.com/?kc=qwRVB.

Figure Credits

American Welding Society, 309F03 (AWS B1.1:2000, Figures 1, 2, 3. Reproduced with permission of the American Welding Society (AWS), Miami, FL), 309F04 (AWS B1.1:2000, Figure 30. Reproduced with permission of the American Welding Society (AWS), Miami, FL), 309F05 (AWS B1.1:2000, Figure 20. Reproduced with permission of the American Welding Society (AWS), Miami, FL), 309F06 (AWS B1.1:2000, Figures 3, 24-26, 28. Reproduced with permission of the American Welding Society (AWS), Miami, FL), 309F09 (AWS B1.1:2000, Figures 7, 8. Reproduced with permission of the American Welding Society (AWS), Miami, FL), 309F10 (AWS B1.1:2000, Figure 18. Reproduced with permission of the American Welding Society (AWS), Miami, FL), 309F12 (AWS B1.1:2000, Figure 35. Reproduced with permission of the American Welding Society (AWS), Miami, FL)

Topaz Publications, Inc., 309F07, 309F19, 309F21, 309F23 (photo), 309F27 (photo), 309F38

Machida Borescopes, 309F14 (photo)

Ridge Tool Company (RIDGID®), 309F15

G.A.L. Gage Co., 309F16, 309F17

Curtis Casey, 309F20

Magnaflux, A Division of Illinois Tool Works, Inc., 309F22 (photo)

Olympus NDT, 309F24 (photo)

Albuquerque Industrial, 309F31 (photo)

Anvil International, 309F32

Ashcroft Inc., 309F33

Girard Industries, 309F35, 309F36

Piping Technology & Products, Inc. (SWECO FAB), 309F37 (photos)

Gardner Denver Water Jetting Systems, Inc., 309F39

Sealfast Products, 309F43

Wheeler Manufacturing, A Division of Rex International USA, 309F44

Maxim Silencers, Inc., 309F49

NCCER makes every effort to keep these textbooks up-to-date and free of technical errors. We appreciate your help in this process. If you have an idea for improving this textbook, or if you find an error, a typographical mistake, or an inaccuracy in NCCER's Contren® textbooks, please write us, using this form or a photocopy. Be sure to include the exact module number, page number, a detailed description, and the correction, if applicable. Your input will be brought to the attention of the Technical Review Committee. Thank you for your assistance.

Instructors – If you found that additional materials were necessary in order to teach this module effectively, please let us know so that we may include them in the Equipment/Materials list in the Annotated Instructor's Guide.

Write: Product Development and Revision
National Center for Construction Education and Research
3600 NW 43rd St, Bldg G, Gainesville, FL 32606

Fax: 352-334-0932

E-mail: curriculum@nccer.org

Craft   Module Name

_____

Copyright Date   Module Number   Page Number(s)

_____

Description

_____

_____

_____

_____

(Optional) Correction

_____

_____

_____

(Optional) Your Name and Address

_____

_____

_____

Glossary of Trade Terms

Acceptable weld: A weld that meets all the requirements and acceptance criteria prescribed by the welding specifications.

Addenda: Supplementary information typically used to describe corrections or revisions to documents.

Adjacent side: The side of a right triangle that is next to the reference angle.

Aggregate: A collection of soil grains, sands, crushed stone, and other particles.

Anneal: To soften a metal by heat treatment.

Antisyphon: A protective device used to prevent erratic movement of a pressure gauge on the pipeline being tested.

Anti-two-blocking devices: Devices that provide warnings and prevent two-blocking from occurring. Two-blocking occurs when the lower load block or hook comes into contact with the upper load block, boom point, or boom point machinery. The likely result is failure of the rope or release of the load or hook block.

Bird caging: A deformation of wire rope that causes the strands or lays to separate and balloon outward like the vertical bars of a bird cage.

Blocking: Pieces of hardwood used to support and brace equipment.

Bridle: A common line used for manifolding two or more instruments to one pair of connections on a vessel.

Cavitation: The result of pressure loss in liquid, producing bubbles (cavities) of vapor in liquid.

Center of gravity: The point at which an object is perpendicular to and balanced in relation to the earth's gravitational field.

Code: A set of regulations covering permissible materials, service limitations, fabrication, inspection, and testing procedures for piping systems.

Cold load: A piping system before it is filled with product.

Component: A single part in a system.

Concavity: The maximum distance from the face of a concave fillet weld perpendicular to a line joining the toes of the weld.

Cornice: The crowning, overhanging lip at the top of a wall.

Corrosion: The result of wearing away slowly, especially by chemical action or abrasive wear through friction.

Cosine: Trigonometric ratio between the adjacent side and the hypotenuse, written as adjacent divided by the hypotenuse.

Counterweight: A device that counterbalances an opposing load.

Crack: A fracture-type discontinuity characterized by a sharp tip and high ratio of length and width to the opening displacement.

Crater: A depression at the termination point of a weld bead.

Cribbing: Timbers stacked in alternate tiers. Used to support heavy loads.

Defects: Discontinuities that render a part or product unable to meet minimum applicable acceptance standards or specifications.

Design pressure: The maximum specified pressure that all the components of a piping system are designed to withstand.

Discontinuity: An interruption or gap in the typical structure of a weldment, such as a lack of consistency in the mechanical, metallurgical, or physical characteristics of the base material or weldment.

Dunnage: Hardwood blocking and pallets placed underneath materials to keep them off the ground and to allow access for forklifts or for placement of chokers and slings.

Equalizer beam: A beam used to distribute weight on multi-crane lifts.

Fabricate: Putting together component parts to form an assembly.

Field routing: Describes a method for installing piping systems without having drawings to give coordinates and specifications. Field routing is normally performed with small-bore pipe.

Fixed block: The upper block of a block and tackle. The block that is attached to the support.

Gasket: A device that is used to make a pressure-tight connection and that is usually in the form of a sheet or a ring.

Grasshopper: A two-wheel device that is used to transport pipe and piping components.

Hauling line: The line of a lifting device that is pulled by hand to raise the load.

Head pressure: The pressure that water applies to a system from above the system. The head pressure for water is 0.4335 psi per foot regardless of the volume of the cylinder. To find total head pressure at a given point, multiply the height of the water above that point (in feet) times 0.4335.

Head room: The space between an object and the ceiling.

Hot load: The actual load of the pipeline in service.

Hydraulic: Operated by fluid pressure.

Hydraulic: Operated by fluid pressure.

Hydrostatic test: A test performed by filling a piping system with water and then applying pressure to the water to check for leaks.

Hypotenuse: The longest side of a right triangle. It is always located opposite the right angle.

Inclusions: Particles of impurities that are retained in the weld metal after the metal solidifies and can occur anywhere in the weld bead.

Kinking: Bending a rope so severely that the bend is permanent and individual wires or fibers are damaged.

Lanyard: A rope or line used to fasten a safety harness to an existing structure.

Load column: Part of a support assembly that is used for adjustment purposes.

Load rating: The usable lifting capacity of a piece of hardware, including the safety factor.

Load rating: The usable weight capacity that a piece of hardware can support.

NDE: Nondestructive examination of piping and related components.

Operating pressure: The pressure at which a system is intended to operate. The operating pressure is lower than the design pressure of the system.

Opposite side: The side of a right triangle that is located directly across from the reference angle.

Outrigger: A steel beam or supporting leg extending from a crane or other piece of equipment to provide stability by widening the base.

Parapet wall: A low retaining wall at the edge of a roof.

Parts of line: The number of ropes between the upper and lower blocks of a block and tackle. These lines carry the load. Parts of line are also called falls.

Pig: A plug that is sent through a pipeline to remove internal debris, such as scale, rust, and welding rods. Types of pigs include steel mandrel, polly pig, and detector.

Pipe riser: A vertical pipe extending from an opening in a floor or ceiling.

Porosity: Cavity-type discontinuities formed by gas being entrapped in the weld during solidification.

Precompress: To place pressure on a spring before it is installed in a spring can.

Pressure differential: The difference in pressure between two points in a flow system. It is usually caused by frictional resistance to flow in the system.

Pressure vessel: A metal container that can withstand high pressures.

Ratio: A comparison of one value to another value.

Reference angle: The angle to which the sides are related as adjacent and opposite.

Root opening: The separation or gap between the two members to be joined by welding. The root opening between two pipes being butt welded is normally 1/16 to 1/8 inch.

Root penetration: The depth that a weld extends into the root of a joint measured on the center line of the root cross section.

Run: A length of pipe that is made up of more than one piece of pipe.

Safety factor: The load or tension at which a piece of rigging equipment will fail divided by the actual load to be placed on the equipment. The safety factor for rigging equipment should never be less than four to one.

Seizing: A binding to prevent rope from untwisting after it is cut.

Severe service: A high-pressure, high-temperature piping system.

Sine: Ratio between the opposite side and the hypotenuse, the opposite divided by the hypotenuse.

Sling angle: The angle formed by the legs of a sling with respect to the horizontal when tension is put upon the load.

Snubber: (1) A protective device that uses weights or hydraulics to dampen the shock placed on a pressure gauge on the pipeline being tested.

Glossary of Trade Terms

Snubber: (2) A hydraulic device that slows sudden up and down movement in pipe systems to prevent damage to supports or piping.

Steam target: A piece of soft metal that is inserted into a steam line during a steam blow test and used to detect the amount of debris contained within the line.

Structural member: A supporting part of a building or structure.

Sway brace: Either a spring or hydraulic shock absorber, or a set of cables, designed to prevent dramatic side-to-side or longitudinal movement.

Tangent: Ratio between the opposite side and the adjacent side, opposite divided by the adjacent.

Test boundaries: An ISO or group of ISOs that define the pipelines, spools, or systems to be tested.

Test medium: The fluid used to test a piping system. Most of the time, this is distilled or process water, but always check the job specifications to determine the test medium to be used.

Testing: Verifying that an item can meet specified requirements by subjecting the item to a set of physical, chemical, environmental, or operating conditions.

Toe crack: A crack in the base metal occurring at the toe of the weld.

Toe of the weld: The intersection between the face of the weld and the base metal.

Turbulence: The motion of fluids or gases in which velocities and pressures change irregularly.

Underbead crack: A crack in the heat-affected zone of the base metal that generally does not extend to the surface of the base metal.

Undercut: A groove melted into the base metal adjacent to the toe or root of a weld and left unfilled by weld metal.

Vessel trim: Piping, instruments, and valves connected to vessels.

Weld root: The points at which the back of the weld intersects the base metal surfaces.

Weldment: An assembly in which the component parts are joined by welding.

Wire rope: A rope formed of twisting strands of wire.

Index

Index

A

Accidents
   burns, 2.19
   crushing, 2.16
   electrocution, 2.18
   explosion, 2.19, 8.22
   hydroblasting, 9.34
   pinching, 2.17, 2.38, 6.25
   tipping, 2.20, 2.21, 2.29
Acetone, 6.15
Acrylic fiber, 6.13
Addenda sheets, 3.3, 3.17
Aggregate, 5.18
Air, 9.35, 9.40, 9.44
Aluminum, 6.15, 9.45, 9.46
American National Standards Institute (ANSI), 3.3, 3.5, 6.7,
      6.11, 6.20
American Petroleum Institute (API), 3.3
American Society for Testing and Materials (ASTM
      International), 2.3–2.13, 2.39, 3.11–3.12, 6.9
American Society of Mechanical Engineers (ASME), 3.3–3.5
American Welding Society (AWS), 3.5, 9.3
Amerilon®, 6.15
Anchors, for pipe support, 8.9, 8.10, 8.16, 8.18
Angle
   bevel, 9.2, 9.7
   bisection with divider and straightedge, 6.18
   conversion of sine value to, 4.14
   determination when side lengths are known, 4.15–4.16
   interpolation, 4.16–4.17
   obtuse, 4.14
   odd, 4.18
   of offset, 2.31, 4.7, 4.15–4.16
   reference, 4.4, 4.22
   right, 4.2, 4.6
Anneal, 1.13, 1.26
ANSI. *See* American National Standards Institute
Antisyphon, 9.30, 9.50
Anti-two-blocking device, 2.23, 2.43
API. See American Petroleum Institute
Arc strike, 9.8, 9.23
Asbestos, 6.13, 6.14, 6.15
ASME. *See* American Society of Mechanical Engineers
ASTM International. *See* American Society for Testing and
      Materials
Auger, 5.14
AWS. *See* American Welding Society

B

Baffle, 7.12
Barricades and barriers
   field routing area, 7.2
   manlift, 5.5
   pipe installation area, 6.29
   rigging operation, 2.15, 2.16
   testing area, 9.12, 9.44, 9.46
Basket, personnel, 2.23, 2.24, 2.27
Beam
   equalizer, 1.9, 1.26
   I-, 5.9, 5.10
   lifting, 2.17
   outrigger, 5.11
   spreader, 1.9, 2.39
Bird caging, 1.13, 1.26
Blind
   slip, 9.41, 9.42
   test, 7.6–7.8
Block and tackle system, 1.17
Blocking (load support), 1.17, 1.21, 1.26, 2.16–2.17, 2.36, 2.43
Blowoff, 8.36
Blowout preventer, 6.14
Boiler, 3.3–3.4
Bolt holes
   alignment, 6.21
   layout, 6.17–6.18
   numbering system, 6.23, 6.24
   on spring can, 8.29, 8.30
   standards for location, 6.20
Bolts
   down-limit, 8.34, 8.35
   flange, 6.22–6.24, 7.7, 7.8, 9.32
   re-torquing, 6.12
   tightening sequence, 6.23, 6.24
   toggle, 8.19–8.20
   U-, 8.6–8.7, 8.8
Boom
   angle and length, 2.28, 2.29, 2.30, 2.38
   effects of wind, 2.20–2.21
   lattice, 2.30, 2.31
   lightning strike, 2.20–2.21
   load capacity, 2.28
   minimum clearance for power line, 2.18
   movement, hand signals to direct, 2.6–2.9, 2.12–2.13
   telescoping, 2.12, 5.2, 5.4, 5.5, 5.6–5.8
Boom point elevation, 2.28, 2.29, 2.30

Certified Occupational Therapy Assistant COTA® Certification Examination

Official

NBCOT Study Guide

Serving the Public Interest

National Board for Certification in Occupational Therapy, Inc.
12 South Summit Avenue, Suite 100
Gaithersburg, MD 20877-4150

www.nbcot.org

Our Mission...

Above all else, the mission of the National Board for Certification in Occupational Therapy, Inc. (NBCOT®) is to serve the public interest. We provide a world-class standard for certification of occupational therapy practitioners. NBCOT will develop, administer, and continually review a certification process based on current and valid standards that provide reliable indicators of competence for the practice of occupational therapy.

National Board for Certification in Occupational Therapy, Inc.
12 South Summit Avenue, Suite 100
Gaithersburg, MD 20877-4150
http://www.nbcot.org

Printed in the United States of America.

ISBN 0-9785178-2-2